# TECTONICS AND THE FORMATION OF MAGMAS

**Yu. M. Sheinmann**
O. Yu. Schmidt Institute of Physics of the Earth
Academy of Sciences of the USSR
Moscow, USSR

Translated from Russian by
**J. Paul Fitzsimmons**
Department of Geology
University of New Mexico

CONSULTANTS BUREAU · NEW YORK-LONDON · 1971

*Yu. M. Sheinmann* was born in Kiev in 1901. In 1927 he was graduated from the Leningrad Mining Institute after having begun his scientific work as a paleontologist in 1923. Beginning in 1926 he conducted field investigations, first in Transbaikalia and the Baikal region, then in the Tien Shan, along the Yenisei, the extreme northern part of Siberia, Tuva, and elsewhere. Field work led to tectonic problems, at first regional, then more general. To his other interests he has added magmatic petrology and the problem of the relationship between magmatism and tectonics. He was the first to uncover the marine Mesozoic geosynclinal deposits of Transbaikalia and the great province of alkalic rocks in Siberia. In recent years he has worked on the investigation of rocks under high temperatures and pressures. He has worked at a number of geological institutions of the USSR, notably in the Geological Survey. At present he is working at the O. Yu. Schmidt Institute of Physics of the Earth, Academy of Sciences of the USSR, in Moscow.

The original Russian text, published by Nedra Press in Moscow in 1968, has been corrected by the author for the present edition. The English translation is published under an agreement with Mezhdunarodnaya Kniga, the Soviet book export agency.

Ю. М. Шейнманн

Очерки глубинной геологии

(о связи тектоники с возникновением магм).

Library of Congress Catalog Card Number 70-136981

SBN 306-10859-3

ISBN-13: 978-1-4684-1586-5          e-ISBN-13: 978-1-4684-1584-1
DOI:10.1007/ 978-1-4684-1584-1

© 1971 Consultants Bureau, New York
A Division of Plenum Publishing Corporation
227 West 17th Street, New York, N.Y. 10011

United Kingdom edition published by Consultants Bureau, London
A Division of Plenum Publishing Company, Ltd.
Davis House (4th Floor), 8 Scrubs Lane, Harlesden, NW10 6SE, **London**, England

# PREFACE TO THE ENGLISH EDITION

It is a pleasure to any author to enlarge the circle of his readers. Naturally this is a pleasure to me. At the same time, however, misgivings arise: whether these pages will convey to the circle of new readers the thoughts that excite the author. Science is advancing rapidly in our day. It is already apparent that many things in the book should have been stated differently. I have tried to make additions to the English-language edition in such a way that they do not require great alteration of the text. I ask my new readers to remember that this book was written primarily for my fellow countrymen and that, because of this, some chapters contain descriptions of regions little known to you and information on the USSR almost unknown from other publications. I hope that, despite this, geologists who read the English-language edition will find something of interest to themselves in it and that their labor will not have been in vain.

Yu. M. Sheinmann
Institute of Physics of the Earth
Moscow

# PREFACE

The study of endogenetic processes has long pushed us into investigation of ever deeper parts of the earth. Not long ago all attention was focused on depths where ore deposits appear, where metamorphic and igneous rocks, which later become exposed at the surface, are formed, where granites originate. As we now know, the depths of these zones do not exceed, or only slightly exceed, the depth of maximum erosional dissection. Consequently, all these processes may be studied directly by geologic observation.

Until recently all attempts to penetrate deeper proved futile. They amounted to skill in making a more or less interesting portrayal of a purely speculative picture. The best grounded of these views proved to be that concerning the origin of basalts in the lower half of the earth's crust, since geophysicists established the existence of the "basaltic layer" there. Very incomplete and disconnected geophysical data have yielded too little for us to build firmly on this foundation. The behavior of silicates under conditions corresponding to the depths, pressures, and temperatures involved is still unknown. And geology itself has not been ready for a "leap" into the depths of the mantle.

This possibility has arisen very recently, and the necessity of coordinating the corresponding investigations has lead to establishment of the Upper mantle Project (V. V. Belousov), which has greatly improved our knowledge of the earth's interior. In conjunction with this powerful attack on deep-seated phenomena, the possibility and necessity of a new discipline have become manifest: abyssal geology. Its framework and structure are still far from clear.

There is no doubt that the method of investigating depths not directly accessible to observation or even to deep drilling must be new, since the geological methods we have long used require direct visual observation and "probing" of rocks and processes.

Geophysicists were the first to probe the depths of the earth. Their data are of utmost importance since only geophysicists know how to make observations at depth by means of their instruments. The problem of describing geologic processes taking place at great depth is therefore still one of the objectives of geophysical research. Success in studies of physics and chemistry of the earth merely eases the approach to this problem, makes the approach possible, but in no way substitutes for the geological solution of the problem.

The present work is devoted to abyssal geology: the connection between tectonic phenomena and the formation of magmas. This problem is perhaps the most acute. Deep-seated tectonics are generally touched upon only in passing, when other matters are being discussed. Almost no light has been shed on the nature of the discontinuities noted by geophysicists.

A great difficulty in setting up and developing problems of abyssal geology is the fact that it encroaches on the fields of tectonics, petrology, geophysics, and the physical chemistry of silicates. At times one even hears that it is inadmissible to use data of petrology or physical chemistry in tectonics, or data of tectonics in petrology. The author, whose specialty is tectonics, fully realizes the difficulty of this problem. But, if we are denied use of available neighboring disciplines, it becomes almost impossible to do anything at the boundaries of scientific fields. To deny oneself such ventures is to erect a path of but limited possibilities, i.e., to proceed along already well-worn paths.

A great difficulty in attempts to investigate the problem is the extraordinarily rapid accumulation of new data, especially in regard to the behavior of silicates at depth. This means that it may well be true that some of the material discussed in this work will be to some extent out-of-date by the time the book appears in print.

The author permitted himself the inclusion of brief descriptions of several petrographic provinces, the peculiarities of which have been alluded to in discussions of more general questions. These are provinces that have received no attention or have been inadequately described in Russian literature. If these descriptions are of no interest to the reader they may be ignored, and this will lead to no difficulty in reading the rest of the book.

The author recalls with thanks his conversations with M. E. Artem'ev, E. V. Artyushkov, V. V. Belousov, V. V. Zhdanov, I. P. Kosminskaya, Yu. A. Kuznetsov, E. N. Lyustikh, V. A. Magnitskii, E. V. Pavlovskii, B. A. Petrushevskii, E. M. Rudich, A. A. Sorskii, A. Ya. Saltykovskii, E. K. Ustiev, N. I. Khitarov, and V. I. Chernysheva, after which it has been easy to comprehend the problems facing him. He expresses his profound thanks to T. M. Kuznetsova and V. V. Kozhanova for their aid in putting this work into its completed form.

# CONTENTS

# CHAPTER 1

# INTRODUCTION

From the time that the existence of tectonic phenomena became clear to geologists, and the presence of magmas in the earth could no longer be explained as the result of the burning of coal, the idea that there was probably some connection between the two phenomena appeared. Infantile ideas persisted a long time. It was necessary first to discover much about the life of magma and about the transformation of magma. It was necessary also to find some basic patterns in the movement of material in the earth. Only after correlation of the data in the two fields of geology was it possible for the science to develop. For us, therefore, the first name will be G. Steinmann, who first formulated the probable connection between volcanic phenomena and the development of folding. Steinmann [1905, 1926] believed that in the early epochs of development of fold zones, during the accumulation of sediments in geosynclines, ophiolitic vulcanism prevailed. Later, basic magmas yielded to acidic magmas. Steinmann's scheme, proposed for Europe, proved to be exceptionally long-lived. Its ideas have been developed by a number of scientists, chiefly in Germany. In the twenties and thirties, ideas along this trend were advanced by S. von Bubnoff, H. Cloos, P. Eskola, K. Bergmann, V. M. Goldschmidt, R. Daly, A. Ritman, G. W. Tyrrell, and others. The greatest effect on further development of this scheme has come from the work of H. Stille [$1940_{1,2}$]. The founder of tectonics as a separate science, at least in its present form, Stille has had a tremendous influence on development of the problem of the connection between magmatism and tectonics.

According to Stille, all magmatic activity may be separated into four stages (terminology of Stille, 1938-40):

1. The stage of geosynclinal subsidence — initial magmatism. Chiefly basic magmas, ophiolites; commonly magmas of intermediate and even acidic composition. The magmatism is confined to orthogeosynclines. Volcanic rocks predominate, but intrusions occur.

2. The stage of folding (orogenesis) — synorogenic magmatism. Sialic magmas of orogenic phases, during greatest development of folding, or at the end of it.

3. The stage of the "quasicraton," that is the initial stage of craton formation — subsequent magmatism. Chiefly sialic magmas. Similar to synorogenic in origin and time.

4. Stage of the craton — terminal magmatism. Predominantly basic magmas, rising in both subsiding and uplifted zones.

The scheme of Stille summarized and refined views already existing in the science. These views reduce chiefly to a distinction of early (geosynclinal) ophiolitic magmatism, acidic intrusions in the stage of folding; and post-folding magmatism. These concepts, not yet in their final form, were thus outlined in the works of the petrologists and tectonic geologists we have mentioned. Stille himself, although but slightly acquainted with petrology, was able to combine organically the changes in character of magmatism with the evolution of geosynclines. However, almost everyone else who developed a theory of tectonics left the problem of magmatism to one side. At best an author might repeat the concept concerning the connection between ophiolites and subsidence and between granites and folding. It was so with our leading tectonic geologists, regardless of which school they belonged to. A. D. Arkhangel'skii, A. A. Borisyak, E. M. Milanovskii, D. I. Mushketov, M. M. Tetyaev, and N. S. Shatskii actually separated magmatic processes from tectonics. The study of magma did not enter the circle of their interests. The most that volcanic processes have been considered in these schemes has been merely in reference to more or less completely discussed views of Stille, which is like making an official visit to the zone of magma, while leaving one's visiting card in the lobby. And petrology itself has not been sufficiently prepared for this correlation with the development of tectonics. This position has persisted almost to the present day. Disregard of magmatic phenomena, sometimes formal discussion based on ignorance of this branch of geology, may be found in the papers of almost all our tectonic geologists of the forties and fifties.

The paper of A. V. Peive and V. M. Sinitsyn [1950] is the "most magmatic" of those for that period. In his later works V. V. Belousov has also related magma closely to the tectonic process [1962, 1966$_{1,2}$]. This problem arose at the same time among petrologists. The advancing study of rocks and minerals has led to establishment of new objectives for petrologists (in addition to the objectives of studying and classifying rocks): profound study of the origin and characteristics of rocks as geological bodies (that is, a study of magmatic formations and the site of magmatic phenomena in the life of the earth). For this latter, a critical problem is the connection between tectonics and magma. In our country, these new tasks of petrology were first clearly pronounced in the works of F. Y. Loewinson-Lessing and, shortly later, the works of A. N. Zavaritskii and V. N. Lodochnikov. The connection between tectonic and magmatic processes as an independent study, however, came to be discussed somewhat later and more fully in the works of Yu. A. Bilibin [1948, 1955] and then of Yu. A. Kuznetsov (a number of papers in the last fifteen years). With most petrologists as well as tectonic geologists, the problem has been restricted only by the development of geosynclines. For tectonic geologists, as we have already pointed out, this has reduced more or less to a discussion of the views of Stille, whereas petrologists have repeatedly distinguished nongeosynclinal magmatic phenomena, which do not belong in the series with geosynclinal phenomena. Probably the clearest demonstration of this relationship was the work of Tyrrell [1937] on the geologic position of basalts.

At the end of the forties and, especially, in the fifties, a number of petrologist—geologists appeared for whom the tectonics—magma problem became important, if not central. The works of Yu. A. Kuznetsov may be cited especially in this regard. Much was accomplished in seeking a solution to this problem by E. K. Ustiev (chiefly in the discrimination and study of volcanic—plutonic formations and belts) and others. Some aspect of the connection between tectonics and magmatism is found in the works of all, or almost all, petrologists. It has become not only necessary but also "fashionable."

A similar evolution has taken place also in foreign countries. Despite all this, the problem as a whole is still far from solution. On one hand, it long ago outgrew particular questions such as the type of structure granitic bodies are restricted to: faults, anticlines, and so forth. On the other, we still do not see any clear pattern in the fields of physical chemistry, geophysics, or geochemistry on the basis of which we may construct a general picture of the process. With each passing year the question concerning the nature of magma, temperature and pressure conditions during genesis of magma, and the site and time of manifestation of such conditions in the earth becomes ever more urgent, and attempts to solve it are many.

# CHAPTER 2

# WHAT PREVENTS DIRECT COMPARISON OF PETROLOGIC AND TECTONIC DATA?

It seems to the author that this is the first question requiring consideration, since only by knowing where the principal difficulties lie can we continue in search of an answer. Let us therefore examine some of the basic data that we must use in our approach to the problem of correlation between magmatism and tectonics.

The range of questions embracing this problem is undoubtedly very broad. In addition to tectonic and petrologic factors proper, the already mentioned problems of physical chemistry are of no little significance in solution of the major problem, such as the limits of pressure and temperature, as a preliminary approximation, existing in the initial material from which a melt of an observed composition was formed, and such as the conditions under which secondary magmatic rocks might be formed from some primary melt, as well as the conditions under which any particular melt yields the observed differentiation series, and so forth. Such questions are very numerous, especially when we consider the fate of individual minerals. Similar significance belongs also to geophysical data, because it is by no means a matter of indifference what physical conditions exist in the depths of the earth and where they occur, how movement of material takes place in these zones, what state the material exists in, and what its chemical composition may be. If we add to this questions concerning the behavior of individual elements at depth and some other questions on geochemistry, data on planetary astronomy, and other information, the problem appears to be practically insoluble in the near future.

In order to avoid these difficulties, we must first focus our attention on a narrower range of questions and must, above all, single out the central question, the solution of which will permit us to approach the problem as a whole. This question may be formulated in the following manner. With what tectonic processes and, consequently, what structures at depth is the rise of primary magmas associated? In stating this question we first of all abstain from consideration of secondary processes that lead to the differentiation of magmas into a series of rocks with their complex interrelations, and also abstain from all confusing aspects of the origin of hybrid rocks.*

---

*Determinations of primary and secondary magmas are discussed in more detail in Chapter 6. Here we note only that we include in primary magmas those melts that, after separation from their parental material, rise to higher levels without changing in composition or change only by settling out of some minerals. Secondary magmas may be of two types: a) liquids obtained as a result of differentiation of a primary magma in a secondary reservoir (such as during the formation of complex intrusive or extrusive complexes), and b) magmas of hybrid origin

The principal factors will be directly observed phenomena and straightforward conclusions from observations, not explanation of these features, i.e., not a translation to the language of the more common concepts of physics and physical chemistry. In other words, we shall consider the paramount data of geology (in the broadest sense of the word) and geophysics. Of course we can in no way ignore material obtained from experiments. Such data may commonly aid us in determining conditions at depth if we know the geological phenomena originating there. We shall not affirm what geological phenomena originate there on the basis of values postulated for depth however. The proposed statement of the question presents appreciable difficulties. The chief of these is the possibility of direct comparison of tectonic and petrologic data. However astonishing it may be, this question has received but minimal attention heretofore. First of all, we may note that the task facing us presupposes a study of the phenomena originating during the appearance of primary magmas, i.e., whenever and wherever they are generated. But we have never observed these phenomena directly.

We know either the rocks, the products of magmatic activity, or magma that has risen to high levels and lost its volatile constituents, lava. However, as experiments have shown, it is possible on the basis of the resulting knowledge to approach some concept of what may be called primary magmas and their variations. These are without doubt the principal features of the liquids generated at depth, and the confused aspects and controversial questions are undoubtedly secondary. It is thus possible to establish types of magma at depth. Furthermore, thanks to geophysics, it is possible in some cases to establish the most probable depths of generation of these magmas. Physicochemical experiment, if it succeeds, may aid in determining conditions at these depths or, perhaps, may furnish an approach for making such determinations.

Another group of data necessary for solution of the stated problem, practically unknown to us, has received less attention in any case. Modern structural geologists describe structures and movements in the earth's crust, more correctly in their upper part. The entire zone of fold structures as a whole is associated with stratified rocks. Faults reach down to greater depths. But even so-called deep fractures generally die out within the continental crust. In any case, analysis of the geological conditions about these features leads to the indicated conclusions. The disappearance of stratified structure at rather greater depths and the relative increase in the role of fractures in the structure of the lower stage is the objective basis for the current widespread enthusiasm for the view that the crust exhibits block structure.

This position is complicated, however, by the fact that tectonic observations have been restricted to very shallow depths. Chiefly these observations have been of the surface and the first few kilometers of depth, at most the first ten. Structures that form at great depths appear only rarely at the surface, and 20 km is apparently the limit of such depths. Deeper structures appear to be entirely outside the range of direct observation. All deeper structures are described by structural geologists with consideration of geophysical data, which are by no means always clear or correctly interpreted.

An example of the difficulty with which we are faced may be seen in the horizontal zoning of the crust and the presence of discontinuities within it. These are very distinct in seismic

---

or contamination by country rocks, leading to a change in composition of the primary magma. The resulting magma, remaining at the cite of its differentiation (as we shall see later), cannot change in composition if the pressure or temperature do not change (or both together), since it is in equilibrium with the remaining minerals of the parent rock. If conditions change, the composition will change, in order to come again to equilibrium in the solid phase; i.e., this becomes a primary magma as before.

investigations, and they have been studied, and will now be studied again, on the basis of tectonic synthesis. The nature of most interfaces is but poorly known, and in many cases we believe that the zones of discontinuity are unrelated to tectonic structures. Underestimation of this fact and confidence in a direct correlation between the discontinuities and tectonics lead to confidence in the view that tectonic features become simpler with depth, that the horizontal occurrence of beds practically disappears at depths of a few kilometers.

On the whole our tectonic geologists are occupied with a description of the structure of only the uppermost film of the crust. Only in very weakly documented guesses do they refer to the zone where magmas are generated. The helplessness of tectonics in this respect may be underlined most effectively by the appearance of extreme views, at times bordering on the aspect of caricature, when the entire tectonic process is considered to be induced from above, at times by direct interference of universal, so-called "heavenly" forces, at times by the evolution of minerals of solar energy accumulated by exogenetic synthesis. It is easy to see that such hypotheses are brought forth as the reverse side of our lack of knowledge of depth, rather of our inability to comprehend the geology of this zone (tectonics in particular).

From the above discussion it follows that the problem of correlation between magmatism and tectonics comes up against the impossibility of combining the accumulated information of two adjoining geological disciplines.

Petrology and some geophysical and physicochemical data have permitted us to formulate a view concerning magmas that originate at depth. Tectonics offers no clear view concerning these depths. It has been concerned only with phenomena that occur near the surface, mostly on the assumption that the movements involved in these phenomena take place at very shallow depths (such movements as occur in geosynclines, superimposed from above on extremely slow, "flabby" movements of platform type). Modern tectonics is therefore in a position to present some view concerning the distribution of magmatic bodies in the upper crust; i.e., it is prepared to establish in some measure a correlation between these bodies and the structure of the upper crust. But it is unable to establish any correlation between the origin of magmas and this structure (or, what is the same thing, movements in the upper crust), except for confused and inaccurate preliminary outlines. These rough hypotheses have long been known and may be summarized in a few lines:

1. The Stille-Steinmann scheme of geosynclines;
2. The discovery that this scheme is inapplicable to nongeosynclinal zones and that basaltic magmas are predominant within them; and
3. The restriction of alkalic rocks to zones of relative tectonic calm.

Thus, to find a solution or, as it were, an approach to a solution to one of the principal problems of modern geology, it is necessary to find methods of studying the tectonics of deep-seated regions. Only comparison of tectonic data for zones in which magma forms with the character of the derived melts may aid us in solving the problem. If we don't do this, we then inevitably come again to conditions under which our hypothesis is practically unrestricted by our knowledge of the environment in which the process was carried out. It is precisely in this kind of situation that views tinged with the imagination of a visionary artist may appear. Such, for example, are the views of Kraus [1959] concerning convection currents at different depths and of different radii, which have practically no basis in facts. And such is the hypothesis concerning magma arising during friction as one mass slides against another. Even the very argumentative hypothesis concerning asthenoliths, advanced first by Willis and Willis [1941] but modified more than once (in this country by Belousov [1966]), is not free of this defect, since it rests in great measure on desired and not observed tectonics of crustal depths.

   Our first objective is thus to examine in what measure the apparatus of modern sciences of the earth permits us to study the tectonics of deep-seated regions and to discover what we may conclude from these observations.

# CHAPTER 3

# THE POSSIBILITY OF UNDERSTANDING THE TECTONICS OF DEEP-SEATED REGIONS AND A METHOD PROPOSED FOR STUDYING IT

We cannot observe and describe structures below the so-called granitic layer of the crust because even on shield areas the crystalline rocks exposed there, down to the granulite facies, could scarcely occur below the upper 20 km. There is no basis for expecting help in the near future, in this respect, from ultradeep drilling.

There has seemed to be no doubt that visual observations might be rather effectively replaced by geophysical data. Actually, reflecting and refracting surfaces of seismic waves have been represented as being directly associated with changes in the rocks and as necessarily revealing deep tectonic structures. However, investigations of this kind, even the most precise (the method of deep-seismic sounding), do not justify this expectation at this time. As a matter of fact, observed differences in seismic velocity of transmitted waves cannot attest uniquely to the cause of these differences. Apart from a transition from one kind of rock to another, the differences may indicate small variations in degree of metamorphism or in density of rocks of a particular type. As a result, boundaries observed by seismic means by no means always correspond to geologic boundaries, and they are constantly being "smoothed," since the method itself does not permit one to detect small kinks or inflections. With modern methods of investigation we cannot establish structures by seismic methods. It is strongly possible that most frequently we obtain information concerning density changes caused by increase in pressure and only in special cases concerning changes in composition. Such a change has been suggested, with more or less uncertainty, for the Mohorovičić discontinuity. All this discussion refers to almost horizontal boundaries. Detection of vertical boundaries by means of seismic waves is made with difficulty, and thus far it has been possible only for rather shallow depths. Even the sharpest boundaries of this type, faults, are not excepted. It is even more difficult to establish the nature of these boundaries.

On the whole it becomes necessary to recognize that the most promising methods of geophysical investigation (seismic and seismological) are of little aid in comprehending deep-seated structure. At present they are not adequate to replace ordinary visual observations of deep-seated features. Of the remaining geophysical methods, the most powerful for studying structure is the gravimetric method.

The distribution of heavy and light masses at depth permits one in principle to make some approximation in studying structural features. It is well known, however, that such information is extremely vague and at best permits one to distinguish regions exhibiting certain anomalies. An especially weak feature of this method is the great uncertainty in determining depth and form of disturbing masses at depth.

In summing up, it may be stated that geophysical methods of investigating deep-seated zones (if we are not speaking of exploration geophysics at shallow depth or about simple structure as a whole) cannot yet, except in very minor degree, replace direct observation of structure as made in tectonic studies. Moreover, it may be stated that, at least in the near future, not one of these methods will be able to supply anything new in tectonics or to produce a picture of the structure at those depths where magmas are generated.

We have thus established the fact that the modern array of methods for studying the earth (for a solution to our problem) is not suitable. We cannot study deep-seated structures with any degree of completeness. Nor can we discover the distribution of magmatic rocks at depth. Geophysics is inadequate to do this for us. In other words, we are not able to compare structure and magmatic rocks at depth, and the whole problem hangs in the air.

In fact, however, the situation is not so hopeless, and it appears that we may be able to find a way to solve the problem.

Whereas tectonic structure in the lower part of the crust and especially the upper part of the mantle may be vague to us, we cannot say the same regarding recent movements there, i.e., regarding neotectonics at depth. Those movements that show the greatest contrast have been clearly recorded on seismograms and are consequently available for study. We may also observe less active movements, not causing any appreciable seismic disturbance. Such displacements may be evaluated by recent and moderately recent movements at the surface (i.e., phenomena studied in neotectonics and geodesy). It is true that data from these latter cannot serve for direct study of deep-seated regions, but they help us to approach closer to some concept concerning conditions at depth than structures observed at the surface do. Lastly, recent upwellings of magma are known for a number of places on the earth, and other regions have magmatic bodies of the recent past. The principal conditions for an approach to the problem of correlation between magmatism and tectonics are thus satisfied: both types of phenomena occur at the same time geologically and in the same locality (at least there is logical expectation of finding this situation). In other words, it is necessary to attempt a study of the interrelations not of already developed structures and magmatic bodies but of tectonic movements and of still mobile or developing (generating) magmas, i.e., to study these phenomena, as chemists express it, in *status nascendi*. However, we should first test the possibility of this in regard to tectonic observations; the actuality of the occurrence of young magmas is beyond doubt.

The most precise information concerning movements at depth may naturally be obtained from earthquakes. They have three very important characteristics: they may be accurately located; the energy of the process may be measured; and, lastly, the direction of movement may be measured. All other movements at depth, not accompanied by shearing and not generating seismic waves, fail to allow us to obtain such information. They do not make it possible even to determine depth with any precision. Our attention should be turned chiefly to the data of seismology, therefore.

It is not necessary to resort to figures and formulas to understand the difference between movements accompanied by earthquakes and those not accompanied by earthquakes. If the contrast and velocity are low, the probability is less for expecting the movements to create stresses in the rocks in excess of ultimate strength of the rocks, thus leading to the formation of dislocations (slippage) and to more or less instantaneous extinction of the stresses. On the other hand, the greater the speed of movement and the greater the contrast, within some range (these two characteristics generally vary in parallel and interdependently), the greater the probability that, other conditions remaining the same, the stresses with exceed the possibility of creep and the ultimate strength of the rock and that an earthous bewith occur.

From the above discussion it follows that, during identical types of movement, it is generally most likely that shear dislocations will occur where the rocks possess previously formed and unhealed fractures. However, the very appearance of these fractures is associated with the increased brittleness of the rocks, and the long life of an unrenewable fault is associated with low creep. With increase in creep rate the self-healing of fractures increases. Thus, in the final consideration, the number of earthquakes increases with decline in creep, other factors remaining the same. This indicates that earthquakes take place chiefly at shallow depths, in the zone of low pressure. The deeper a point is within the earth, the greater the stress required to produce dislocation; i.e., it is necessary to increase the contrast and rate of movement for this. This pattern is clearly seen at focal depths of about 20 km. At greater depths earthquakes become rare.

We have no grounds for stating that there is a noticeable change in the state of matter from one geographic locality to another. Therefore, as a first approximation, it may be said that, in regions where earthquakes are frequent at a given depth, tectonic forces are more intense than in regions where earthquakes are lacking at that depth. Later we shall see that zones of strong and frequent earthquakes are generally characterized also by somewhat higher temperatures. The limits of viscosity must therefore be greater in zones of tectonic activity than in zones of comparative quiet at the same depth. Consequently, the appearance of shear dislocations in such zones becomes increasingly more difficult, and, correspondingly, the increase in tectonic activity must be still greater.

The data we have discussed permit us to approach a subdivision of tectonic (or neotectonic, if we wish) depth zones. In the future, comparison of observed results with structure of the near-surface layer should strengthen and supplement this scheme.

It is possible that earthquakes at depth do not take place because of shear dislocation but because of spontaneous change in volume during structural change of matter. This idea leads us to state that shear dislocations are generally impossible at these depths (as a result of considerable rise in temperature, perhaps due to the effect of increasing pressure and a corresponding decline in viscosity). However, we may readily conclude that this change in the proposed mechanism of earthquakes changes nothing in our conclusions.

The manifestation of earthquakes as a result of change in the volume of matter is possible only if this change takes place so rapidly that matter cannot adapt itself by overflow. By introducing the concept of this kind of mechanism for deep-focus earthquakes and by considering the fact that dislocations are impossible at these depths, we are even more compelled to recognize that the change in volume takes place in a very short interval of time and to reject the idea of slow rearrangement of internal structure, which might occur during gradual change in conditions (i.e., there must be a rapid change in conditions in the zone of future earthquakes). But these are the conditions of intense tectonic activity, and we cannot imagine such a scheme without corresponding movements.

Thus, without active tectonic movements it is difficult to imagine conditions under which there will be a "lag" in the reconstruction of matter, not going beyond changes in pressure and temperature. Therefore, in case of earthquakes of this type, their presence will indicate the existence of a zone of higher tectonic activity. It should be noted also that the observed pattern of orientation of the stress ellipsoid at foci of deep-focus earthquakes [Balakina et al., 1967; we shall return to this below] leads us to state that shear dislocations represent the cause of these earthquakes. When change in volume is the cause of an earthquake, this proper orientation is unlikely, and special subsidiary hypotheses must be proposed.

CHAPTER 4

# REGIONAL TECTONIC SUBDIVISIONS
# ACCORDING TO DEPTH

## Types of Seismic Regions

In approaching the problem of making a regional tectonic subdivision according to depth (I think it is already clear to the reader that by "depth" we primarily mean the subcrustal zone), we find it possible in practice to use only data for deep-focus earthquakes. The one-sidedness of this information, its obvious incompleteness is the result of inadequate study. To a certain degree the information is random, the result of our lack of information concerning the distribution and change of intensity in frequency of earthquakes of the past. Lastly, we lack knowledge of what quantitative characteristics define in nature delimiting types of regions. It would appear that all these deficiencies compel us to reject any simple regional classification according to intensities of seismic activity. A more fundamental approach would be to distinguish regions according to qualitative change in seismicity with depth. This would essentially lead us to a subdivision of regions according to focal depth of earthquakes. In this case, we distinguish regions according to presence or absence of earthquakes at particular depth intervals. The intervals themselves are determined by composite geologic and, in part, geophysical data.

Thus, in order that our classification not be arbitrary, we must combine data on earthquake distribution according to depth with data that we know concerning the structure and surface of the earth and with the geologic history of the region. As a basis for this conclusion, we assume that the geologic development and the surface structure are directly related to deep-seated development, and similar structures above correspond to definite types of deep-seated processes, that a particular deep-seated process determines, in different parts of the earth, approximately identical structures in the upper structural zone. Of course, we cannot extend this view in regard to detailed developments of structure, nor for phenomena of clearly exogenetic origin. It is true that many of the latter are strongly depth-dependent, although indirectly (such as the formation of sedimentary rocks).

The adopted position is hypothetical, since we still know too little concerning occurrences at depth. However, direct knowledge compels us to adopt this view since it reduces to a postulate that "identical causes give rise to identical results, and different causes give rise to different results." It is the simplest of hypotheses. If it appears, in fact, that a solution may arise in which identical near-surface structures are genetically heterogeneous, we must introduce a proper correction into the scheme. This complication must first be proved, however. We must not forget that it is easy to adopt a position in which it is precisely the inadequacy of our knowledge concerning deep-seated phenomena that compels us to believe that the various near-surface structures have formed under a single set of deep-seated conditions. Therefore, before we adopt this view, we must check whether the similarity of deep-seated processes is a result merely of a scarcity of information. We must be just as careful in

11

maintaining that different deep-seated phenomena create like structures above. If a difference in deep-seated environment is obvious, it is necessary to make another check of this history and type of near-surface structures.

Thus, on the basis of two groups of phenomena, apparently in some degree independent of each other, we find it possible to construct a scheme for subdividing tectonic regions according to depth that will actually approximate nature and may aid us in studying natural phenomena.

We assume it proper to distinguish the following three groups of seismic regions in depth zones:

1. Regions where shear dislocations are not observed either at depth or at the surface. Such regions are characterized by very low contrast and, mostly, low rates of movement (although small displacement is by no means an associated feature). Movements in one direction and at approximately a uniform rate embrace large areas. As a result there is complete lack of expression at depth, and quiet conditions prevail even at the surface. This type occupies the greater part of the earth's surface: platforms, oceanic regions (except for midocean ridges), and, to a considerable extent, regions of culminated folding.

2. Regions where numerous shear dislocations take place in the upper part of the crust (in continental regions), throughout the entire thickness of the crust, and even in the upper part of the mantle (in ocean basins). Earthquakes are practically absent at great depths. Movements are contrasting and, on the whole, stronger than in group 1. The absence of deep-focus earthquakes, however, indicates that either the contrast of movement is insufficient to lead to earthquakes in zones where creep is intensified, or that the intensity (including contrast) of movement declines with depth. This is clearly a heterogeneous group. It includes (if we consider near-surface zones according to developmental history) regions of different origin, such as zones of accelerated activity, Alpine fold zones, zones of recent subsidence, zones of "continuous-structure faults," large grabens, and midocean ridges. Correspondingly, when working out a subdivision scheme, group 2 must be divided up.

3. Regions where shear dislocations and very strong earthquakes occur in the near-surface zone, and also earthquakes at great depths, some with foci hundreds of kilometers deep. Deep-focus earthquakes indicate great contrast (and intensity) of movement. These conditions are preserved to depths of hundreds of kilometers, depths many times those in other zones. Approximately half the thickness of the upper mantle is seismic here. This group includes island arcs and some parts of young fold belts (Burma, Kamchatka, and elsewhere).

## Comparison of Seismic Regions with Tectonic Zones

In principle it is not difficult to compare the three groups of seismic regions distinguished above with what we know of the tectonics of the surface layer. However, it is first necessary to rise above the extreme detailed character of systematic tectonics, which is clearly excessive from the viewpoint of possible study of deep-seated phenomena. It is easy to see that for depth subdivision of regions there is no fundamental difference between regions of group 1 and many regions of group 2. The entire difference reduces to the presence or absence of seismic activity at shallow depths. At depths as shallow as 30-40 km all these regions behave alike. Therefore, as a first approximation when studying deep-seated tectonics, we must make no distinctions between the two groups. In turning to tectonic data we must correspondingly refer platforms and zones of culminated folding to a single group (we shall except only the Alpides). The Alpides and other zones typical of group 1 are practically aseismic. However, we must also refer zones of recent activation to this group. We must note only that this term does not also include such features as the African rift or recent subsidence in the region of the Mediterranean Sea or in the Sea of Japan. Apparently the regional significance of the term tectonic activation involves renewal of appreciable movements in ancient structures that may be said to have become consolidated. Activation involves both folded zones and platforms. It may lead to structure on a new plan, but most commonly following to some degree the trends or lines of the earlier plan. We must distinguish from this type of tectonics relief-producing movements in young fold zones. This type of movement is most commonly reflected directly in the epoch of landscape development at the end of the folding, or it belongs to the landscape development itself. In either case, the movements represent the last stages of geosynclinal transformation in the folded region.

From phenomena of activation it is also necessary to distinguish newly formed oceans and geosynclinal subsidence, extending the zone of geosynclinal segments preserved during the epoch of folding. An example of this is recent subsidence in the Mediterranean Sea.

We may thus describe three clearly expressed types of zones of deep-seated tectonics.

1. Zones characterized by the absence or rarity of earthquake foci below the near-surface "seismic" layer, i.e., below approximately 20 km. It must be assumed that movements at depth show little contrast and are slow. In only the near-surface layer may the marked decline in rock creep and the increase in brittle behavior yield zones, where the depths are somewhat more active, of appreciable or strong seismicity. At the surface all these zones correspond primarily to platforms, zones of completed folding, or activation zones. It is possible that we should also refer to this group, zones of subsidence that lead to the formation of ocean basins with shores of Atlantic type.

2. Zones corresponding to part of group 2. These are not yet reliably differentiated. To them we refer regions confined chiefly to the central crests of midocean ridges and to other especially active fracture zones (the Hawaiian rise?). We still know too little concerning the nature of these regions. Only further accumulation of data will therefore allow us to make a more fundamental characterization of this second type of deep-seated tectonics. There is no doubt that continuations of midocean ridges upon continents should be referred to this same type. This is not entirely clear for the Baikal system of grabens, however, although it is similar to the African rift zone. The absence of rather deep earthquake foci within it and its isolation from midocean ridges compel us to exercise care. It must be emphasized that in this second type of zone the stress of recent movements is insufficient for the manifestation of shear dislocations in zones with high values of creep. Therefore, if further studies do not furnish fundamentally new data we have grounds for combining the first two types and for setting them up in contrast, in some measure, to the third.*

3. Zones of deep-seated tectonics, for which deep-focus earthquakes are characteristic. A comparison of these zones with the tectonic character of the upper stage shows that this type cannot be limited only to some island arcs and the region of the Andes. Along with zones where shocks are common at depths of 500-700 km, regions are known where earthquake foci occur at appreciably shallower depths. As already pointed out repeatedly, it is possible to find a practically continuous series of earthquakes according to depth, from 700 km deep and more under the Indonesian arc to approximately 100 km deep under the South Sandwich arc. It is fundamental to note here that in all cases these regions correspond at the surface to island arcs; i.e., they are of a single type tectonically.

We must thus refer zones under which seismic activity may be traced to depths exceeding 100 km (approximately) to the third type of deep-seated tectonic region. Their near-surface

_____
*Recent data on midocean ridges lead us to believe these are completely separate structures with very large energies at depth.

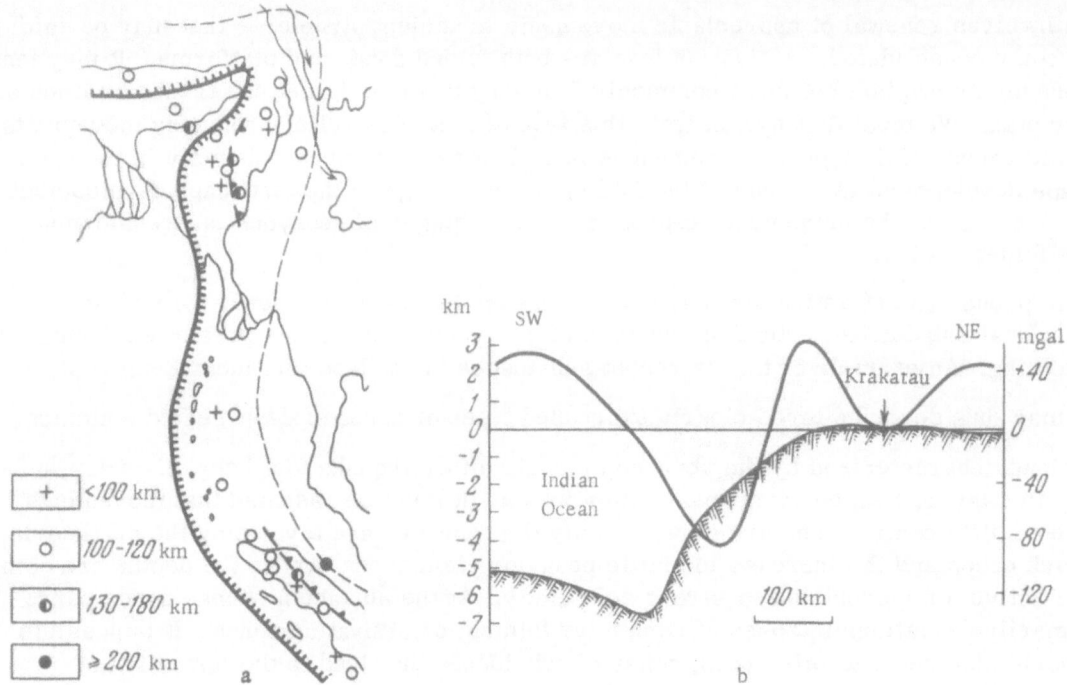

Fig. 1. a) Transition of the Indonesian arc into the fold belt of Burma: the outer mar-
gin of the arc, the western boundary of deep-focus earthquakes, and focal depths. b)
Isostatic anomalies of the region (after Artem'ev): profile from Krakatau into the
Indian Ocean.

characteristics must also be somewhat expanded. It is well known that some island arcs gradually change along the strike to young fold belts. A number of such transitions are accompanied
by decline in depth of seismic activity. The transition of the Indonesian arc into the Burma
fold belt (Fig. 1) may be the best example of this type.

It is also indicative that there are regions difficult to refer to young fold zones or to
island arcs, since these zones are transitional. Such are the southern part of the Appenine –
Sicilian and the Cretan arcs (Figs. 2 and 3). Both of these are characterized by deep earthquake foci. The third type of deep-seated region must therefore include both island arcs and,
even if in part, young fold ranges.

It is necessary to keep in mind, however, that zones of deep-focus earthquakes are
found also beneath some Alpine fold ranges. It is true that the depths of these foci are not
very great, but they are greater than 100 km and are fully comparable to the deep foci under
the Antillean arc or the South Sandwich arc and are even more like those under the Burma fold
belt. There are no grounds for separating these zones from each other. The principal difference here is not the focal depth of the earthquakes but the fact that the zone of deep foci is
commonly concentrated in a very small district. This is the picture for the Carpathians and,
in part, for Hindukush, although the zone of epicenters for the latter extends along the range.
It should be noted that even for some island arcs similar "point-like" foci are observed. In
this case, however, the foci are at very great depths. For example, a focus in the Tyrrhenian
Sea lies at a depth greater than 450 km, and with this are associated shallower earthquakes of
the Southern Italian arc. These relations were described by Petterschmitt [1956] and Blot
[1964]. Somewhat similar relations have been described also for the New Hebrides and Tonga
arcs in the Pacific Ocean. There, a similar focus occurs in the vicinity of Fiji at a depth of about
600 km.

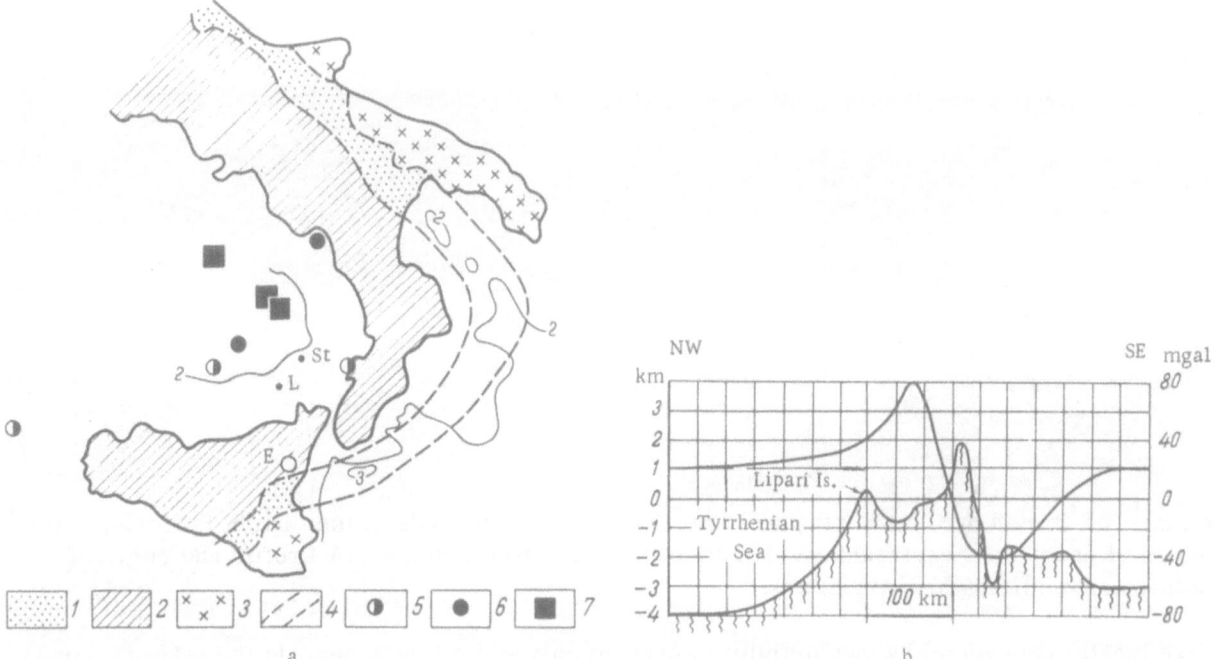

Fig. 2. Southern Appenine—Sicilian arc. a) Earthquake foci (isobaths in thousands of meters: St, Stromboli; L, Lipari; E, Etna). 1) Foredeep; 2) fold arc; 3) frame of fold zone; 4) probable continuation of foredeep into the sea; 5-7) epicenters of deep-focus earthquakes: 5) 100-200 km; 6) about 200 km; 7) 250-300 km. Boundaries on land after Wunderlich (Tectonophysics, I, 6, 1965). b) Profile and curve of isostatic anomalies (after Artem'ev).

It should be noted that the difference between foci of the Carpathian type and the latter two is apparently very substantial. It is a matter not only of depth but also of the geometric pattern of earthquake distribution. Beneath the Carpathians we have to do with a single and very small zone (Vrancha earthquakes). All these shear dislocations take place at approximately a single depth, though they probably give birth to secondary earthquakes at shallower depths above this focal zone. In southern Italy, the New Hebrides, and Tonga (according to Blot), and in some other earthquake zones, earthquakes with deep "point" foci are accompanied by a series of later and shallower quakes. Attention should be drawn to the fact that here the earthquakes appear to be located, as it were, on a single conical surface (Fig. 4). If the characteristic noted by Blot is confirmed, it would be of interest to ascertain to what extent it is universal and to check that we are not dealing with extended fronts of deep-focus earthquakes along island arcs as a result of a summed series of similar "cones."

The above discussion should underline the profound difference between earthquake zones beneath island arcs and foci of the Vrancha type. But the presence of transitions from island arcs to fold belts, the decrease in focal depth in this transition, and the undoubted absence of a "cone" or "fan" in the second case furnish ample grounds for rejecting the view that the phenomena occurring beneath island arcs and beneath the Carpathians differ in nature. We undoubtedly have to do with different phases of a single phenomenon. At one end of the series we find young island arcs; at the other, the Carpathians. All intermediate stages may be noted, as it is possible to observe gradual transitions from the tectonics of island arcs to the tectonics of fully completed fold structures. However we treat these transitions, there is no doubt that they are associated with gradual weakening of seismicity with depth. Whereas in an island arc, such as the Kurile—Kamchatka, the deep zone beneath it is active throughout its

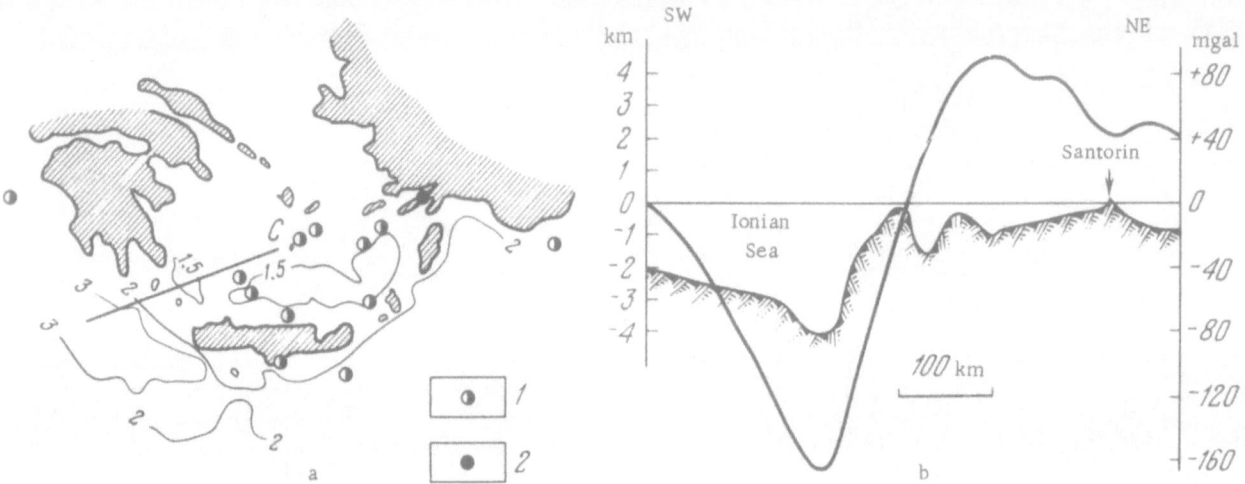

Fig. 3. The Cretan arc. a) Earthquake foci (isobaths in thousands of meters; S, Santorin); epicenters of deep-focus earthquakes: 1) 130–180 km; 2) about 200 km. b) Profile and curve of isostatic anomalies (after Artem'ev).

entire length; beneath fold zones activity is present only in limited zones. In the extreme form (the Carpathians), earthquakes take place in only one small district. This picture is most easily represented as a result of the fact that activity at depth is so weak that earthquakes occur only where a potential seismic zone at depth intersects a zone of unhealed fracture.

Where activity at depth proves to be still lower, all manifestations of deep-seated seismic effects fall off entirely, and we have to do with ordinary fold zones, such as the Caucasus, the Northern Carpathians, the Alps, and other such ranges. It is entirely inadmissible to assume that the absence of earthquakes at depth in this case may be grounds for classifying such zones as of a different type. In this case the leading possibility proves to be a tectonic relationship (which, generally speaking, should hardly surprise us). Moreover we have weighty reasons for believing that older fold zones are similar in origin to Alpine structures; they also were at one time tectonically active at depth. But this is also a problem of a different order, and we shall return to it later, as well as to a discussion of whether the island arc–fold zone series is the result of varying intensity of deep-seated processes or the result of stages of a single process differing in time.

Thus, we have decided to include in the third type of deep-seated regions those zones in which earthquakes at depths of hundreds of kilometers (and as shallow as 100km) normally occur beneath island arcs. We also include the continuations of such zones in the form of young fold chains, in part being accompanied by earthquakes at depth, in part being free of such earthquakes.

In the following chapter we shall attempt to analyze how we obtain data on direction of movement and on the orientation of the stress ellipsoid for deep-focus earthquakes.

For the moment let us note that it is scarcely possible to consider the indicated subdivision into three types as exhaustive. This is the crudest of classifications, the only one possible in the initial stage of tectonic subdivision of deep-seated regions. When we learn to read in recent tectonic movements not only their surface characteristics but also the depth at which they originate, and perhaps also their aspects at depth, then we may begin to add details to the scheme. We are not justified yet in attempting such detailed classification.

Furthermore, some types of manifestation of deep-seated tectonic activity may be lacking today, and this does not permit us to introduce such types into our scheme. With this kind of phenomenon, however, our conclusions must be drawn with extra care, since even the certain identification of these phenomena on the surface of the earth is very difficult, and the interpretation of their connection with depth processes is almost impossible. Of the type of phenomenon established beyond doubt but not manifested today, special interest must be aroused by the formation of basalt (trap) fields. But we cannot distinguish them as a fourth type of region, firstly because there are no real grounds for separating them from other basaltic zones (oceanic) and secondly because we have absolutely no knowledge of the deep environment beneath them during the period of their formation. The only present-day analogs — Iceland and Hawaii — cannot serve as an example, the first because it is not clear here where the area of continental type of vulcanism ends (traps) and where phenomena of the midocean ridge begins, the second because it is generally associated entirely with the ocean and it is difficult to draw analogies. We shall thus content ourselves with three types, and for brevity we shall label them:

1. Extrageosynclinal (a very tentative label),
2. Midocean ridges,
3. Geosynclinal (since it is directly associated either with still-existing geosynclines or with fold zones recently developed from them).

## Concerning Isostatic Anomalies

We may find basic support for the views just formulated in the distributional pattern of isostatic gravity anomalies. A recent book by Artem'ev [1966] was devoted to this question. Isostatic anomalies show a picture of now-completed movements, and they are therefore valuable for our purpose. Artem'ev in his book arrived at the following conclusions:

1. There are two types of anomaly fields: one, geosynclinal, corresponds both to island arcs and to young fold belts; the second, termed the "platform" field, lies outside zones of geosynclines and young fold belts. The first is characterized by elongated outlines; the second by more or less equant forms.

2. Within the region of an island arc, positive anomalies predominate on the back side. The belt of largest negative anomalies extends into the lower part of the slope of the deepwater trench. A belt of positive anomalies appears again on the outer slope of the trench and in the adjacent part of the ocean basin. The greatest change in anomalies per unit distance across the trend (maximum gradient) is found on the slope of the arc facing the basin.

3. The anomalies have the same character in young fold mountain ranges: the belt of positive anomalies is confined to the inner slope; near the crest of the range lies the zero anomaly line; the outer slope is the zone of highest gradient; and the greatest negative anomalies occur in the lower part of the inner slope of the foredeep.

4. A second belt of negative anomalies (lower values) is found beyond the zone of maximums and farther than it from the crest of the range or the outer island arc. We find a volcanic belt in this zone of minimums; in island arcs it forms a volcanic arc. Earthquake foci under this zone lie at depths of about 100-200 km.

5. A comparison of the anomalies of island arcs and those of young fold zones shows a great similarity. Secondary differences are most likely due to the fact that these two types of structure correspond to different stages in the development of a single structural type. Transitions may be traced along the trend from island-arc anomalies to anomalies of the fold zone.

We may note that the isostatic-anomaly curve along cross-profiles is clearly preserved even for fully culminated fold zones, but the amplitudes of the anomalies here may be less and the curves gentler. For example, in the Alps and Carpathians anomalies in the foredeeps are small, almost zero or slightly negative, but positive anomalies over the ranges (over the inner slope) are appreciable. This is attested to by weakened subsidence in the deep and by continued elevation of the range. For us, this analogy with island-arc anomalies is fundamental in the sense that it confirms our view concerning the existence of a series: from island arc through final formation of a fold zone to completed folding. During the second, intermediate, stage, there still exist individual deep foci; these are missing in the last stage.

The distribution of isostatic anomalies thus confirms our view concerning the existence of a geosynclinal type of deep-seated tectonics. Artem'ev [1966, pp. 27-29, 43-48, 71-75, and 85-90] came to a very similar conclusion (the existence of geosynclinal and "platform" isostatic anomalies, and the existence of a single series from island arc to fold zone) independently of the author of the present work and, to a considerable degree, on the basis of other data.

Judging from data obtained in the American Cordilleras, the same distribution of anomalies may be observed in a very weak form in the Mesozoic structures, but this is hardly true for all Mesozoic structures.

We still have too few data on isostatic anomalies to form a proper judgment concerning midocean ridges. This type of deep-seated neotectonics cannot yet be verified gravimetrically.

CHAPTER 5

# ORIENTATION OF STRESSES AND THE GEOMETRY
# OF DEEP-FOCUS EARTHQUAKES

It is well known that the overwhelming number of earthquakes take place as a result of slip (more accurately shear in the sense of mechanics but by no means in the tectonic sense). In this process, slippage at the focus frees accumulated energy and leads to dissemination of this energy. The dissemination of energy must perform work in overcoming forces preventing movement, chiefly friction along the walls of fractures and cohesion of particles not yet broken by rock fractures.

Many hundreds of earthquakes have now been studied from the viewpoint of direction of the effective stresses. The collected data are therefore sufficiently representative, and we may draw some conclusions, even if preliminary, on the basis of these. Unfortunately, the number of deep-focus earthquakes studied from this point of view is not yet large, and we cannot consider them separately from shallow quakes. A comparison was made for some regions however, permitting us to discover in what measure the orientation of stresses at depth corresponds to surface orientations. For our purpose we used the recent summary on the study of stresses at earthquake foci [Balakina, Vvedenskaya, Misharina, and Shirokova, 1967]. Because of the indicated insufficiency of data on deep-focus earthquakes, we shall describe briefly the pattern established by the authors for the entire number of earthquakes. As a rule, rather strong shocks were studied, with magnitude of 6 or greater. Only for the Caucasus, the Arctic region and the Atlantic Ocean, and the Baikal—Mongolia region were shocks of lower magnitude considered (down to approximately 5). The above-named authors noted that the number of earthquakes studied from our point of view amounts to about 49% of all those recorded during the period chosen by them for shocks of appropriate magnitude (the period for the different regions varied, ranging from 3 to 13 years). The number of studied earthquakes amounted to only 20% of those recorded for the Arctic—Atlantic Ocean region (thus, the percentage amounted to 65% for the other regions). Furthermore, these data, obtained in the USSR, are in agreement with results obtained by other investigators for most foreign seismic regions.

The designated seismic belts of the world are shown in Fig. 5. If we approached this question from the geologic viewpoint, we will make an entirely different subdivision than that made by seismologists. Instead of (I) the Circum-Pacific belt, (II) the Eurasian belt (from the Mediterranean Sea to Asia, extending through Burma into Indonesia, and with a branch from Central Asia into the region of the Baikal grabens), (III) the Atlantic—Arctic belt (including the Verkhoyansk zone and the Cherskogo Range), and (IV) the East African belt, the following would be considered (the figures in parentheses correspond to arabic numerals in Fig. 5):

a. The Circum-Pacific belt (3),
b. The Asiatic belt (from the Black Sea to Indonesia), fusing on the east with the Circum-Pacific belt (1),
c. The Mediterranean zone (4),

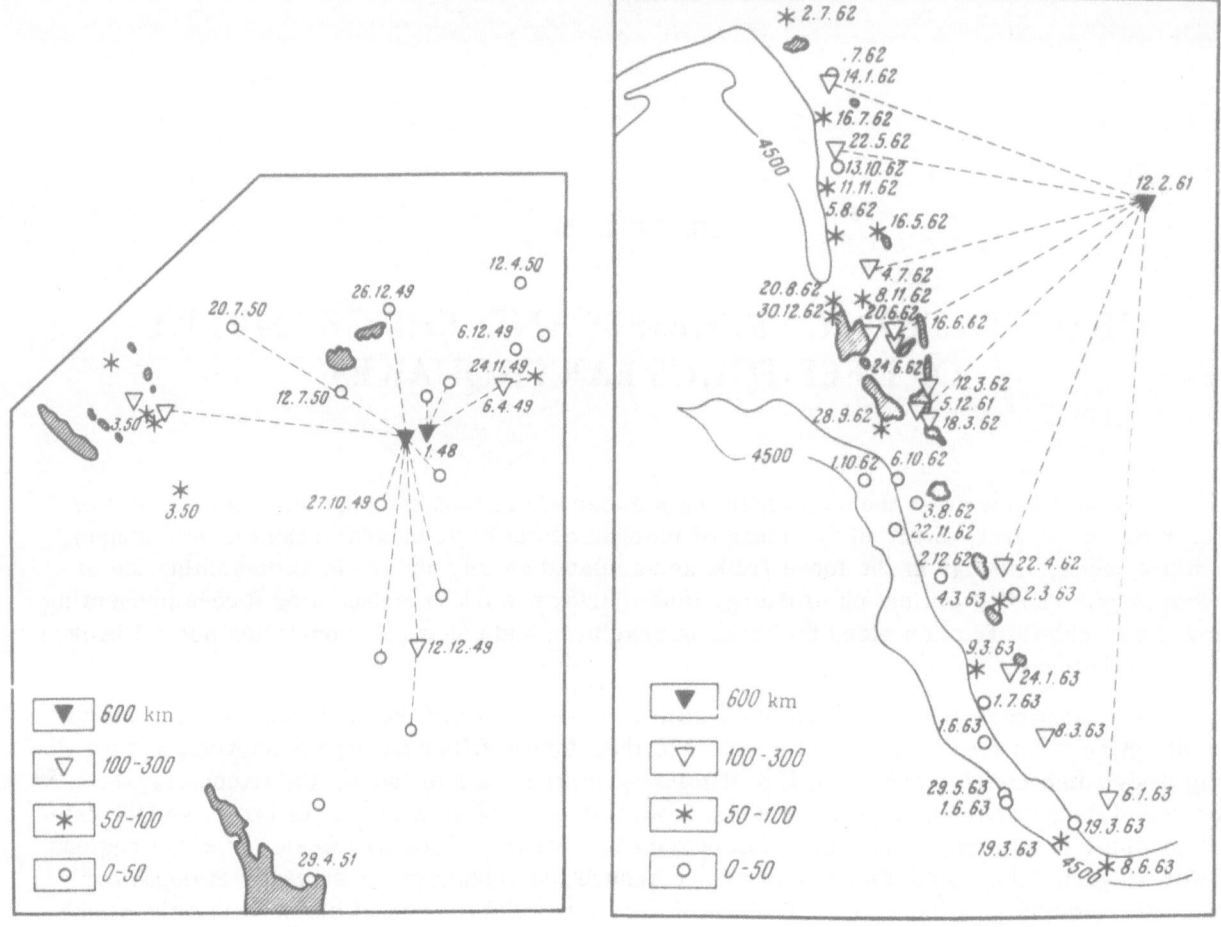

a                                                                    b

Fig. 4. Distribution of earthquake foci on a conical surface with apex at the depth of focus (after Blot; focal depths are shown, and the dates of corresponding earthquakes are given ). a) Zone of the deep-focus Fijian earthquakes; b) the same for the New Hebrides. (Dates on figure are given as day-month-year.

 d.  The Central Asiatic zone (2),
 e.  The Atlantic—Verkhoyansk zones (5 and 5a),
 f.  The East African belt (6), and
 g.  The Baikalian zone (7).

The basis for making this subdivision will be clear from the discussion below.

A very important and entirely indisputable conclusion may be drawn concerning the systematic pattern of orientation in earthquakes: in each region most of the earthquakes exhibit their greatest stress near the horizontal (or slightly inclined), oriented across the trend of the structure. For some regions this stress is compressional; for others, tensional.

For most of the regions we may mark out a more or less broad belt that coincides with the structural trend of that region: the Pacific Ocean (3), Asiatic (1), Atlantic—Verkhoyansk (5-5a), East African (6), and Baikalian (7). In contrast to these, some regions exhibit a great diversity of directions, corresponding to the complex structural pattern. The direction for

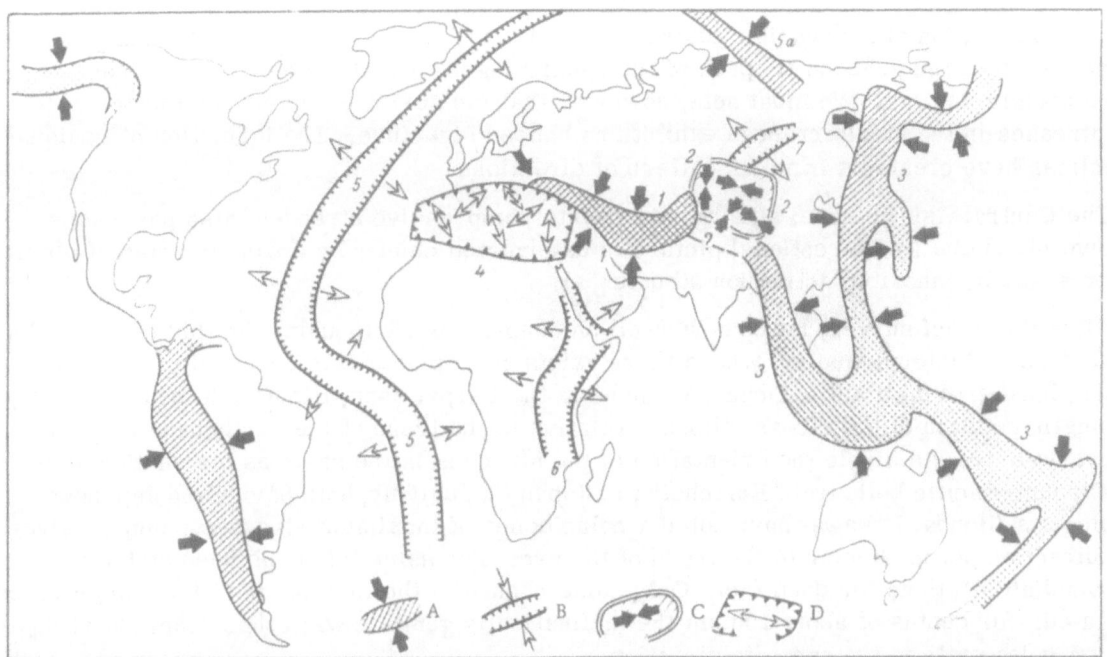

Fig. 5. Orientation of the principal stresses at earthquake foci (after Balakina et al. [1967]). Seismic belts and zones: 1) Asiatic; 2) Central Asiatic zone; 3) Pacific Ocean; 4) Mediterranean zone; 5) Atlantic—Verkhoyansk; 6) East African; 7) Baikalian zone. Principal stresses: A) compression, across the belt; B) tension, across the belt; C) compression, variously oriented; D) tension, variously oriented.

each segment of such a zone so far fails to coincide with the trend of neighboring belts that, without considerable stretching of the facts, it is impossible to consider this segment a component part or a continuation of any neighboring belt. There are two such zones among those outlined on Fig. 5: the Central Asiatic (2) and the Mediterranean (4).

Compressional stresses are confined to the framework of the Pacific Ocean, the Asiatic belt, the Central Asiatic zone, and the Verkhoyansk segment. All these are young fold zones, activated older fold zones, and present-day geosynclines (in the stage of initial uplift and folding).

The better known zones of tensional stresses are the zone of the Atlantic—Arctic mid-ocean ridge, East Africa, the Baikal region, and the Mediterranean.

Even if we call to mind that on Fig. 5 the seismic zones of the Mid-Indian Ridge and the Eastern Pacific Rise have been left off entirely, we still gain a rather clear picture: compression across the structural trend is characteristic for geosynclinal zones and fold zones, but tensional forces characterize the great grabens or rifts. Against this background the Mediterranean zone arouses considerable interest. A typical zone of young folding should have been characterized by maximal compressive stresses. This does not appear. It seems to us that the following may be an explanation: the European Alpine structure has completed its development, and only within the Calabrian, Cretan, and Carpathian arcs does the process continue in a weak form. In such segments, compressive stresses should be expected. Throughout this structure as a whole, a sequential process is impressed: at first a fresh subsidence, followed by the formation of new, residual geosynclines, the so-called "Late Quaternary" stage (cycle). And this process is bound to a different geometry of stresses in the upper stage than the process of converting the geosynclines to a fold structure. The process in the Mediterranean is, in part, the formation of new basins, future geosynclinal troughs, in some measure similar to the formation of large grabens (I speak of the upper stage). The distribution of stresses is correspondingly similar. We must note, however, that the series with a predominance of tensional stresses in the Mediterranean exhibits no banded structure. The formation of residual geosynclines here creates a complex pattern of directions.

The Central Asiatic zone, although it exhibits compressive stresses, also possesses multiform structure (the directional picture is determined chiefly by a complex plan of older structures and by inherited activation structures).

After this brief survey, let us now dwell on deep-focus earthquakes (in this case we shall combine deep and intermediate). Where there exists appropriate data concerning them (the northwest border of the Pacific Ocean, Vrancha in the Carpathians, Hindukush, and Indonesia) a systematic position of the stress ellipsoid relative to the trend of the earthquake zone is noted everywhere. As a rule the orientation of the stresses is the same as for shallow foci. In the Circum-Pacific belt, from Kamchatka to Japan, inclusively, both Soviet and Japanese seismologists (Honda, Isikawa) have noted a coincidence of maximum stresses (compressive) with a direction perpendicular to the trend of the arc. The same thing has been stated for foci of intermediate depth as for deep foci. Only some change in the inclination of the compression is indicated: for depths of about 100 km the inclination is gentle toward the ocean, but at depths of 300 km it is gentle in the opposite direction.

For earthquakes of intermediate depth in the Hindukush region, compression (again the greatest stress) takes place at right angles to the structure; the slope of the line is appreciable. In the Vrancha zone, the stresses in deep-focus earthquakes are oriented the same as for shallow quakes (compression at right angles to the arc and approximately horizontal).

Something of an exception is found in the deep-focus earthquakes of Indonesia. For near-surface earthquakes here, as everywhere in geosynclines and fold zones, stresses are chiefly compressive and are oriented approximately in a horizontal direction at right angles to the arc. The orientation for the deep-focus earthquakes is different, although it is also systematic. According to Ritsema and Weldkamp, the principal stress here proves to be tensional, also directed at right angles to the trend and along the horizontal. In this zone dominant compression at the top thus appears to give way to tension at depth.

From the above discussion we may make the following conclusions: a) the general rule is almost complete correspondence in orientation and nature of the principal stress for shallow and deep-focus earthquakes in a particular zone (but this is only the rule); b) the law (to the degree with which it confirms existing data) is uniform orientation of stresses relative to the trend of the principal structure at deep foci. It seems unlikely that such conditions could be caused by stresses manifested during adaptation of material to new conditions of pressure and temperature, i.e., during restructuring of lattices. It is perfectly natural to expect the relations, as a consequence of the mechanism giving rise to earthquakes, that exist at shallow depths.

# CHAPTER 6

# PRIMARY MAGMAS

We must recognize the fact that huge masses of acidic rocks cannot be separated directly from basic melts. Their appearance requires special conditions, a special productive melt of substances, and even the most extreme proponents of metasomatic granites must admit that the process is accompanied by the outpouring of acidic lavas. Apparently the problem of ultrabasic magma may be solved with this same degree of accuracy. And here geologic data indicate practical independence of this magma from basaltic magma, although there exists a genetic deep-seated connection in this case. Controversy over the primary character of ultrabasic magma lasted a long time, but it has now died down.

There can be hardly any doubt that particular types of magma differ in composition at different places and different times and that the three principal varieties indicated above by no means exhaust all possibilities. It is thus shown that there exists a "hierarchy," and that acidic, basic, and ultrabasic magmas correspond to certain types of magma that combine numerous kinds of magma, differing from each other, as well as to actual independent magmas.

This view is the direct consequence of our failure to find in the earth any huge and common magmatic reservoir, attesting to a long past stage of a molten and still uncovered earth crust. Establishment of this fact became at the same time one of the reasons for rejecting the Laplace hypothesis and a cause of the fundamental change in views concerning magma and its origin.

The absence of a constantly existing melt in the solid earth (except for the liquid outer core, the role of which is negligible for magmatic geology at best) points to the undoubted generation of magma from solid matter. And this, in turn, makes it possible to maintain that melts of different composition may be produced. Their classification is already a matter of concrete study.

Before we begin a discussion of differences in primary magmas, let us attempt to define, even if in only a few words, what we mean by the term primary magma.

Any magma may travel a complex path and may change its composition markedly during this time, and the boundary between primary and secondary magmas may prove to be tentative. We may consider the following division a natural one, however. Magmatic fluid that appears at depth may be mixed with solid phases (the result of partial fusion of matter). It forms a magmatic mush which can hardly be considered magma in the direct sense of the word. This mush will be more or less mobile, depending on the ratio of solid to liquid. When the liquid content is rather large, a distinctive deep-seated "quicksand" appears, easily displaced by relatively small forces. In any form the magmatic mush is free to move (particularly to form intrusions) if the pressure gradient is sufficiently large. We have no grounds for denying that this mush is a magma, but it should be distinguished from magma proper (liquid).

At the place where the liquid is generated, or at some distance from it, separation of the liquid from the solid phase usually takes place (we should note that from this time, if we have to do with a closed system, and up to the time when crystals begin to precipitate out, the magma ceases to change in composition, since reaction between it and the residual solid phase has stopped). We have to do with primary magma proper from the moment the liquid is separated out.

During the rise of the liquid it comes under new temperature and pressure conditions, and it may discard some of its component parts. Crystals may separate out, and the magma may change in composition. If this takes place at depth, however, in the zone where the magma originated, or near it, it is scarcely proper to consider this secondary magma. For basic magmas this depth is greater than 30-40 km (i.e., within the mantle under continents and near the upper edge of the low-velocity channel under the oceans). For acid magmas this depth is that of the place where they are generated and the zone of migmatite development. This type of magma may be termed primary derivative.

Thus, among primary magmas, we shall distinguish primary magmas proper, primary magmatic mush, and derivative primary magmas.

It will be convenient to call those magmas secondary that 1) form as a result of differentiation in secondary magma chambers (most commonly near the surface or at a place where magmatic bodies of complex composition form) or that 2) form as a result of contamination of country rock or as a result of hybridization (mixing of several melts).

The view that there are different primary magmas was arrived at by two independent paths: the first followed from geophysical investigations, the second from considerations of the geologic environment. The last allows us to distinguish three types of magma and to derive individual magmatic varieties from them. An astonishing feature comes to light here: on the whole the appearance of magmas of any particular type is not connected with the tectonic peculiarities of the region. Even such fundamental differences between continent and ocean would appear to be feeble. The differences in some measure point to the composition of the magma, but any magmatic difference is, so to speak, of a second order: different contents of secondary elements (Ti, K) and not of principal elements. The differences in magmatic composition have no effect on determination of the magmatic type. Only silicic magmas are excluded. They are clearly associated with the continental crust or some intermediate crust, nowhere appearing in regions of oceanic crust.

The separation into geosynclinal and extrageosynclinal regions also has no effect on the appearance of magma of any particular type. Again acidic magmas are excluded. Their special place in all cases is determined by their generation only within the crust and by their low temperature and pressure conditions. Consequently, the appearance of any particular type of magma is determined primarily not by the tectonic conditions but by the composition of the initial material and by the physicochemical conditions. For our purposes, therefore, the distribution of different types of magmas across the face of the earth signifies very little. We must view the distribution of magmas of different varieties within major types. While not striving for great detail, we may still note the following subdivision of magmatic types (Table 1):

Magmas of alkalic (chiefly silicic) complexes are not mentioned in Table 1 since none of them can be considered primary. This type of magma is always derivative, the result of development of primary melts at times of tectonic calm. A difficulty here is that the process of alkali enrichment is undoubtedly gradual and cannot lead to a clear boundary between "normal" and alkalic complexes.

TABLE 1

| Acidic magmas<br>Granitic | Basic magmas<br>Basalt-tholeiitic | Ultrabasic magmas<br>Periodic (Alpine<br>type of ultra-basics) |
|---|---|---|
| Intermediate acidic<br>(granodioritic) | Olivine-basaltic<br>Andesite-basaltic<br>to andesitic | Alkalic-ultrabasic |

We will not touch here on the question of the oldest granites ("Archean"). We do not know the conditions that existed at the time of their formation. If we wish to form some judgment concerning the nature of their magmas, therefore, we are forced to vague uncertainty, whereby we must believe (touching on magmas of the recent past) that so many groundless hypotheses will be introduced that consideration of the problem becomes impossible or that consideration of it would require a great amount of special labor. The author is of the view that the oldest granites were most likely formed in approximately the same manner as younger granites, and that their direct separation from basalt (i.e., juvenile origin) is highly improbable.

Some uncertainty arises in the classification itself. This uncertainty comes from the impossibility of "verifying the primary quality" of all the indicated magmas. Some of them raise no doubt in this respect. The appearance of soda-granite and granite melts from solid matter of the crust is beyond doubt, and these may be observed in deep sections of the crust. There also appears to be no doubt concerning the primary character of tholeiite, andesite-basalt, and peridotite magmas (we shall say more about the last in Chapter 10). The problem of olivine-basalt and alkalic-ultrabasic magmas is more complex. These are characterized by high alkali content, and, consequently, they may be considered a first stage in the formation of alkalic melts proper, i.e., a result of alteration of primary magmas under certain physico-chemical conditions. Both these magmas deserve special attention, therefore. We must also consider andesite-basalt magmas on the basis of other considerations, however.

Let us examine some controversial questions arising during the classification of magmas:

1. The nature of alkalic-ultrabasic magma has been the subject of much controversy, and the controversy continues to the present day. The present writer has considered it an independent type of primary magma [Sheinmann, 1955; in a report of the Mineralogical Society at the beginning of 1949]. Its primary character and independence have raised no doubts. The problem will be considered in more detail in Chapter 9.

2. Tholeiite and olivine-basalt magmas. Until very recently perfectly good grounds were found for believing in the existence of two independently forming basaltic magmas. One of these, the tholeiitic, is especially characteristic of geosynclines and is widespread outside geosynclines on the continents; the second, olivine-basaltic or the similar trachybasaltic, characterizes all oceanic regions (except regions of modern geosynclines — island arcs) and is rather common on continents (outside geosynclines). This division is demonstrated by the almost complete absence of tholeiites in the oceans and by the dominance of such basalts in geosynclines. The work of some American geologists [Engel and Engel, 1963, 1964] has shown that the view that there is a type of oceanic basalts is incorrect. It has been impossible to find olivine basalts on the floor of the ocean; only the upper parts of high oceanic

volcanoes consist of olivine basalt. The remaining, larger part of such mountains and their bases are formed of tholeiite. Until the development of deep dredging we observed only, or almost only, the olivine-basaltic volcano summits — islands.

A second view of the Engels, arising from the first, is that olivine-basalt magmas in oceanic regions are derived from tholeiite magma. This relationship has been observed directly in Hawaii, where not only olivine-basalt caps rise above the ocean surface but the tholeiitic bodies of the volcanoes as well.

From the scheme of the Engels it follows that 1) basalts of the oceans are not fundamentally distinguishable from basalts of the continents, and 2) olivine-basalt magma is not primary and should not be placed in the table of primary magmas.

These conclusions cannot be considered definitely proved, however. As a matter of fact, it is still necessary to prove that 1) there are no provinces in the ocean where olivine-basalt series are widespread and tholeiites are absent, and that 2) trachybasalt or olivine-basalt provinces on continents are of secondary magmas, arising from tholeiite. The first question is touched on directly by the statements of Engel and Engel. If olivine-basalt provinces are actually missing in the oceans, this points to a difference between oceanic magmas and continental magmas with their trachybasalts. If such provinces are discovered in the ocean, further investigation then becomes necessary in order to explain to what degree this magma may be considered primary.

It should be noted that the problem of primary or secondary character of trachybasalt (or olivine-basalt) melts on the continent is far from simple. It would seem that there is almost no doubt that trachybasalts, occasionally observed among tholeiites in regions of basaltic lava fields, are formed from derivative melts generated from tholeiite magma. On the other hand, huge masses of olivine-basalt lava practically without any accompanying tholeiite are known. It appears likely that the most representative, but not the only, example of this kind is that of the Ethiopian lavas. It is probably not without interest to dwell briefly on this and some other examples.

## Lavas of Ethiopia, Syria, and Lake Kivu*

Ethiopia. We now know a great deal about the lavas of this country, chiefly through the work of Mohr [1960].

These lavas were long ago divided into two series: the older Trap Series and the younger Aden Series. It was suggested that a long break occurred between them, even an unconformity. More recent investigations have shown, however, that, although this division is generally proper, it is very difficult to draw a boundary in some places, and elsewhere it is extremely tentative. The principal difference between the two series is their positions relative to the rift valleys.

The Trap Series was formed by outpourings of pre-rift age and, correspondingly, it forms the Plateau of Ethiopia and Somaliland. The rocks commonly exhibit appreciable weathering. The lavas cover large areas and the sequence is well stratified, although it is impossible to make direct correlation of sections from different regions. Exposures of these rocks may be found even within the rift valleys, however, and, if they are not cut by younger lavas, recognition of the Series may be very difficult.

The Aden Series came into existence after subsidence occurred along young faults. It covers a much smaller area, forming the floor of the rift valleys. The rocks are commonly scoriaceous. Columnar jointing, which is characteristic of the Trap Series, is weakly

--------

*Tables of chemical analyses are given on pp. 139-166.

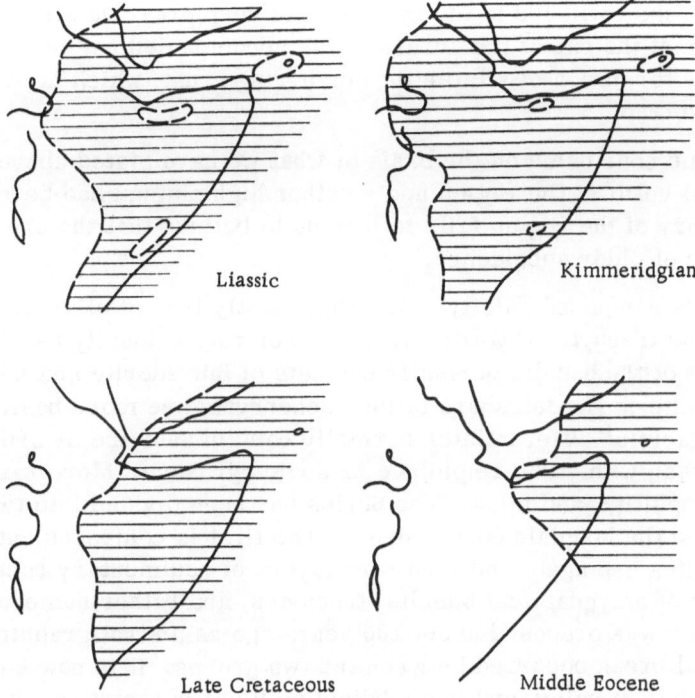

Fig. 6. Paleogeography of Ethiopia —Somaliland (after Mohr [1960]).

developed or absent entirely. The flows occupy a smaller area, and stratification of the sequence is poorer. Numerous volcanoes and explosion craters are known within the area where
this series occurs, probably because of the youthfulness of the eruptions. This may indicate
that the Aden Series was not formed solely by fissure eruptions as the Trap Series apparently
was. Aden vulcanism is not yet completely extinct, and several active volcanoes are known.
We should add that the Aden Series appears in small patches even on the Plateau. Therefore,
the principal feature, the position relative to the rifts, by which we distinguish the two series,
is not an absolute characteristic. We must apply the criterion with some care.

The Trap Series. This series occupies a very large area and covers the entire region of
Early Tertiary uplift in northeastern Africa. The lavas poured out immediately after uplift
(perhaps accompanied it in some measure?) (see Fig. 6).

The lavas of Ethiopia are thus a beautiful illustration of the danger of restricting the
region of outpouring to downwarps and the zone of intrusions to uplifts. We have already found
it necessary to point out the inadmissibility of this view. The marine basin that existed during
the second half of the Mesozoic by no means defined the zone of synclinal subsidence in the
present uplands. As seen by the maps taken from Mohr (Fig. 6), we are dealing with subsidence of the margin of the continent, beginning in the Rhaetian and reaching a maximum in the
Jurassic. Regression began in the Tithonian, and the sea gradually retreated to the east (if we
don't call attention to the very small transgression at the beginning of the Cretaceous). The
region of present Ethiopia became land at the end of the Cretaceous, and during the Eocene
only the very end of the "Horn of Africa" was still under water. The exposed land was beveled
during the Cretaceous and the first half of the Eocene, and it became a low plain with shallow
stream valleys. In the Late Eocene the land was uplifted rapidly, an event that Mohr tentatively connects with subsidence of the adjacent parts of the Indian Ocean. Faults apparently
developed at that time that determine the present configuration of the coast line, creating the

Gulf of Aden. Later Eocene uplift led to the formation of deep pre-lava stream valleys. Interruption in sedimentation at the end of the Jurassic was accompanied not only by erosion and planation but also by the formation of a lateritic weathering crust, which is now preserved in many places beneath the trap.

We may draw some conclusion on the basis of what we have stated above: 1) the lavas poured out after general uplift of the region and a rather high plateau had been formed, and 2) all the previous history of the region fails to lead us to believe that the present lava field corresponds to the zone of older subsidence.

The Trap Series is composed chiefly of basalts, mostly free of olivine. In the upper part of the sequence are found trachyte, rhyolite (?), and other rocks, locally as significant components. The groundmass of the basalts normally consists of labradorite and iron-rich pyroxene (clinoenstatite or pigeonite). The feldspars of the phenocrysts are more basic, and are generally zoned. The "acidic" differentiates normally contain sanidine or orthoclase, perhaps nepheline, sodalite, nosean with sodic amphibole or sodic pyroxene. More rarely one finds rocks with leucite, kaliophilite, and mica. The Series has been divided into two groups: the Ashangi Group below and the Magdala Group above. The first is composed entirely of basalts with amygdules of zeolites and agate and with rare layers of sedimentary rocks. The second consists of alternations of amygdaloidal basalts, trachytes, and rather numerous layers of sediments. This division was proposed about 100 years ago as a stratigraphic one. It was thought that an erosional break occurred between the two groups. It is now known, however, that the break is commonly missing, and the subdivision itself is tentative. It is impossible to correlate the groups by age in different parts of the country. It should be noted that almost everywhere only basalts were poured out during the first stage; later basalts alternated with differentiates richer in silica.

The character of the sequence varies from region to region. The sequence is best developed in central Ethiopia, where it generally exhibits the typical two-member structure. The thickness is extremely variable, ranging from hundreds of meters to 3500 m (Simen Mountains). Besides trachytes, the upper group also contains trachyandesites and rhyolites (?). The lavas have been but poorly studied in the southwestern part of the country. Here they consist of 90% basalts, with only the upper 10% consisting chiefly of trachytes. Nepheline rocks are also encountered (bostonite, even phonolite), and augitite. Remains of Burdigalian mammals have been found here in a layer of tuff near the base of the series.

In the eastern part of the country the thickness of the sequence is less, but it still reaches 1500 m. In Eritrea the rocks are almost exclusively basalts. Farther south, in Tigre, a layer of limburgites occurs in the middle of the series, above basalts but below the sequence of basalts, trachytes, and underlying sediments. In other parts of the same province (Aduwa and Aksum) silica-rich alkalic rocks are abundantly developed: selvsbergite, grorudite, quartz bostonite, and others. They were preceded by tinguaites.

The Series also makes up the Plateau of Somaliland, where olivine basalts are dominant. Silica-rich lavas are few. The thickness of the series reaches 2500 m.

The age of the Trap Series has been determined in great measure by its tectonic relations. Rather numerous discoveries of fauna and flora have been made in the inner sedimentary layers; they cannot give the limiting or boundary ages. The most important fossils are the above-mentioned Burdigalian fauna. Judging from them, the lavas had already poured out in the vicinity of Lake Rudolf at the beginning of the Miocene. In the Kenya region, vulcanism reached its greatest development in the Miocene. In the provinces farther north, however, vulcanism apparently began somewhat earlier, and it seems that the Trap Series may be dated most likely as Oligocene—Miocene. Mohr even believes that the greatest extrusion of these lavas took place in the Oligocene.

Tectonic criteria lead us to believe that vulcanism did not occur in the Eocene, since the rather well-dated Late Eocene uplift clearly preceded the first outpourings of lava. The upper limit of the series is determined by the fact that the rift in Ethiopia has been cut into the plateau composed of these lavas. Thus, if vulcanism also took place at the end of the Miocene and in the Pliocene it must have been rare and weak.

The Aden Series. The discrimination between this series and the Trap Series is by no means always easy. It is especially difficult where the Trap Series ends and the Aden Series begins with silicic alkalic lavas. Sedimentary layers in the series are rare, and age determinations are not very precise for this reason. There is no doubt that the lower age limit is the time of formation of the rifts, i.e., sometime after the end of the Miocene. A more accurate age is defined by the underwater lavas of Alid Volcano, which are interbedded with Upper Pliocene sediments. The Aden Series is distributed over a much smaller area than the Trap Series, and its thickness is many times less than that of the latter. As a rule it begins with silicic alkalic lavas and ends with olivine basalts. Among the first are found pantellerite, sodic rhyolite, comendite, trachyte and sodic trachyte, and dacite. Many volcanoes of this epoch have been preserved, differing rather strongly from each other by the association of lavas. From a short description of them it is sometimes difficult to give account of the sequence and connection among the individual lavas. For Odmat Volcano, for example, "rhyolite" and phonolite are mentioned as preceding basalt flows. It is possible that we are here dealing with terms used in a manner different from ordinary usage. We have not been able to find a detailed description of these rocks.

Perhaps the greatest assemblage of rocks characterizes the large Alid Volcano and its environs (Eritrea). Here, along with olivine basalts, occur amphibole felsodacites, hyalodacites, rhyolitic obsidian, anorthoclase rhyolite, rhyolite, and pantelleritic pumice. Dankalite, especially rich in alkalies, falling in the field of typical phonolites on the qz diagram, is found among the trachyandesites of this region. Approximately the same complex of rocks is found also in other regions. Everywhere the silicic alkalic rocks are most characteristic: pantellerite, trachyte, and, in association, rhyolite.

Both complexes of lavas in Ethiopia, as may be seen even from a list of the rocks, undoubtedly belongs to the alkalic-basalt association. This is especially clear on the qz diagram for the Trap Series. The qz diagram was proposed by the author [Sheinmann, $1965_2$] and is now used by many geologists in the USSR for distinguishing basalt series with alkalic and tholeiitic trend. Fundamentally it represents the position of the qz coefficient of Niggli. It characterizes excess or deficiency of silica in a rock. The value of qz with a positive sign defines the amount of free quartz that should appear in the rock during crystallization under ideal conditions. When the sign of qz is negative, this defines how much $SiO_2$ would need to be added to saturate the rock. Thus, qz determines the difference between the amount of silica present and the amount that may be bound up with bases. It is convenient to express qz in the following way:

$$qz = \frac{SiO_{2\,present} - SiO_{2\,bound}}{total\ bases} \cdot 100.$$

If, for the molecular quantities, we set $Na_2O + K_2O = a$, $Al_2O_3 = al$, $CaO = ca$, and $FeO + MnO + MgO = fm$, then,

for most rocks

$$qz = \frac{SiO_2 - 5a + al + ca + fm}{a + al + ca + fm} \cdot 100;$$

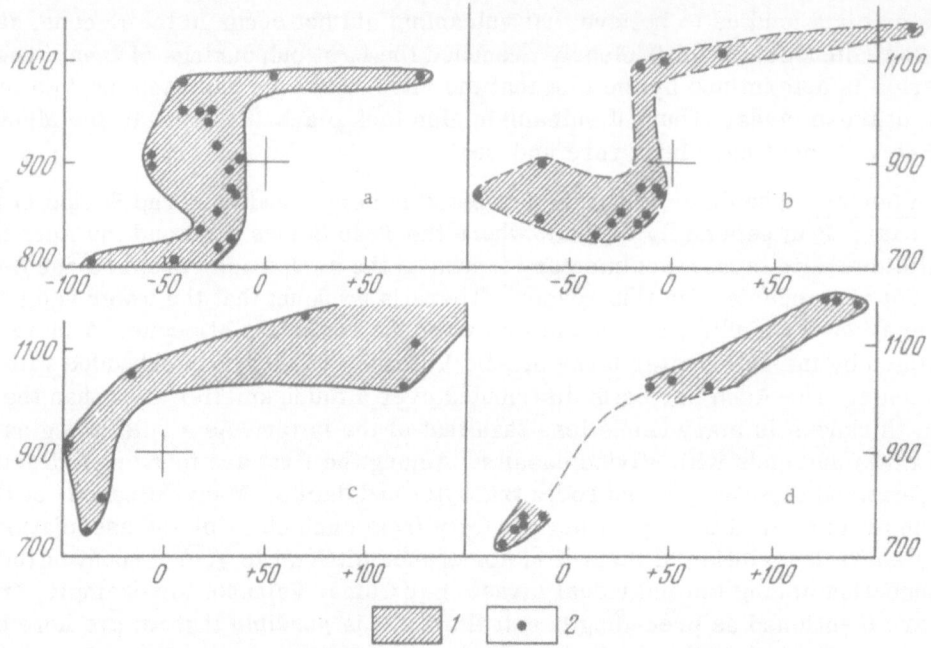

Fig. 7. qz diagrams for lavas of Ethiopia (Trap Series). a) Southwestern
Ethiopia; b) Somaliland Plateau; c) Tigre and Eritrea; d) Central Ethiopia.
1) Field of analyses; 2) individual analyses.

but if $a > al$, then

$$qz = \frac{SiO_2 - 4al + 2a + ca + fm}{a + al + ca + fm} \cdot 100.$$

On the diagram, $SiO_2$ (molecular quantity) is placed on the vertical axis; qz is plotted on
the horizontal axis. It appears that points for any series saturated with silica forms a straight
inclined belt (Figs. 19, 27), but the series of alkalic rocks exhibits a curved belt (Figs. 7, 8).
If we measure the slope of the middle part of these lines, we may introduce the coefficient

$$K = \frac{\Delta qz}{SiO_2{}_{[100]}}$$

where $\Delta qz$ is the change in qz on the diagram with the change in $SiO_2$ per hundred units.

This coefficient is characteristic for $SiO_2$ contents between 700 and 1000 (i.e., between
42 and 60% by weight). It amounts to $K \leq +10$ for alkalic series, even reaching negative values,
but for series saturated with $SiO_2$ the value is not less than +20 and may reach +50.

In Fig. 7 one may clearly see the typical curve of the alkalic-basalt belt for rocks in the
vicinities of Wallega and Ilubabor in southwestern Ethiopia. The same type of curve, but much
sharper, dropping its "knee" in the zone of ultra-alkalic rocks (such as ijolite or phonolite),
has been plotted for the rocks of Eritrea and Tigre. The following rocks (for which we have
analyses) occur in these associations: for Wallega — olivine basalt, basalt, kenyte (a "syenite"
rich in nepheline and glassy anorthoclase), trachyandesite, bostonite, alkalic trachyte, and
rhyolite; for Ilubabor — "Gore rock" (apparently a variety of basalt), highly siliceous basalt,
trachyte, and comendite; for Eritrea and Tigre — olivine basalt, tinguaite, selvsbergite,

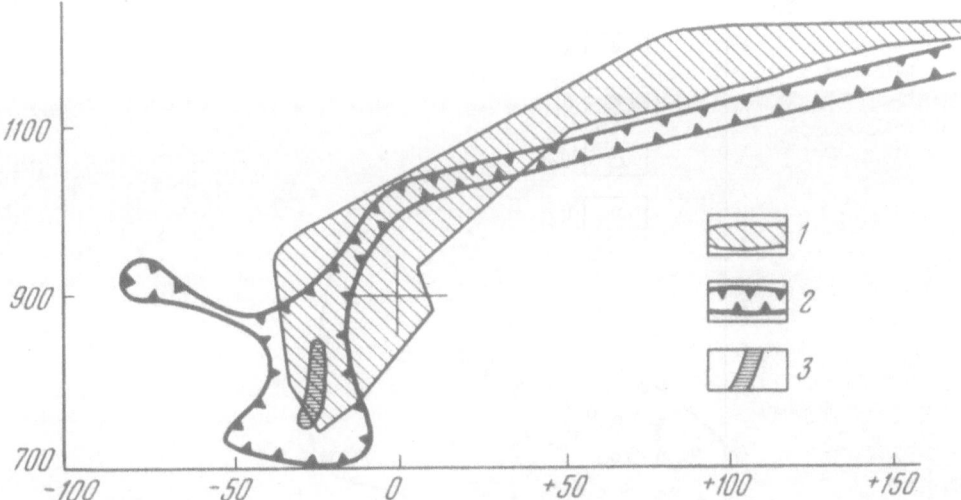

Fig. 8. qz diagram for lavas of the Aden Series in Ethiopia (post-rift lavas).
1) Rift of Ethiopia; 2) Quaternary lavas of Lake Tana; 3) the same for the field
south of Lake Rudolf.

bostonite, and obsidian. The diagram for lavas of central Ethiopia is not so clear. For this
association there is no information on the intermediate members of the series (trachyandesite
and the like). As a result, the series consists of basalts and olivine basalts, anorthoclase
trachytes, comendite, and pantellerite. However, the nature of these silica-rich differentiates
indicates that there can be no andesites or andesite-basalts here as intermediate members,
and that in this case the association is typically alkalic-basaltic. The value of K for these
series apparently ranges from −25 to +5.

For the Aden Series we are lacking analyses of moderately basic rocks (trachyandesites,
perhaps trachybasalts). There is thus a jump to rather silica-rich varieties. However,
analyses of alkalic trachyte (Adama Volcano) and trachyte (Fantale Volcano), both in the province of Shoa, undoubtedly indicate an alkalic-basalt band in the curve. This is confirmed by
analyses of recent olivine basalts and trachybasalts from the lava field south of Lake Rudolf
(Fig. 8). In this case K = 6; for the lavas of Eritrea—Shoa, K may lie between 1 and 10.

Thus, it is impossible to find tholeiitic associations in any of the lavas of Ethiopia.
These lavas are apparently entirely alkalic basalts, and in this are sharply distinguished from
typical tholeiitic trap formations.

S y r i a . Analyses of Syrian basalts were sent me by A. A. Krasnov, and they should permit this province to be described in very general outlines. The lava field here is directly related to the northernmost segment of the African Rift zone. In age the basalts correspond
more or less accurately to the Aden Series of the regions to the south. Slight differentiation
makes it very difficult to determine types. All groups, within which Pliocene and Lower,
Middle, and Upper Quaternary lavas may be found, contain only basalts, and only basic basalts.
The basalt richest in silica contains at most 46.3% silica; the most basic contain 41.5%. However, the marked silica deficiency of all associations and the presence of alkalic and even
nepheline basalts permit us to assign this sequence to the alkalic-basalt association to which
the Ethiopian traps are assigned (Fig. 9).

L a k e  K i v u . (Southern field) (after Bowen [1938] and Holmes [1940]). This field of
basaltic lavas lies about the southern end of Lake Kivu (western rift of East Africa, north of

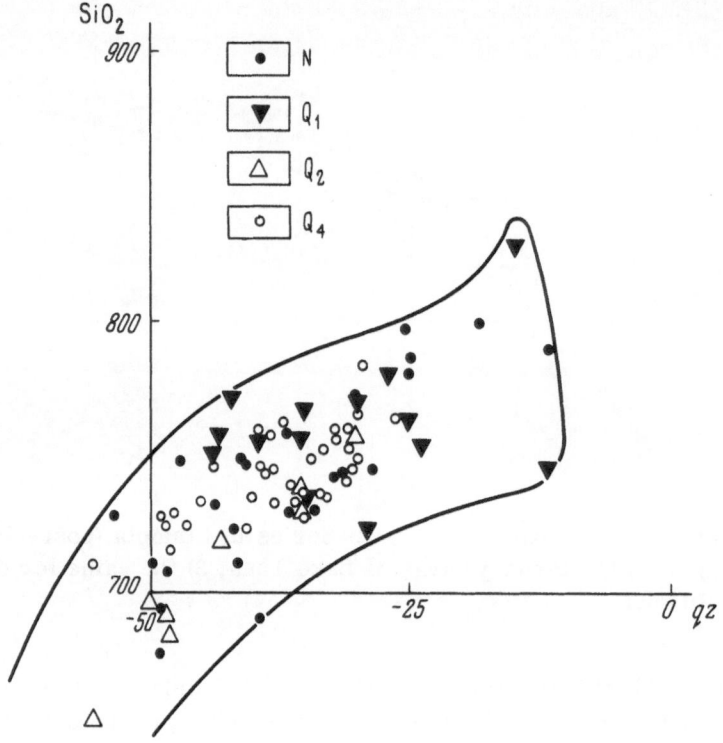

Fig. 9. qz diagram for the basalts of Syria.

Lake Tanganyika). The field (essentially two contiguous fields) occurs in the region where the rift line changes rather sharply from a west-northwest to an east-northeast trend. This break corresponds to a change from faults of the Tanganyika system to faults of the Lake Albert system, both systems extending in the form of weakly expressed valleys, which were occupied by lavas beyond the intersection of the systems. Eruptions in the northern field (Lake Kivu field) were confined to two large volcanoes, now almost completely eroded. The northern, the Kahusi, is a neck 2 km in diameter. Farther south occurs a group of fractures, which join and give rise to the subvolcanic Biega granosyenite body. The lavas are almost entirely olivine basalts (over an area up to 5500 km$^2$). Trachyte occurs in small amounts in only three places. The southern field (Mwenga—Gandu) is made up entirely of olivine basalts and olivine-free basalts.

The youngest rocks of the region are rhyolites. According to the descriptions, they are characterized by very appreciable enrichment in potassium and silica, a feature that distinguishes them from the rhyolites of other regions. The age of the eruptions in the Kivu field is determined by the fact that in the valley of the Ruzizi River both basalts and trachytes are covered by Lower Pleistocene sediments, and the lavas must therefore be assigned to the Pliocene. The basalts of the southern field (Mwenga—Gandu) are considered somewhat older than the Kivu lavas (Kasmichev). Rhyolites were formed after the main bulk of the lavas began to be eroded and after stream valleys were developed. In particular, rhyolites cut Lower Pleistocene sediments.

In all, five analyses are at our disposal. Of these, two are for basalts: olivine basalts from the field of Mwenga, Gandu, and Mukaba near Kahusi. The two basalts are similar to each other and belong to the Sorochin group of "ophitic" olivine-poor basalts, distinguished in this region. A third analysis corresponds to the second group: "olivine basalts." As first pointed out by Bowen, these are characterized by the presence of small quantities of alkalic

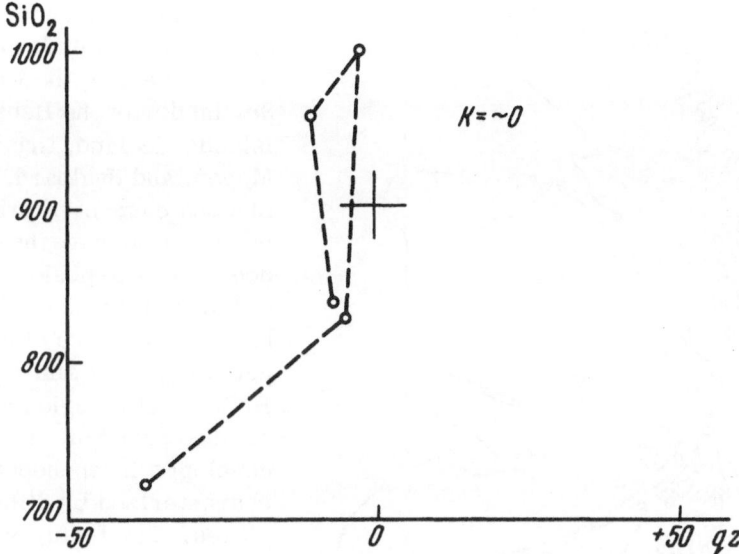

Fig. 10. qz diagram for lavas of the southern Kivu field (from
analyses of Bowen and Holmes).

feldspar in the interstices. Holmes has called these rocks essexitic basalt. The fourth and
fifth analyses are of trachytes. According to Holmes, the group of trachytes in these lava
fields consists not only of trachytes but of other rocks as well, from trachyandesites to
phonolites.

Both the petrographic characteristics and the qz diagram very clearly show that the
Kivu basalts belong to the olivine-basalt association (Fig. 10). The later rhyolites (if they
are differentiates of magma and not the product of fusion of sediments) indicate that the process ran to the very end, and after trachytes there occurred a second turn of the curve toward
normal silicic melts.

Whereas for Ethiopia and the other lava fields of East Africa there are essentially no
grounds for seeking any mysterious primary tholeiite magma, since it has nowhere risen to
the surface or else has passed entirely into olivine- basaltic magma, the question for the
Brito-Arctic province is not clear. The presence of a series of large fields, comparatively
equal in size, of basalts of both types leads us to seek for a single source for the magmas of
the region and, consequently, to doubt the origin of olivine basalts from tholeiitic magma. However, we have less right to suggest the independence of an olivine-basalt melt. Furthermore,
it was here that a mechanism was first described for supplying tholeiitic magma that at the
end of volcanic activity was replaced by olivine basalts with a distinct alkalic trend.

## The Tertiary Brito-Arctic Province*

This region is practically without description in our literature. Its characteristics are
derived almost exclusively from the interesting work of Bailey and others [1924], serving as a
model of keen geologic—petrologic analysis, and also from the works of Black [1952], Kapp
[1960], Noe-Nygaard and Rasmussen [1957], Patterson [1952], Patterson and Swaine [1955], Richey
[1935], Tyrrell [1926], Walker [1930], Walker and Davidson [1936], Wager and Deer [1939], and
Willie and Drever [1961-62].

---

*Tables of chemical analyses are given on pp. 139-166.

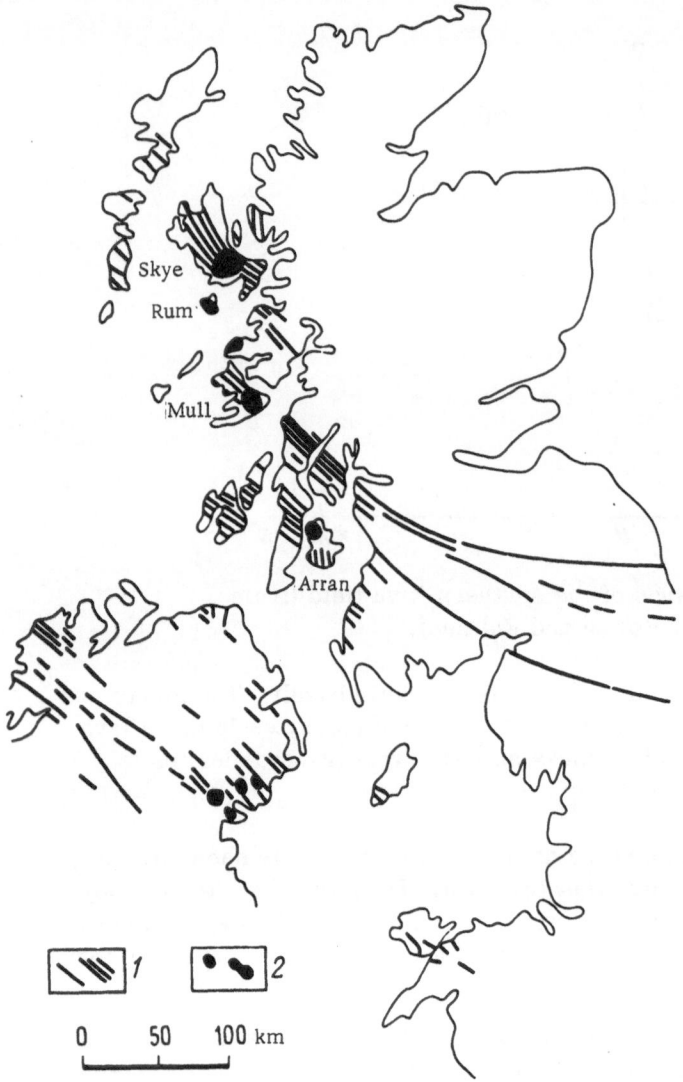

Fig. 11. Distribution of Tertiary basic dikes and central complexes in northern England, Scotland, and northern Ireland (after Richey and others [1930]). 1) Dikes and dike series of basic rocks; 2) central intrusive complexes.

This tremendous area of Tertiary vulcanism embraces the northern part of Ireland, the western part of Scotland with the Hebrides, the Faeroe Islands, Iceland, Greenland, Jan Mayen, and Svalbard. We should note that the concept of province in this case does not have the same meaning it does when we speak of the provinces of Siberian traps or the Deccan basalts. In place of a single field of generally very similar lavas, in the Brito-Arctic region we encounter a series of volcanic regions that developed independently and are characterized by different rock complexes. The features that unite these complexes are age (all the rocks are Tertiary lavas) and the basic character of the magma. It would probably be difficult to point out structural features common for the entire province: it includes segments of ancient platforms (within Greenland), a midocean ridge (Iceland, Jan Mayen), and activated Caledonide zones (Ireland, Scotland, Svalbard, and, probably, the Faeroe Islands). Of course, with this variety of tectonic base we cannot deny the possibility of some unity in the system of superimposed structures. The fact that the vulcanism is everywhere dated as Tertiary and that nowhere did it appear earlier than approximately the middle of the Eocene leads us to suggest the existence of some fundamental tectonic phenomena, embracing the entire province in the Eocene-Oligocene and not characterizing neighboring regions.

First, we shall attempt to limit the zone of these, as yet hypothetical, movements or at least, their manifestation in sufficient strength so that volcanic centers might arise. This zone is clearly bounded on the south: Tertiary vulcanism did not reach the southern part of Ireland or the southern half of England. There is no record of Tertiary vulcanism along the eastern coast of England or Scotland, in the Orkney Islands, or in Scandinavia. Young basalts are found on both the eastern and western coasts of Greenland (probably also beneath the ice in the center of the island). But they do not extend southward. Thus, Tertiary basalts are found over a tremendous area, embracing both sides of the Atlantic, its near-polar part, and a considerable part of northwest Europe. It would appear that the most direct solution to the problem would be a comparison of this vulcanism and of the tectonic movements causing it

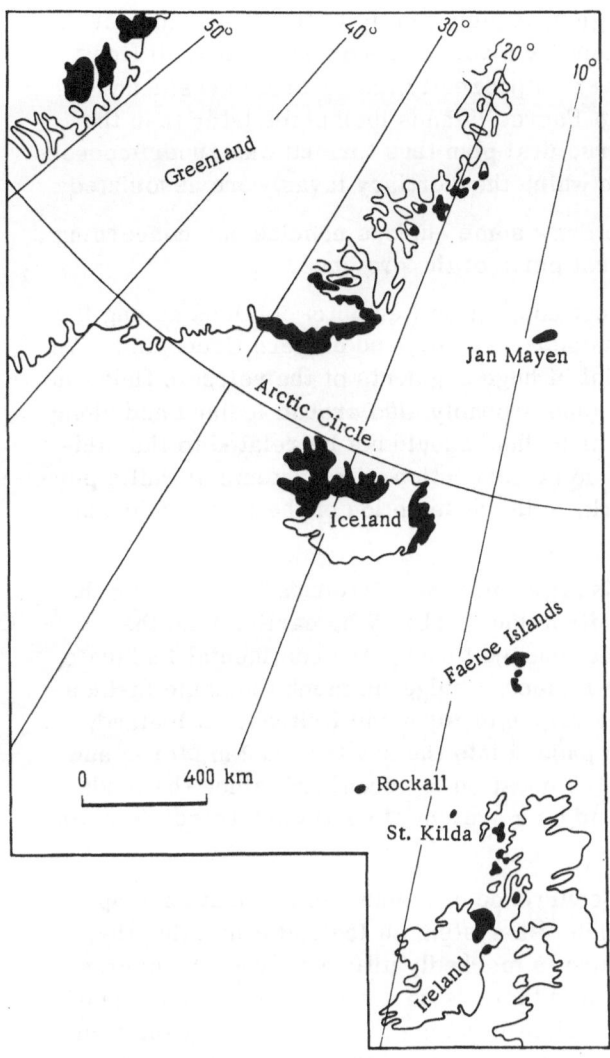

Fig. 12. Brito-Arctic province of Tertiary igneous rocks (after Richey [1935]).

with subsidence in the region of the present Atlantic Ocean. However, such direct comparison is hardly sufficient.

Let us turn first to that part of the province most thoroughly studied: Scotland. In Fig. 11 we see a very interesting picture of Tertiary faults in this region. The system of fractures, now occupied by dikes of basic rocks, has a well-defined northwestern trend on the whole. Large central intrusive complexes are subordinate to this system. If we transfer the outlines of this system to a general scheme of the province (Fig. 12), it then appears that the system should be extended to the Rockall bank and to the islands of St. Kilda. The trend of this system would seem to indicate a possible continuation to Iceland and, farther, to the basalts of eastern Greenland. The arc-like picture of the eastern half of this cluster leads us to suggest its continuation in the region of the Faeroe Islands. Thus, that which is directly known concerning the structures with which the Tertiary basalts are associated in no way confirms any connection between these structures and the border of the ocean. It might be more properly said that the present ocean basin disrupts the volcanic structures.

If we turn to the nature of the sedimentary rocks, we find that the Cretaceous sediments are marine. The Eocene, immediately underlying the volcanic rocks, is everywhere continental. And the character of the lavas themselves in the lower part of the sequences (where these lower units are accessible to observation) indicates subaerial eruptions. In any case, there are no grounds for seeking any great ocean depths near the present Hebrides or the Faeroe Islands before the beginning or at the beginning of volcanic activity. This was a continental region, which stretched across the present Atlantic to Greenland. The author wrote about this several years ago [Sheinmann, 1961].

There are thus no grounds for believing that the system of deep fractures, later filled with basaltic magma, was formed under conditions of the present distribution of deep marine basins and continental uplifts. It was formed in a continental zone, perhaps partly occupied by shallow seas. We have no data for connecting this system to any other tectonic phenomena. Perhaps it was in some measure a reaction to phenomena taking place southward in the Atlantic Ocean or in the central part of the Arctic Ocean. It is obviously impossible to say anything more definite about it.

The onset of extensive subsidence in this region, leading finally to the appearance of the present near-polar part of the Atlantic Ocean, is clearly a young phenomenon. It is difficult to date this phenomenon with any precision, however. Apparently, large-scale subsidence in the region northwest of the Rockall band and of the Faeroe Islands took place later than the Eocene, perhaps later than the Oligocene. The structural plan thus formed was superimposed on the previously formed system of fractures with which the Tertiary lavas were associated.

If these statements are true, they lead us to draw some curious conclusions concerning connections between volcanic phenomena in different parts of the province.

During the Tertiary, as we have seen, a direct connection (tectonic) obtained among the volcanic regions of Ireland-Scotland, the Faeroe Islands, Iceland, and eastern Greenland. We should note that very young subsidence led to burial of huge segments of the volcanic fields in the region of the Hebrides and the Faeroe Islands (and probably also around Iceland and along the coast of Greenland). The Tertiary vulcanism of Iceland should not be related to the mid-ocean ridge. More likely we have to do here with some connection with systems of faults perpendicular to the present ridge. Later in this work, in the description of the lavas of Iceland, this position will be established more fully.

The later formation of an oceanic basin in this region furnishes grounds for believing that the midocean ridge was extended recently in this direction, probably no earlier than the Neogene. It may be suggested, naturally, that faults passed through the continental Icelandic bridge, in continuation of the central crests of the midocean ridge, in much the same fashion that such faults now emerge on the continent in the region of Aden and Eritrea. In Iceland, after appearing on the land, they might have again passed into the sea toward Jan Mayen and farther to the north. Whether this is true or not, the question is secondary, since the fundamental dependence of Tertiary vulcanism in Iceland on a system of northwest-trending faults seems very probable.

The existence of a midocean ridge and the occurrence of Iceland on it might have appeared distinctly later, particularly in the formation of a system on Iceland continuing the central trough of this ridge (to the point, this feature is markedly different from the narrow submarine troughs farther south in the Atlantic). And there is no doubt that the existence of the midocean ridge is related to the fact that vulcanism in Iceland did not cease at the time it died out in other parts of the province but that it has continued to the present day. The question is merely: to what degree are the young and present-day lavas typical of the mid-ocean ridge and to what degree have they inherited aspects of the Tertiary continental-type lavas ? Available data are in favor of such inheritance, and the role of the structure of the midocean ridge reduces more or less to a mere focus of submarine activity.

Much less is known concerning connections with tectonics of the ocean and with other volcanic regions for Svalbard and, apparently, Franz Josef Land (Fridtjof Nansen Land), which belongs to the same province. Of course, there is no doubt that neither Svalbard nor Franz Josef Land are connected with the system of faults we spoke of above. But, at the same time, the phenomena lead us to seek a different, less concrete relationship with some still insufficiently clear phenomena within the entire province during the epoch of intense subsidence of the ocean basins.

Below we shall dwell on only a few regions, sufficient for obtaining a general view of the Brito-Arctic province (northwestern Sçotland, the Hebrides, northeastern Ireland, the Faeroe Islands, Iceland, Jan Mayen, and eastern Greenland).

Western Scotland and the Hebrides. This part of the Brito-Arctic province has been especially well studied, and it is of great interest because of the complexity of its structure. This is the region that is characterized by the existence of olivine-basalt and

Fig. 13. The Tertiary volcanic region of western Scotland (after Richey [1935]). 1) Silicic intrusions; 2) basic intrusions; 3) basalts; 4) pre-Tertiary rocks.

tholeiite series and where we know of volcanic processes with very complex and prolonged histories. The principal regions of volcanic activity here are the island of Mull and its environs, Skye, the Small Isles, and the Shiant Isles, all in the Inner Hebrides. The ridge of the Outer Hebrides is not volcanic, but beyond it, in the open sea, lavas and hypabyssal intrusions are known on Rockall Island and the St. Kilda group as well as around them on the floor of the surrounding shallow-water zone. The continuation of the region to the southwest is found in the Antrim plateau and other volcanic phenomena in north-eastern Ireland (Fig. 13).

The most interesting is the district of Mull (the island itself, the small islands to the west of it, and the Morven coast). Here a thick basaltic sequence (up to 1000 m thick), consisting of olivine basalts, rests on pre-Tertiary rocks. In the upper third of the sequence are found a "horizon" of mugearites and, stratigraphically near this horizon, flows of basalt with large plagioclase phenocrysts. For a long time it was considered that both rocks are sills, but it was later shown that they are flows, alternating with flows of olivine basalt.

It appears most likely that fissure eruptions occurred here, at least at first. However, during rise of the mugearites and porphyritic basalts, central eruptions were already taking place. It must be thought that, on the whole, olivine basalts formed a very gently sloping shield volcano of impressive size, at least 30-40 km across. Later after some break, the olivine-basalt lavas gave way chiefly to nonporphyritic basalts free of or with but small quantities of olivine (the thickness of this unit reached 800 m). The first rocks are characterized by columnar jointing and are of continental origin. The nonporphyritic flows poured out under water during the first stages, giving rise to pillow lavas. Then terrestrial flows again became dominant. The underwater eruptions were not associated with transgressions of the sea. They took place in a lake that occupied the caldera that was beginning to form. It was the development of this basin that leads us to suggest that a break occurred after termination of the olivine-basalt eruptions (it is true that no clear indications of such a break are given in the sketch of Richey or in the monograph of Bailey and others, and it may be that the assumption of a break is incorrect).

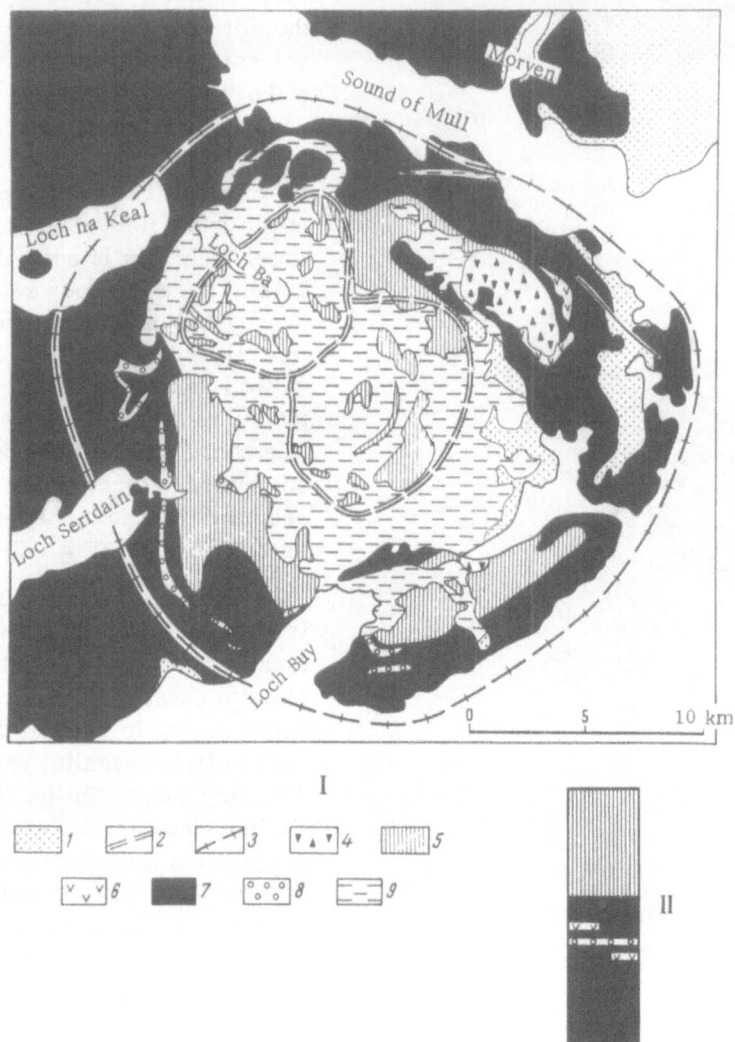

Fig. 14. Map of volcano on the island of Mull (I) and a diagram showing the sequence of lavas (II) (after Bailey and
others [1924]). 1) Pre-Tertiary rocks; 2) boundary of
caldera; 3) limit of pneumatolysis; 4) agglomerate; 5)
"central-type" lava; 6) basalt with large feldspars; 7)
"plateau-type" lava; 8) mugearite; 9) intrusive rocks, un-
differentiated.

The eruption of nonporphyritic basalts ("central-type basalts" according to Bailey) was
accompanied by intrusions of basic rocks and by further sinking of the center, particular along
a ring fracture (see Fig. 14, the center). Somewhat later a second ring fracture developed,
joining the first on the northwest and having its center near the southeastern end of Loch Ba.
Both rings were in existence some time; movement then ceased along the first.

The appearance of the ring fractures approximately coincided in time with the rise of
silicic magma that filled the complex series of ring and cone fractures, forming stocks and
plugs in the vents and accumulations of agglomerates in lower parts of the landscape. These
phenomena may be divided into three stages of activity of the Mull volcano. The first magma
filling the ring and cone fractures was basic, but this gave way to acidic magma. Following

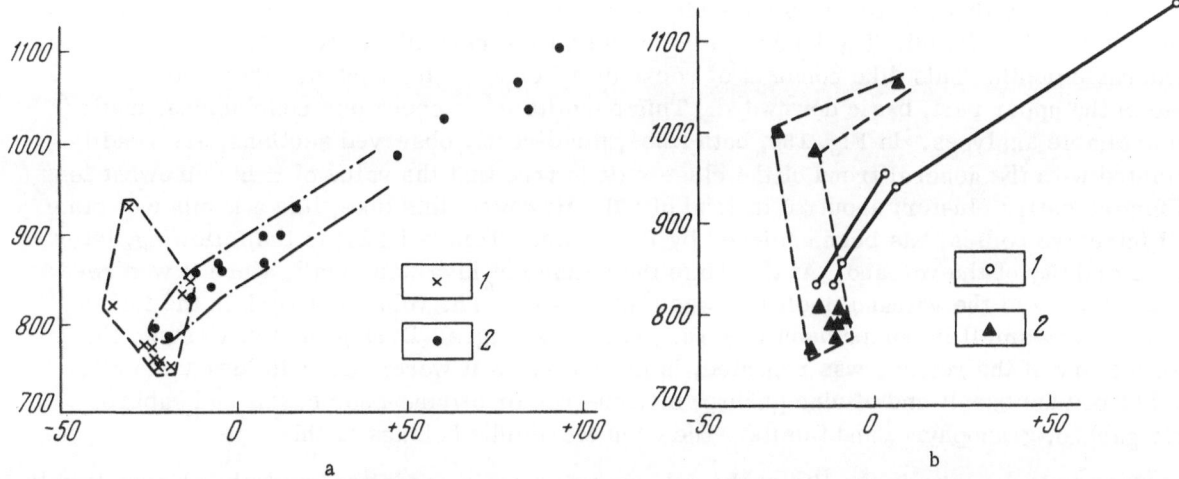

Fig. 15. qz diagram for the Mull volcano and adjoining regions. For volcanic rocks of the first stage (a): 1) olivine basalt and mugearite; 2) nonporphyritic basalt and quartz diorite. For later volcanic rocks (b): 1) Glen More dike, two sections; 2) volcanic rocks of the later stage.

this, dikes of intermediate composition were formed: quartz diorite, ranging from completely crystalline to glassy varieties (craignurite, inninmorite, leidleite), and still later to felsites, granophyres, and rhyolites. This entire series of rocks is directly connected with the basalts and forms with them a single line of differentiation.

Still later, basic dike rocks again appeared (the fourth stage in the life of the volcano). According to the literature, it is rather difficult to distinguish these from other basic intrusions. Apparently they are chemically more nearly similar to basalts of the first stage (olivine basalts) than to the later basalts. This part of the magma, being differentiated, formed syenites and trachytes.

Thus, on the basis of the map of Bailey and Richey, we may summarize the history of this extraordinarily long-lived volcano in the following way. After deposition of Eocene sediments (fresh-water), perhaps still in the Eocene, fissure eruptions of basalt began; the rocks were basic (44-46% $SiO_2$), rich in olivine. Before the end of this epoch, apparently after the formation of a volcano from a central eruption and after plugging of the fissures with olivine-basaltic magma, a mugearitic melt very rich in alkalies (up to 10% of the total) was separated. Compared with the mass of lava poured out, the volume of the mugearites is negligibly small.

The character of the magma later changed. The nonporphyritic basalts are tholeiites, and their differentiates appear on the qz diagram on the tholeiitic line with K = 27 (Fig. 15a). If it were not for the mugearites, one might, with some stretch of the imagination, consider that the olivine basalts also belong to this same group. But the development of olivine basalt-mugearite is perfectly obvious. It is therefore necessary to seek an explanation in some change in the character of the magma. We may assume that tholeiitic basalts are the product of an olivine-basaltic magma, from which, after separation of the mugearite, excesses of alkalies disappeared and melts saturated in silica appeared. It appears, however, that this process can no longer be observed anywhere. Another possibility is change in composition of the melt, i.e., the appearance of new conditions at depth. The first part (with mugearites) was produced at moderate depths by the separation of enstatite from the rising magma. There was then no entrapment at moderate depths, and primary deep-seated tholeiitic magma reached the surface.

It is curious that on Mull it is possible to observe directly the result of gravitative separation of basaltic liquid. The landscape is such that the ring dike of Glen More is exposed to considerable depth. This dike consists of rocks of different silica content: they are light and silicic in the upper part, basic downward. This permits us to check our conclusions, made from available analyses. In Fig. 15b, both lines, for directly observed sections, are readily correlated with the general trend of the cluster (it is true that the value of K is somewhat less than for the entire cluster; about 20 instead of 27). However, this dike, like a number of other small intrusive bodies, has been assigned by Bailey and Thomas [1924] to the following, later stage of activity of the volcano. At this time the volume of lava was small, since it was restricted chiefly to the volcano itself and was thus eroded. The volume of rock in the fourth stage was very small in comparison with the preceding stages. During the fourth stage the entire history of the volcano was repeated, in miniature as it were: there followed one after the other olivine basalt and olivine gabbro, olivine-free or olivine-poor basalt and gabbro, quartz gabbro, granophyre, and felsite. The Glen More dike belongs to this time.

The final moments in the life of the volcano were again associated with the olivine-basalt lineage, with syenite and trachyte as differentiates. The basic rocks contain very much $Al_2O_3$ (up to 23-26%). These high-alumina gabbros and dolerites, in contrast to those described by Kuno for Japan, are not transitional from olivine basalt to tholeiitic basalt but appear as a result of changes in the tholeiitic magma even after its eruption.

The question concerning the last part of the magma we shall touch only briefly, from necessity. We may state merely that it was either residual magma after the removal of the nonporphyritic basalts from the magmatic reservoir, and underwent no appreciable contamination, or it was contaminated by the country rock during its separation and subsequent rise through these rocks. Strong enrichment in aluminum would seem to indicate the second view. But invariability in the Na/K ratio does not permit us to assume contamination of the rocks by a silicic crust. In any case it came from the magma of nonporphyritic basalt. It is impossible to assume any independent origin. We cannot determine precisely the time when volcanic activity ceased on the island of Mull. It was certainly long before glaciation, since erosion had succeeded in destroying the volcano and in removing a considerable thickness of basalts before glaciation began.

Having acquainted ourselves with the Mull volcano, it should be possible to understand readily the character of other volcanic fields. The second region is the Isle of Skye with its large volcanic field and subvolcanic body in the southern part. The several analyses at our disposal do not permit us to understand the chemistry of this complex (these analyses give but a very fragmentary picture of the series). The geologic characteristics are given in the work of Richey [1935]. The plateau of the island is deeply eroded, and the later products of vulcanism are thus poorly preserved. The present plateau is composed chiefly of flows of olivine basalt, the same type as that found on Mull. Olivine-free basalts (the second stage on Mull) are few, probably because of erosion. As on Mull, mugearites and basalts with large plagioclase phenocrysts are present within the sequence. In the northern part of the Cuillin Hills silicic rocks up to 600 m thick are exposed. These are chiefly rhyolites and corresponding tuffs. Trachytes and andesites are found in the lower part of the section. We have no analyses of these rocks, and the interrelations are not altogether clear. It is therefore impossible to shed light on the connections between the different kinds of rocks.

Two phases are distinguished in the subvolcanic complex on the Isle of Skye. The first (and principal) is represented by large bodies. They indicate a clear sequence from ultrabasic rocks through different varieties of gabbro to granophyre; i.e., this series corresponds to the second and third stages on Mull, where the larger intrusions are assigned to this same phase. The second intrusive phase on Skye is represented by small intrusions. The sequence in these

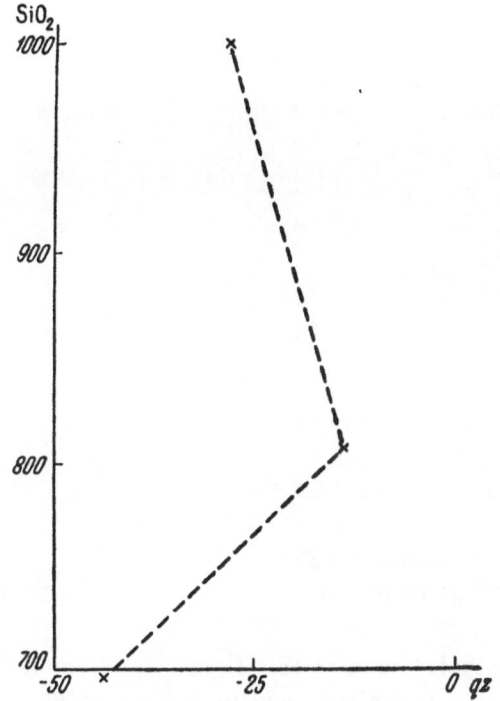

Fig. 16. qz diagram for the sill of Shiant Isles (in the Hebrides).

is complex, indicated by the following series: sills of complex composition → sills of basic rocks → acidic intrusions → dikes of basalt and dikes and sills of basic rocks → small ultrabasic bodies → trachyte and trachyandesite bodies → volcanic glass (dikes) → dikes of basic rocks. This phase clearly corresponds to the last stage in the life of the Mull volcano.

Thus, to the extent we find possible without a proper set of analyses, it appears valid to state that fundamentally the volcanoes of Mull and Skye were identical, were fed by identical or very closely related magmas. The Skye volcano began its activity with olivine-basaltic magma (with mugearite); the character of the melt then changed and a tholeiitic series was formed (on Skye these rocks are chiefly intrusives; the lavas have been eroded). During the third stage, rocks of the olivine-basalt series again appeared. On Skye the gabbros are especially rich in aluminum. With volcanoes approximately 100 km apart, this means that we may consider this feature to be not random for the entire district.

Between Mull and Skye lies the Inverness-shire group (the islands Canna, Rum, and Egg) with the partly preserved volcanic center of Rum and the Ardnamurchan Peninsula, also with subvolcanic bodies. The subvolcanic complex of Rum is the most interesting [Black, 1952]. Ultrabasic rocks are extensively developed in it (not peridotites, but the gabbro family): allivalite (anorthite-bearing rock with augite), harrisite (intermediate between peridotite and anorthosite), and, further, eucrite, gabbro, and granophyre with granite. Two phases have been distinguished. The rocks of the first have just been listed. The second is represented by radial dikes and cone sheets of basic rocks. It is of interest to note a well-exposed sequence composed of alternating intrusive sheets of allivalite and rock near peridotite in composition.

On the islands of Canna and Egg, near Rum, basaltic lavas are found, and with them mugearites. There are thus grounds for believing that the Rum volcano passed through the same stages as the two previously described.

Northwest of Skye, near the Outer Hebrides, lies the small group of Shiant Isles [Walker, 1930]. A horizontal sill from these islands has been described. It reaches a thickness of 150 m at one place, but it changes in makeup from one district to another, and the incompleteness of individual sections leads us to believe that the total thickness is much greater. The following generalized section is cited [Turner and Verhoogen, 1960]:

5. Teschenite . . . . . . . . . . . . . . . . . . . . . . . . . . . . . . . . . . . . . . . about 60 m
4. Teschenite with very small amounts of analcime . . . . . . . . . . . . 100-120 m
3. Olivine diabase . . . . . . . . . . . . . . . . . . . . . . . . . . . . . . . . . . . .      30 m
2. Picrite . . . . . . . . . . . . . . . . . . . . . . . . . . . . . . . . . . . . . . . . . .     3-5 m
1. Basal (chilled ?) facies of picrite, thickness not given

Three analyses have been cited: an average for the teschenites, for the picrite, and an individual one for syenitic rock. The picture on the qz diagram is shown in Fig. 16.

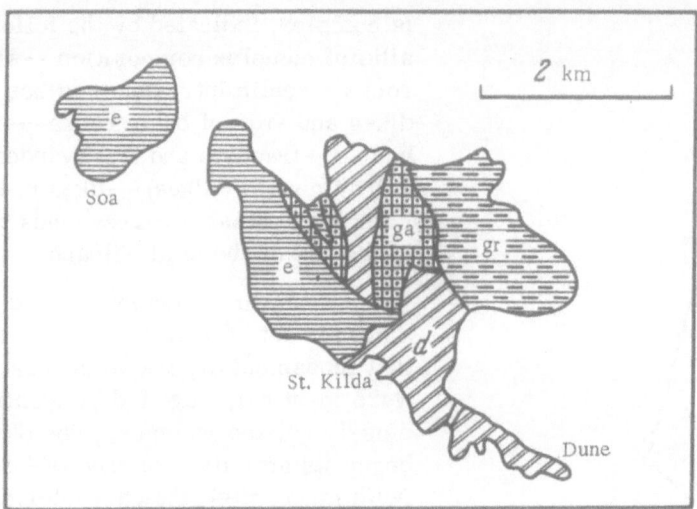

Fig. 17. St. Kilda Island group (after Cockburn [1932]):
e) eucrite; ga) gabbro; d) dolerite; gr) granophyre.

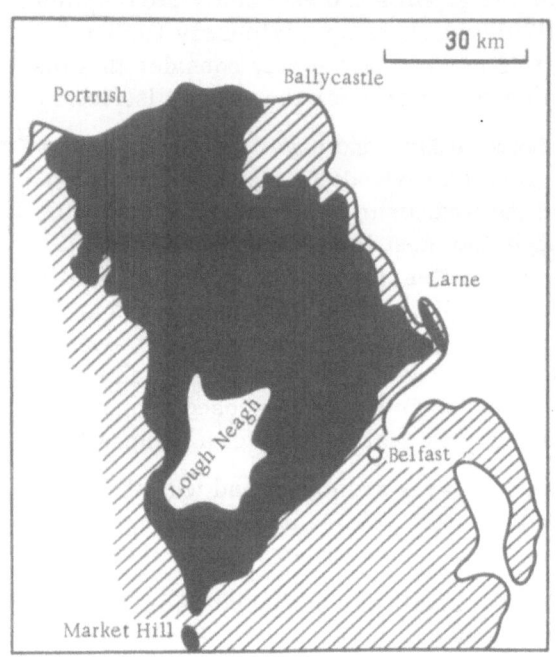

Fig. 18. The basaltic Plateau of Antrim, north-eastern Ireland (after Patterson). Basalt shown by black.

The "alkali" bend on this figure is extremely sharp, reaching negative values of K. The insufficiency of data means that we cannot state whether this sill represents a single stage of vulcanism in this region, or whether three stages may be present, as in the preceding examples. It is equally difficult, adopting a general scheme of vulcanism for the entire region, to refer this sill to either the first or the third stage. Only the large size of the body lends credibility to the view that it belongs to the first stage.

Basalt fields appear again beyond the Outer Hebrides, but they are covered almost entirely by the sea. Basic intrusive rocks are found on Rockall Island. Basalts of the same type as those in Scotland have been raised from the Rockall bank, and olivine gabbro has been picked up from the Porcupine bank. Intrusions of eucrite, gabbro, dolerite, and granophyre, i.e., clearly the tholeiitic lineage, have been studied on St. Kilda (Fig. 17) [Willie and Drever, 1961-62].

Northeastern Ireland. The continuation of the western Scotland zone to the south-west is the Antrim Plateau in Northern Ireland (Fig. 18). This plateau is about 100 km across from north to south; 60 km from east to west. It is made up of a series of basalt flows. Quartz trachyte, obsidian, and rhyolite are found in small quantities. The section of rocks is

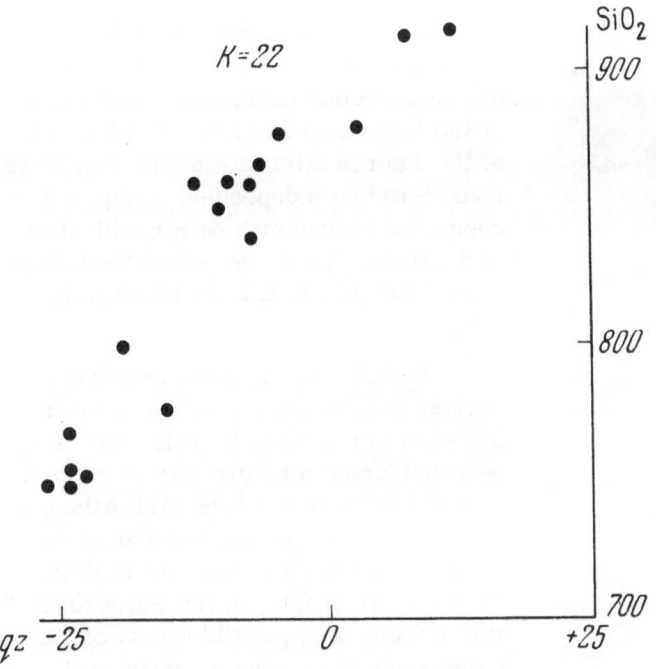

Fig. 19. qz diagram of basalts from the Plateau of
Antrim.

divided into three parts: the lower and upper are composed of olivine basalts of the same time as those in Scotland (the "Hebrides type" in the terminology of S. I. Tomkeieff). No petrographic or chemical differences have been found between these two units. The middle part of the section is represented by typical tholeiitic varieties of basalt, very similar to the nonporphyritic basalts of Mull. Analyses (taken from the papers of Patterson [1952] and Patterson and Swaine [1955]), plotted on a qz diagram (Fig. 19), form a remarkably straight belt, making it difficult to doubt the unity of the members. Analyses of quartz trachytes also lie within this belt. We thus appear to be dealing with a single tholeiitic association (K= 22, which is similar to the K for the tholeiitic volcano of Mull).

The similarity of these rocks to the associations of western Scotland merely serves to emphasize the existing difference: the complete absence of mugearites or their correlatives. There is no tendency toward development of the olivine-basalt series in Antrim. Nor is there a terminal stage with its alkalic rocks and high-silica basalts.

Such comparisons lead but to surmises. Available data do not permit us to confirm or deny. There is no doubt concerning the kinship between the magmas of Antrim and those of Scotland. The similarities are so great as to make the two practically identical. It may be consequently assumed that magma of this type may easily form a tholeiitic association. This is confirmed by the above-cited diagram; this and the diagram for the Scottish complexes show that both groups of local basalts are clearly related. If this is so, why do mugearites appear in Scotland? We can hardly assume that this is the result of a special path of differentiation in some local "trap." The phenomenon has been observed at three centers at least. There remains the possibility that such differentiation might occur under certain specific conditions. The recent data of Ringwood and Green [1966] permit us to gain some concept of these specific conditions, to assume that the magma is held at depths of at least 60 km in case of the olivine-basalt—mugearite lineage. Under conditions of the described region (it is emphasized that this particular region is referred to, not just any region) the proper conditions were created beneath a center of eruption. This interpretation appears possible when we recall that late olivine-basalt series arise here also in a zone of volcanic centers but are absent in Antrim.

The Faeroe Islands. [Walker and Davidson, 1936.] Here young lavas are widely developed; they consist chiefly of olivine basalts (oceanites and other ultramafic rocks are found with them) and porphyritic (plagioclase) basalts. All associations are very basic; their peak on the curve for $SiO_2$ falls at 48% (Figs. 20, 21). The qz diagram gives no clear picture, since it includes no data on relatively silica-rich rocks. If the most silicic of the analyzed rocks mentioned by Noe-Nygaard and Rasmussen [1957] is characteristic, the Faeroe Islands basalt then belongs to the olivine-basalt series. A comparison of these rocks with the Scottish lavas is also difficult, since the porphyritic (feldspar) basalts of the Faeroes are not

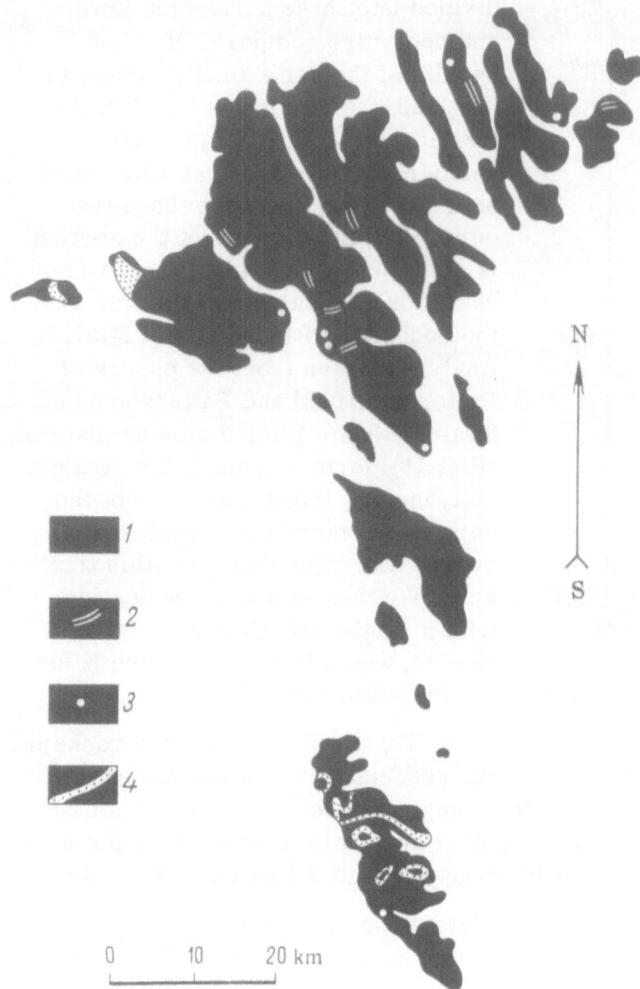

comparable chemically with the non-porphyritic basalts of Scotland. Nor are there rocks correlative with those of the last stage on Mull. The basalts of the Faeroe Islands must be therefore considered an independent group, not connected by common origin with those of Scotland. They are somewhat tentatively assigned to the olivine-basalt series.

Iceland. As already noted earlier [Sheinmann, 1964], Iceland is a difficult problem. Its vulcanism may be considered to be directly connected with the vulcanism of the Mid-Atlantic Ridge and also with the vulcanism of the ancient land mass in this region. As a matter of fact, in the Paleogene (the Eocene and, possibly, part of the Oligocene), there was a continental segment connecting North America (Greenland) with northern Europe. Even a morphological trace of this segment has been preserved in a distinct transverse ridge passing through Iceland. The island is thus situated at the intersection of this ridge and the Mid-Atlantic Ridge.

Some time ago, when there was as yet no summarized information concerning the chemistry of basaltic rocks taken from oceanic depths [Engel and Engel, 1963, 1964₁ 1964₂;

Fig. 20. Map of the Faeroe Islands (after Walker and Davidson). 1) Basalt (lava); 2) sill; 3) neck; 4) sedimentary rocks (with lignite).

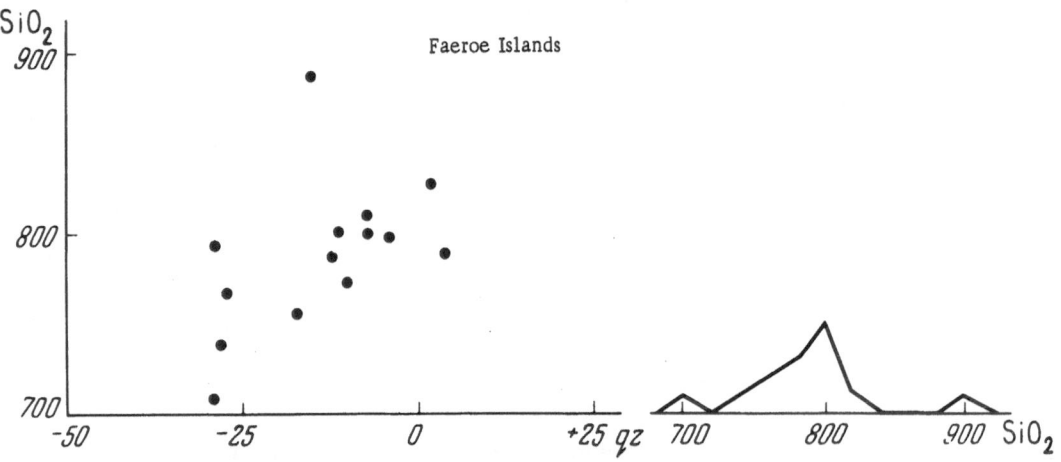

Fig. 21. qz diagram and variation diagram for $SiO_2$ for rocks of the Faeroe Islands.

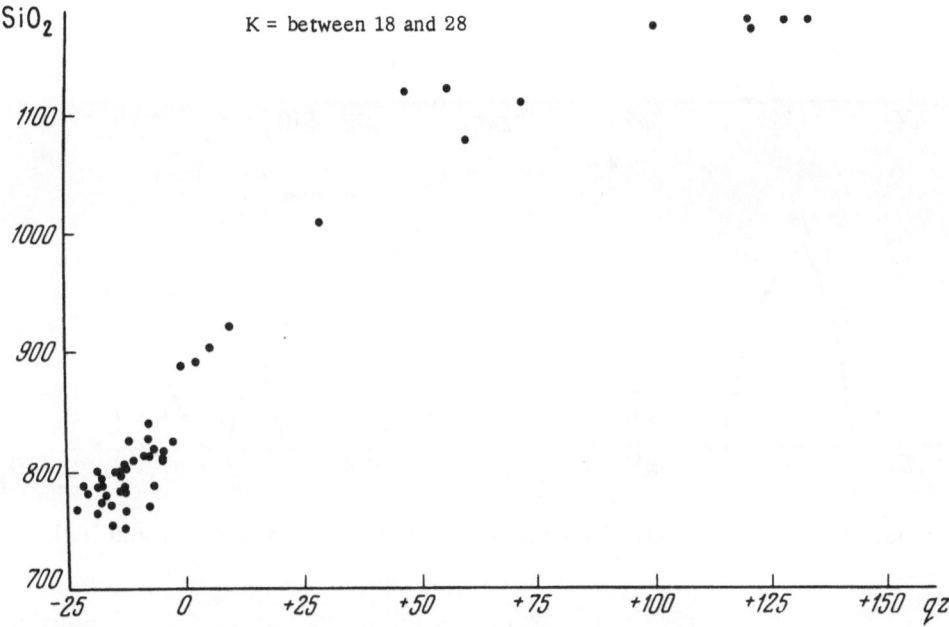

Fig. 22. qz diagram for the rocks of Iceland.

Sheinmann, 1964], it was assumed that the trap character of the Icelandic lavas was direct indication that they are manifestations of deep-seated processes inherited from the continental crust that existed here earlier. However, in light of present data, this simple solution of the problem is inadmissible. It is necessary once again to compare the characteristics of the Icelandic lavas, trap, and oceanic tholeiites. This we will do a little later, but here we pause to consider some of the general aspects of vulcanism in Iceland.

Tholeiites of oceanic islands have been studied only in Hawaii. They are nowhere widely developed (although they occur in notable quantities). In Hawaii the principal stage — tholeiitic — gives way to an olivine-basalt stage, rather minor, the rocks of which form the uppermost part of the volcanoes. Only Mauna Loa and Kilauea have not emerged from the tholeiitic stage. Engel and Engel proposed that all complex olivine-basaltic rocks of oceanic islands are the summits of tholeiitic volcanoes. This hypothesis is very likely, but it requires confirmation from factual data.

There are no olivine basalts on Iceland. All the lavas and intrusive rocks belong to the tholeiitic series. Thus, in assuming that these rocks belong to the oceanic type, we must believe that the process is still in the first stage and that nowhere has it reached the stage of formation of olivine-basalt magma. This can hardly be the most likely solution, since the duration of volcanic activity in Iceland has been at least as great as in Hawaii, and it would therefore be logical to expect advent of the second phase (in Hawaii the first stage has been preserved only on the youngest island, Hawaii; it ended long ago on all the older islands). The existence of the olivine-basalt series on Jan Mayen reinforces our doubt. The qz diagram leaves no doubt concerning the tholeiitic nature of the association (see Fig. 22), with $K \leq 28$ but greater than 20. It has been impossible to find any appreciable difference between the Tertiary and young lavas in the trend of this line. A comparison of the value for Icelandic lavas with other tholeiites shows the following:

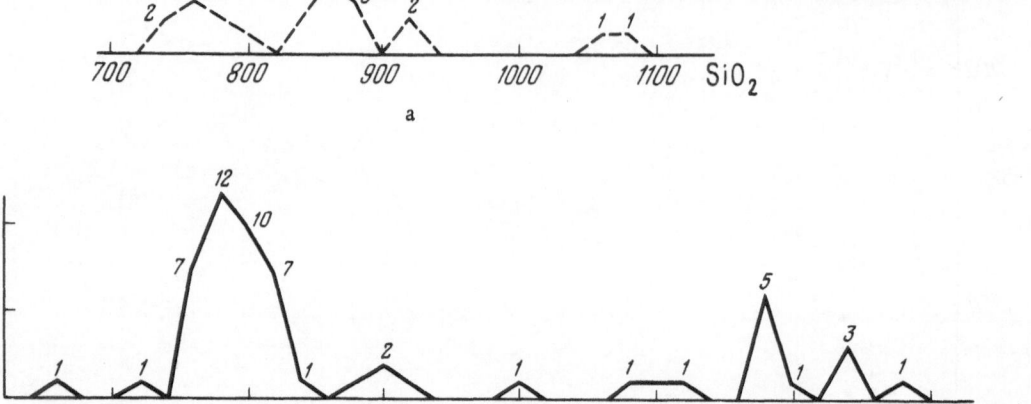

Fig. 23. Variation curves for SiO₂ of rocks on the Plateau of Antrim (a) and Iceland (b).

$$
K \begin{cases}
\leq 28 & \text{Iceland} \\
\approx 29 & \text{"Deep-water" basalts (average of all samples)} \\
= 28 & \text{Hawaiian lavas} \\
30 \text{ to } 50 & \text{Siberian traps} \\
= 40 & \text{Karroo dolerites} \\
= 33 & \text{Deccan traps}
\end{cases}
$$

The value of K for the Icelandic lavas is thus lower than for typical traps and is similar to the value for oceanic tholeiites.

In silica content, the Icelandic lavas are characterized by a curve with a peak typical for traps, very sharply expressed but there are also intermediate rocks, as a result of which that between silicic differentiates and basalts is somewhat blurred (Fig. 23). A comparison of positions of the peak on the curve (undoubtedly characteristic of each association) is shown in the table below (the peak corresponds to SiO₂ content in weight %).

| Location | SiO₂ Wt.% | Comments |
|---|---|---|
| Iceland | 47 | |
| Hawaii | 47-51 | The peak is broad, with two small crests and a shallow saddle between them; the principal crest is at about 50% |
| "Deep-water" basalts | 49 | |
| Deccan | 49 | |
| Karroo | 50-53 | The average value is between 51 and 52% |
| Siberian traps | 48 | |

The Icelandic rocks thus prove to be very basic. The Siberian traps appear to be most like them.

Lastly, it may be of interest to consider the nature of the differentiation. We know that trap complexes typically exhibit extreme siliceous differentiates (rhyolites, granophyres) and practically no intermediate varieties (andesitic composition). We should add, however, that the silicic rocks are most frequently found in intrusive masses, rarely among lavas. However, lavas of this composition are encountered: in the Karroo series (Lebombo) and, as a rarity, in the western part of the Deccan field. For oceanic ("deep-water") tholeiites and the Hawaiian tholeiites, silicic differentiates are unknown (if we ignore one or two very thin dikes on the island of Hawaii). In Iceland we know of both silicic volcanic rocks (such as some of the Hecla lavas, liparite from Hrafntinnusker, dacitic obsidian from Rautnaskrit or Hrafntinnuhryggur, and others) and intrusive granophyres. This distinction cannot be absolute, of course, but the difference between the islands of Hawaii and Iceland strikes the eye, although both regions are found in approximately identical geologic situations (this leads us to consider the probable differentiation of silicic rocks to take place in like manner) and are approximately alike from the point of view of exposures, i.e., the possibilities of finding these differentiates during geologic investigation. We should add to this the fact that the island of Hawaii has probably been studied more thoroughly than Iceland.

From the above discussion we may make the following summary: 1) the Icelandic lavas according to the change in silica content during differentiation (the value of K) clearly belong to the tholeiitic group and are similar to the principal tholeiitic formations of the world (especially to the tholeiites of Hawaii and the ocean floor); 2) the differentiation series of the Icelandic association is not oceanic, but clearly repeats the group of continental tholeiite-trap formations; 3) the rather poorly defined break in the differentiation series is a point in common for the Icelandic lavas and the Karroo dolerites; and, lastly, 4) in silica content the rocks constituting the bulk of Icelandic lavas are more basic than the corresponding rocks of most tholeiitic associations. In this regard the Siberian traps are most nearly like them.

Thus, a simple comparison does not permit us to draw any absolute conclusions concerning whether the Icelandic rocks are typical of midocean ridges or of the continents. But if we compare what we have said with the fact that we have not yet found in Iceland a single rock representative of the olivine-basalt series, we are forced to believe that in Iceland we have to do with rocks similar to tholeiite-trap. In their appearance there are more features relating them to this type than to the oceanic basalt association. It is not difficult to find the cause of this phenomenon. There is practically no doubt that the appearance of silicic differentiates is related to conditions of magmatic life within a continental crust (secondary reservoirs). Apparently, under conditions of the thin oceanic crust, prolonged existence of conditions favoring long-continuing differentiation is not very likely, if it is possible at all. Therefore, the presence of silicic lavas and intrusive rocks in Iceland most likely indicates that a rather thick, thermally insulated crust existed beneath the island. Further, the presence of these silicic rocks would appear to confirm the fact that this type of crust existed here for a long time, most likely from the time the continental bridge was destroyed. But this was then a primary continental crust (or, perhaps, intermediate), and only to the extent the region of Iceland was drawn into an oceanic framework was the crust gradually converted to the oceanic type. Even in this it is anomalous, however: in thickness and in the presence of a low-velocity layer.

Iceland may be therefore included in the region of continental basalts (although it also exhibits processes originating within and under the midocean ridge). The specific influence of the latter is apparently small. We have seen that until now (since the end of the Oligocene, beginning of the Miocene, at least) no olivine-basaltic rocks have formed in Iceland, although they are present in Jan Mayen. We may add that vulcanism in the zone of the midocean ridge, where this latter emerges onto the land, by no means has to be similar to that of Iceland.

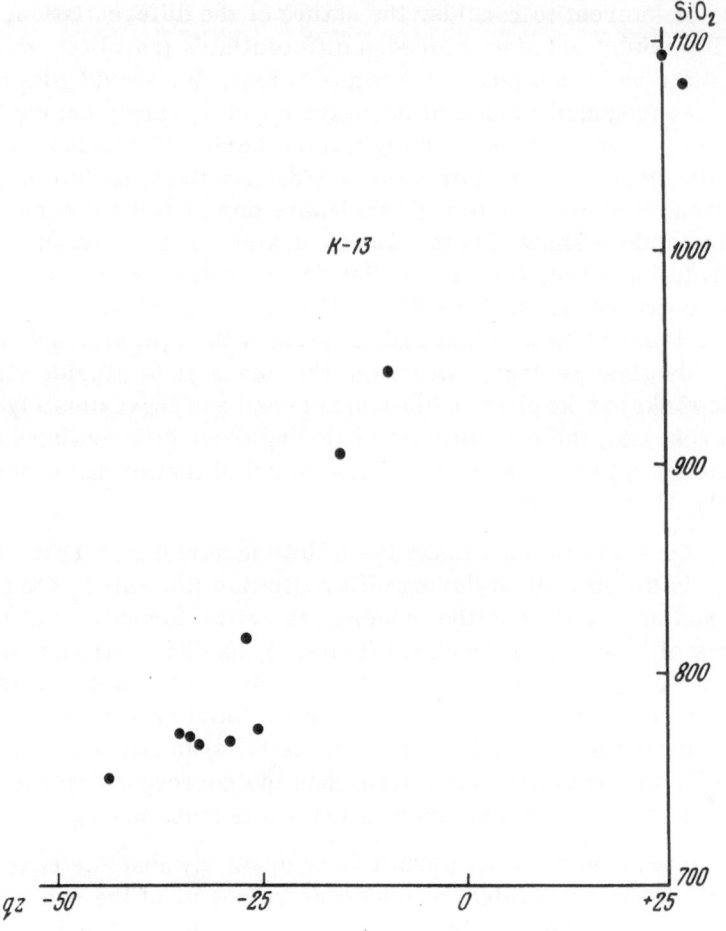

Fig. 24. qz diagram for the rocks of Jan Mayen.

Apart from Iceland, the single undoubted region of this kind is the rift zone of the Red Sea and East Africa, but there we nowhere encounter tholeiites, only the olivine-basalt and alkalic ultrabasic associations.

Thus, in all probability Iceland has survived two epochs of vulcanism: the first when it became part of the Brito-Arctic province, and the second when, being drawn into the oceanic zone, it lay on the continuation of the midocean ridge (the zone may here emerge on the surface of a past land mass). There may or may not have been a break between these two epochs.

Jan Mayen. Judging from the works of Tyrrell [1925-26], Holmes [1918], and Carstens [1961], typical rocks of the olivine-basalt lineage are developed on this island. In these works are described olivine and olivine-free trachybasalts (basic, with a silica content of 45-47%), along with basic olivine trachyandesite-ankaramite, trachyandesite, and trachyte. When plotting a qz diagram from the available 12 analyses (Fig. 24), one's eye is struck by the absence of analyses (and rocks, perhaps) with an $SiO_2$ content of 48-53%; the line of differentiation proves to be inaccurate. Generally in the olivine-basalt series, apart from the characteristic upper bend in this line from trachyandesites to trachytes and beyond, we find a lower bend from basic basalts through varieties richer in silica to trachyandesites. As a result, the curve as a whole is S-shaped. On the diagram of Jan Mayen lavas the lower bend appears to be absent (because of the indicated lack), and the value of K is clearly exaggerated. A formal plot gives K = 13. Actually it should be written K ≤ 13, and if we had these intermediate

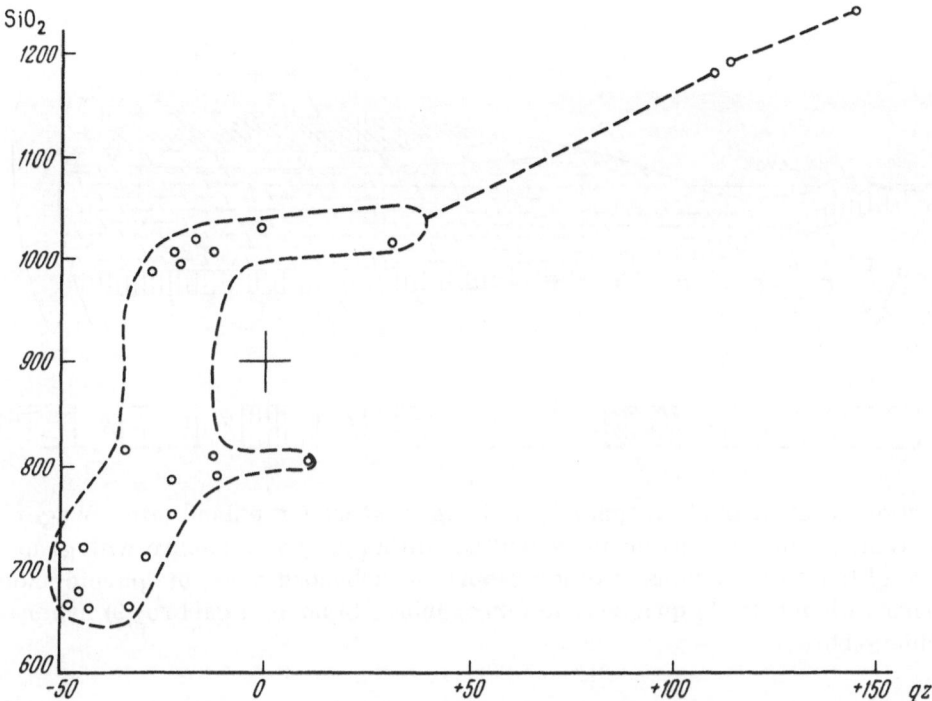

Fig. 25. qz diagram for the intrusions on Scoresby Land, eastern Greenland.

analyses it would prove to be appreciably lower. The peak on the $SiO_2$ curve apparently falls approximately at 47%. As pointed out above, the inclusion of the typically oceanic island Jan Mayen in a continental province must be considered unfortunate.

Eastern Greenland. A group of doleritic intrusions with long-continued differentiation has been described from Scoresby Land at the entrance to King Oscars Fjord [Kapp, 1960]. The intrusions occur along a northeast-trending fracture and are subvolcanic bodies. Their age is Tertiary. They cut horizontal strata ranging in age from Carboniferous to Cretaceous (Campanian). They consist of gabbro, gabbro-diorite, monzonite, syenite, and alkali granite. Petrographically and chemically (Fig. 25) this complex is distinguished by a clear alkalic trend (K = 6). In this region basic magma formed a number of sills and dikes composed of dolerites and diabases. In age they are older, coeval, and younger than the gabbro-syenite bodies.

The absence of analyses of olivine, olivine-free, and porphyritic (feldspar) basalts from the region prevents our drawing any general picture of vulcanism. The presence of the olivine-basalt lineage cannot be doubted, however. We note that the characteristic features of the province as a whole are preserved in eastern Greenland: the olivine-basalt series gives way to the tholeiitic in other regions. An example of this is the well-known Skaergaard intrusion [Wager and Deer, 1939] with its tholeiitic series (Figs. 26, 27). Furthermore, in other fields composed of volcanic rocks, typical tholeiitic series apparently predominate. It is possible that the olivine-basaltic rocks of eastern Greenland will prove to be derivatives of tholeiitic magma. On the western coast of Greenland we again encounter tholeiitic basalts of approximately the same age, and this means that the province extends even farther to the west.

It may be thought that in the Brito-Arctic province we have come up against relations of basaltic magmas not yet considered by us. Some fields, such as the Tertiary lavas of Iceland, have been formed solely by tholeiitic magma, undoubtedly primary. No traces of

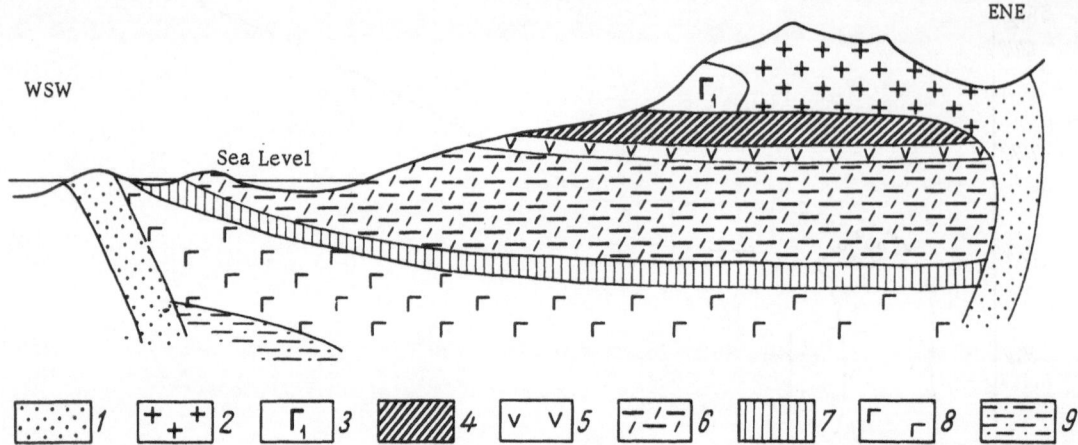

Fig. 26. Section through the Skaergaard intrusion, eastern Greenland (after Wager and Deer). 1) Olivine gabbro with granophyre and xenoliths; 2) quartz gabbro with granophyre and xenoliths; 3) inclusions of older gabbro; 4) unbanded zone; 5) "purple" zone; 6) ferrogabbro with quartz; 7) quartz-free ferrogabbro; 8) normal gabbro; 9) hypersthene-olivine gabbro.

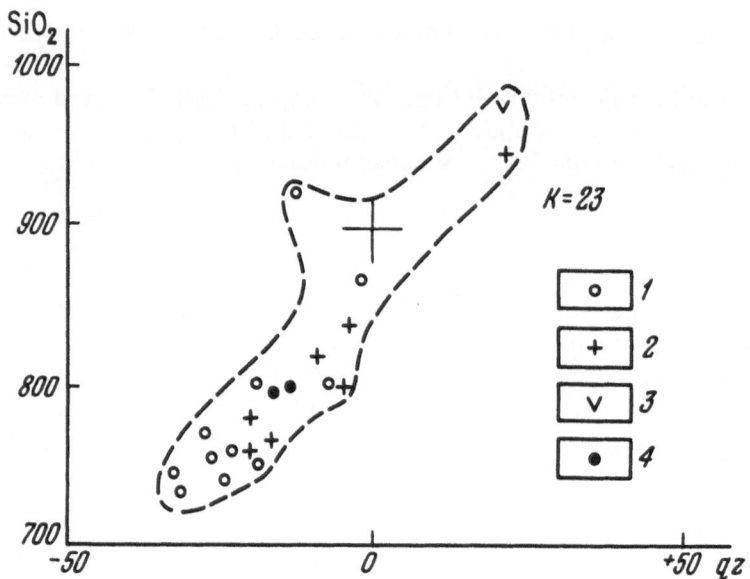

Fig. 27. qz diagram for the Skaergaard intrusion. 1) Rocks of gabbroic body; 2) rocks of marginal intrusion; 3) dike; 4) chilled zone.

olivine-basalt magmas have yet been discovered here. Tholeiitic basalts have continued to pour out almost without break up to the present. On the other hand, fields are known within the same province, comparatively small it is true, in which the only lava is olivine basalt and where there are no reasons for seeking a tholeiitic magma.

Such is true for the Faeroe Islands and, perhaps, Jan Mayen (if it actually belongs in this province), and also Scoresby Land in eastern Greenland.

## TABLE 2

| Analysis No.<br>Component | 1 | 2 | 3 | 4 | 5 | 6 | 7 | 8 |
|---|---|---|---|---|---|---|---|---|
| $SiO_2$ | 49.3 | 49.1 | 49.1 | 49.3 | 49.8 | 50.7 | 50.3 | 48.9 |
| $TiO_2$ | 2.8 | 3.0 | 3.2 | 3.9 | 1.3 | 1.4 | 1.6 | 2.1 |
| $Al_2O_3$ | 13.5 | 13.7 | 13.8 | 13.0 | 16.9 | 17.1 | 16.1 | 13.7 |
| $Fe_2O_3$ | 5.4 | 5.5 | 4.2 | 5.3 | 1.9 | 1.7 | 2.7 | 2.7 |
| FeO | 8.7 | 8.9 | 9.6 | 8.1 | 7.3 | 7.0 | 7.2 | 9.6 |
| MgO | 5.9 | 5.2 | 5.5 | 5.5 | 8.2 | 7.5 | 7.0 | 7.2 |
| CaO | 11.0 | 10.7 | 10.5 | 10.3 | 11.3 | 11.5 | 11.8 | 10.3 |
| $Na_2O$ | 2.5 | 2.8 | 2.5 | 2.7 | 2.8 | 2.8 | 2.8 | 2.6 |
| $K_2O$ | 0.4 | 0.5 | 0.5 | 0.4 | 0.1 | 0.1 | 0.2 | 0.7 |
| $P_2O_5$ | 0.3 | 0.4 | 0.5 | 0.4 | 0.1 | 0.1 | 0.2 | ? |

Note: 1) Faeroe Islands, 2) eastern Iceland, 3) eastern Greenland, 4) western Greenland, 5) Mid-Atlantic Ridge, 6) Mid-Indian Ocean Ridge, 7) East Pacific Rise,
8) young lavas of Iceland (average of 12 analyses, added by us).

Lastly, large volcanoes in this province have been described, the main mass of which comes from both olivine-basalt and tholeiite magmas. At the end of their activity they again yielded lavas of the olivine-basalt series (northwestern Scotland; indeed the term "tholeiite" was first used for the rocks here). The existence of this type of eruptive mechanism outside of ocean basins indicates that tholeiitic volcanoes, crowned with a cap of olivine basalt, may exist under very different tectonic conditions and that the appearance of olivine-basalt magmatic residue in the magma chamber (probably secondary) is by no means related to specific conditions of the ocean.

After the present work had been completed, the author became acquainted with a paper of Noe-Nygaard [1966], in which it is stated that basaltic sequences of Eocene age, perhaps of Cretaceous age also, are present in the Faeroe—Iceland—Greenland zone. These basalts of the Brito-Arctic province occur on the Wyville-Thomson Ridge and can in no way be associated with the basalts of the Mid-Atlantic Ridge. The latter basalts are represented by the young lavas of Iceland and Jan Mayen. It has been noted (by Hawkes) that a break occurs in the middle of the Tertiary between outpourings of these two series in Iceland. There are thus two basaltic belts in Iceland, differing in age and crossing each other. This conclusion of Noe-Nygaard is in complete agreement with what we have just discussed in this work. It is further reported that tholeiites forming the lower third of a 3-km sequence of lavas have been recently discovered in the Faeroe Islands. A comparison of the average composition of lavas along the Wyville-Thomson Ridge with the composition of basalts on midocean ridges shows substantial differences. For example, with approximately the same silica content, basalts of the first locality contain much more titanium, more iron, potassium, and phosphorus, and less aluminum and magnesium than the oceanic rocks. This may be seen in Table 2.

Rocks of the Wyville-Thomson Ridge are thus not oceanic and are clearly enriched in "insubordinate" elements (see Chapter 14). We may note that the average composition of the young Icelandic lavas deviate somewhat toward oceanic basalts (this was suggested by us earlier). On the other hand, the lavas of the East Pacific Rise also diverge somewhat from the lavas of midocean ridges.

Consequently, that which we know about continental basalts indicates that primary basaltic magma with an alkalic trend will very probably appear under certain conditions. The existence of such a magma is apparently possible in regions where only olivine-basalt melts are manifested (Ethiopia, Syria, and elsewhere) and where both basaltic magmas appear at the same

time, or with some interval of time between, each under definite  conditions of temperature
and pressure, i.e., at different depths.  On the other hand, observations in northern Scotland,
and also in Hawaii, show that magmas indistinguishable from primary olivine-basalt magmas
form also as residuals during the development of tholeiitic magmas.  It appears that there are
no real grounds for doubting the possible existence of primary olivine-basalt magma.  In the
list of primary magmas given above, therefore, it was necessary to mention it.

### The Relations among Olivine-Basalt, Andesite-Basalt, and Granite Magmas

We have already spoken of the fact that within geosynclines olivine-basalt magma is
absent or practically so.  However, in place of the typical tholeiite-basalt magma, a new
variety appears here, although it yields, like the other, a normal series of differentiates, i.e.,
not oversaturated in alkalies.  It still deserves to be distinguished as an independent group of
primary magmas.  If we turn to the series of differentiates of tholeiitic magma proper (tholeiitic basalt), a special characteristic proves to be the complete or almost complete absence of
rocks of intermediate composition.  The series consists of basalts (from olivine basalts, some
even very rich in olivine, commonly to andesite-basalt; olivine-free basalt is more characteristic) and silicic rocks, most commonly rhyolites or granites.  Correlatives of granodiorites
may also be encountered.  Andesite and their correlatives are absent or are extremely rare,
and they make up but a negligible volume.

The andesite-basalt magma is characterized by an almost complete series.  Typically
this contains olivine-free basalts, andesite-basalt, andesite, and dacite.  Determination of the
composition of the original magma to the extent it is generally possible gives the composition
of olivine-basalt for tholeiite-basalt magma and andesite-basalt for the andesite-basalt magma,
since it is richer in silica than the first.  Apparently all transitions exist between these two
magmas.  However, discrimination of andesite-basalt magma as an independent group is justified by its restriction to geosynclines, whereas tholeiitic magma predominates outside of geosynclines.  In geosynclines, the only magma that may resemble it may be the spilite-kerato-
phyre series or basalts of the period of subsidence.  However, it is hardly proper to consider
these geosynclinal magmas fully identical to the tholeiite series.

It is well known that the andesite-basalt series of island arcs commonly gives way in
later epochs or perhaps grades along the strike to typical andesites, in places rather rich in
silica (according to E. K. Ustiev these are correlatives of granodiorite).  It is now almost universally acknowledged that chemically andesites correspond to the average composition of the
continental crust or to its upper part.  From this it is commonly concluded that andesites are
generated in the crust, are the fusion product of this crust.  On the other hand, no one acquainted with these data can now maintain that basalt is also the product of crustal fusion.
Still very recently this view was considered self-evident.  Correlation of the "basaltic" layer
with basalt and the melting out of basic magma from it aroused no doubt.  The data of geology,
geophysics, and physical chemistry, however, have shown that these views are incorrect:  of
geology because basalt of the ocean cannot be melted within the thin and cold oceanic crust; of
geophysics because the rise of basaltic magmas from depths many times the thickness of the
crust has been traced by accompanying earthquakes; of physical chemistry because the limits
of temperature and pressure at which basalt may fuse, even for the thick continental crust, are
not generally reached.

The result of the statements above is the assumption that the magmatic center, gradually
shifting upward, at first supplies basaltic magma, which then gives way to andesitic when the
center rises into the crust.  Further rise of the center leads to the formation of granites.  This
surmise can hardly be true, however.

Some time ago, Blot [1964] called attention to the fact that the connection between earth-quakes and the rise of magma before eruption of a volcano has been noted not only for Hawaii (it was first noted there for Kilauea [Powers, 1955]) and the New Hebrides [Blot and Priam, 1963], but also for a number of other localities. Blot himself was not interested in the composition of the erupted lavas; his interest lay merely in the fact of the relationship itself. Among the localities mentioned by him, data were given for Halmahera and Paricutin. For these centers, rise of lava was recorded from depths many times greater than the depth to the Mohorovičić discontinuity, and the composition of the magmas is andesitic. Thus, doubt is sometimes raised concerning the absence of any connection between the crust and andesite, more properly the absence of the connection crust—andesite; the reverse, andesite—crust, appears very probable.

These observations lead us to believe that the transition to andesite-basalt and andesite magma, taking place somewhere in the upper mantle, is systematic and is in no way connected with processes in the crust. Andesite magma also proves to be from the mantle. On the other hand, it is directly related to andesite-basalt magma, which in no way differs from it in composition (completely gradual transitions) and is restricted to the same tectonic environment. Therefore, to separate it from andesite-basalt magma, i.e., to place it in a separate group, is impossible. We must consider that we are confronted with two not yet classified groups: we have just discussed one and have found that we cannot relate it to any processes in the earth's crust; the second is in no way akin to basalt; its magma grades directly into typical silicic magmas; it is a crustal magma and may be somewhat more silicic than the preceding.

CHAPTER 7

# THE TECTONIC DISTRIBUTION OF PRIMARY MAGMAS

In attempting to explain the tectonic conditions for the appearance of any particular variety (but not type) of magma, we must base our conclusions on tectonic peculiarities of depth, since magmas cannot be generated near the surface. Correspondingly, we shall use the division into three types of region that we adopted in Chapter 4: geosynclinal, midocean ridge, and extrageosynclinal. The overall picture is shown in Table 3.

Even a casual glance at the scheme of Table 3 shows us that there is no reflection of the contrast between continents and oceans. Actually there are no magmas that might be considered specific for oceans. The only exception is granitic magma, which is associated with continents. This, however, is explained by the special position of such magmas: they can form in large masses only within a continental crust. Their temperature of formation is low. The picture is different for all varieties of basaltic magma and for ultrabasic magma. The melting out of basaltic liquid from the substance of the mantle, even more the continuous fusion of this substance, is possible only at temperatures much greater than temperatures existing in the crust (even considering the low pressures there). Direct connection with any particular type of crust is therefore impossible in this present consideration. There remains the possibility of some indirect connection, since conditions at depth under oceanic and continental crusts differ. As we know [Magnitskii, 1965; and others], heat flow at the floor of the ocean and at the surface of the continents is approximately the same. On the continents it is generated chiefly by substances in the crust and only in minor degree from the mantle. In the oceans there are no such concentrations of radioactive elements in the near-surface zone, and heat generators are more or less uniformly distributed to considerable depth, though the concentration of these is notably lower than in the continental crust. As a result, in order for the heat flow to be approximately the same in both types of region, a much more rapid increase in temperature with depth is necessary under the ocean. An approximate computation [Magnitskii, 1965, pp. 9–12] shows that what we expected is true. In Fig. 28 it is seen that at depths less

TABLE 3

| Type of magma | Tectonic depth zone | | |
| --- | --- | --- | --- |
| | Geosynclinal | Midocean ridge | Extrageosynclinal |
| Ultrabasic | "Alpine-type" peridotite | "Alpine-type" peridotite | Alkalic-ultrabasic |
| Basaltic | Andesite-basaltic | Tholeiitic | Tholeiitic |
| | | Olivine-basaltic | Olivine-basaltic |
| Granitic | Granitic | Very poor | |

55

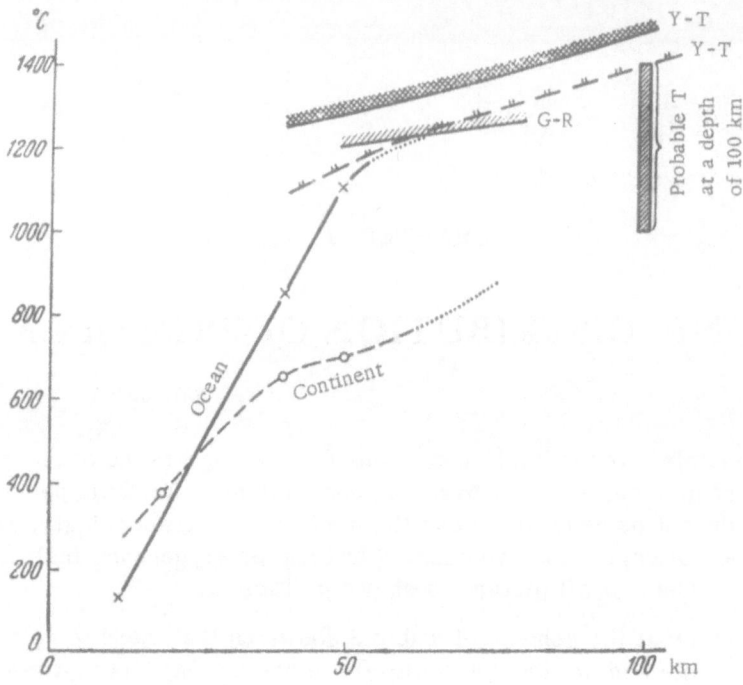

Fig. 28. A comparison of assumed temperatures beneath the
ocean and the continents, the probable temperature at a
depth of 100 km (according to Magnitskii [1965]), and data
on the fusion of basalt (after Yoder and Tilley, Y-T, and
Green and Ringwood, G-R).

than 30 km temperatures beneath the continents and under oceans are comparable, but temperatures at greater depths are higher under the oceans than beneath the continents. Somewhere at depths exceeding 60-80 km (it is still impossible to measure this depth, even approximately), the temperatures again approach each other, and there are no differences at depths
below this.

Magnitskii cites the results of various computations for temperatures at depths of about
100 km, and he shows that all these fall in the interval 1000-1400°C. Apparently no difference
between oceanic and continental zones is found at this depth.

A comparison of these data with the results of experiments by Yoder and Tilley [1962],
also plotted on our diagram, shows that we cannot expect the melting of basalt at any depth less
than 50 km, and that only under the ocean can this take place, under certain conditions, at any
depth notably less than 100 km (we should not forget that the curve for temperature beneath the
ocean bends to the right with increasing depth to 100 km, in the interval 1000-1400°C as shown
in Fig. 28). As we know, the magmatic center for Mauna Loa and Kilauea is assumed to lie at
this depth on the basis of seismic data. Thus, the absence of oceans and continents in Table 3
is understandable. The differences due to these two zones are effective only in the sense that
the upper boundary of possible basalt fusion is somewhat nearer the surface beneath the oceans
than beneath continents. Other conditions are the same, and these determine the possibility of
the appearance of basaltic and ultrabasic magmas.

We may note still another feature, emphasizing so to speak the nonexistence of any distinction on the earth's surface between oceans and continents, from the viewpoint of the formation of magma. We have already seen that it is possible to distinguish three types of regions

according to the character of present-day movements. Two of these—geosynclines and mid-ocean ridges—clearly grade from the ocean to the continent, and the reverse. This is the situation in modern geosynclines and young fold belts, and the same is true in respect to mid-ocean ridges, which, in at least one place, East Africa, emerge upon the continent.

CHAPTER 8

# THE EXTRAGEOSYNCLINAL TYPE OF MAGMATIC DEVELOPMENT

## Basaltic Magmas

Judging from the present, magmas do not appear under conditions of tectonic calm. Any appearance of modern magmatism or magmatism of the recent geologic past has been associated with zones that are to some extent tectonically active. We cannot determine the conditions of massive outpourings of trap on platforms or the penetration of large masses of the same basaltic magma into the body of a platform. However, it is hard to imagine that such vigorous phenomena might appear under calm conditions. Outpourings of plateau basalts relatively near us in time clearly took place against a background of rather active tectonics: movement along steep-angle faults, formation of deep fractures, and the like. At least the eruptions of Iceland and Ethiopia have been of this character. Outside platforms, within zones of culminated folding, there is no doubt about such conditions, but here the volume of erupted lava has generally been less than on platforms.

It might be noted that it is extremely difficult to determine precisely the nature of movements accompanying the outpouring of lava. Many, in attempting to establish some connection with subsidence (rarely with uplift) of districts flooded with lavas, have stated that intrusions are restricted to zones of uplift, and the like. There are no real grounds for maintaining such beliefs, however. Along with zones of subsidence (platform basins – syneclises – of the Tunguska region and the Karroo) we know of zones of uplift (Ethiopia). Nor are their any such systematic relations in the distribution of intrusions (the Bushveld complex can hardly be referred to a zone of uplift).

The connection between folding and zones of approximately simultaneous magmatism is much more substantial We have already had occasion to mention this more than once [Sheinmann, 1956], and, although we are not obliged to seek connections in detail in this case, their presence on the whole cannot be doubted. Such a connection leads us to state that tectonic movements accompanying eruptions and intrusions are echoes, as it were, of a more powerful tectonic process in the neighboring geosyncline. The regional connection between magmatism and folding is a significant indication of tectonic activity at the time and place of the rise of magma. This spatial connection between trap and folding is involved in the fact, as we know, that the trap region is attached to the border of the fold zone, as it were, and, if the platform is rather large, the trap region lies on that part adjacent to the fold zone. Furthermore, we may note a tendency of the trap field to be drawn out perpendicular to the trend of the folding, apparently indicating a corresponding elongation of the disturbed zone (but not individual components of its fractures).

The entire environment of a trap field is apparently rather distinctive, and it possibly differs strongly from other manifestations of magmatism in extrageosynclinal zones. In

59

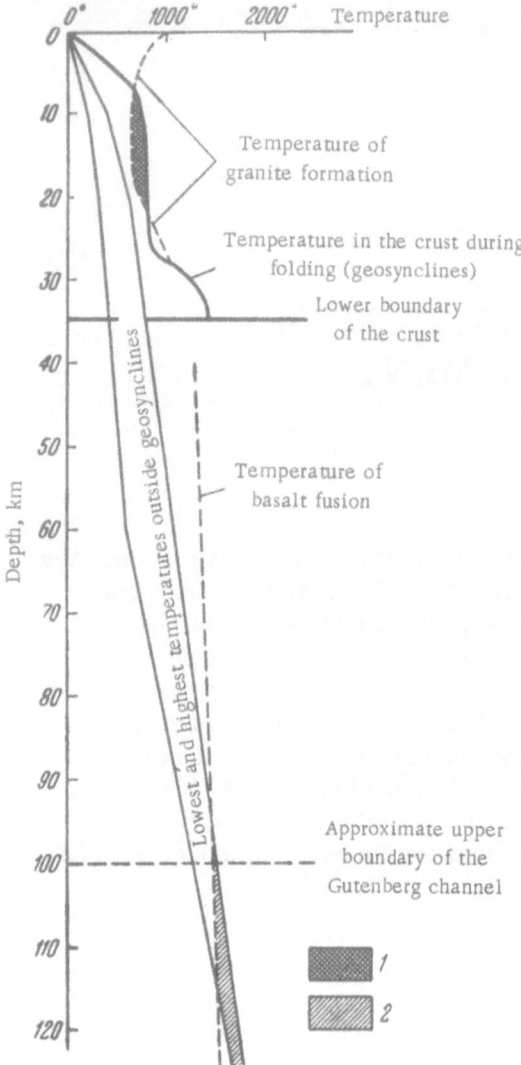

Fig. 29. Diagram of V. V. Belousov. 1) Melting zone of granites; 2) melting zone of basalts.

particular, the huge areas of trap fields astonish one. We are in no position to present a clear picture of the deep-seated tectonics of these regions, however. As a matter of fact, there is not a single modern trap field that might have been used as an example for studying these phenomena. It is possible that trap fields are regions of specific deep-seated tectonics and that should be separated as a special fourth type, or that might prove to be in some measure similar to the midocean ridge type.

We still know very little about the tectonic conditions of vulcanism in the oceans. The guyot fields undoubtedly differ appreciably from those observed on midocean ridges or on the Hawaiian uplift. In any case these regions require further study.

There can be no doubt of the conclusion that any phenomena of magmatism in extrageosynclinal regions are associated with tectonic mobility. This mobility is restricted, however. We may observe it both by the occurrence of deformation and by the distribution of earthquakes. The absence of any deep-seated foci most likely indicates very low contrast of movements at depth. Under these conditions, one cannot expect any increase in the local advent of energy in a region of magmatic centers.

The generation of magma may thus be associated primarily with conditions in situ. It is necessary to seek the cause of magma generation at the site. Data on the low-velocity channel of Gutenberg are of most interest in this respect. Observations of this channel indicate that it (or they, if we use the data of Luk [1966]) develops in its typical form only within extrageosynclinal zones. Data on oceanic ridges [Heezen and others, 1959; Le Pichon and others, 1965; Talwani and others, 1965], for zones of still very young folding, such as the ridges and valleys of the American Cordillera [Pakiser, 1963], and for island arcs [Fedotov and Kuzin, 1963], all show that in all these examples a so-called "crust-mantle mixture" appears. We shall not analyze this matter here; for us it is important to note the diminished velocity of waves beneath the Mohorovičić discontinuity. In the American Cordillera and in the Kurile arc this means that the upper boundary of the low-velocity channel is destroyed. And, since the lower boundary is very diffuse, essentially representing gradual transitions, we are obliged to admit that the low-velocity channel is indistinct in such regions, or is lacking entirely. In midocean ridges the process has apparently not gone so far, and there is no "erosion" of the upper boundary. It appears to be separated from the crust-mantle mixture by a layer with a wave velocity normal for the mantle.

In his last large work, Belousov [1966₁] presented a very interesting diagram (Fig. 29). Using the most significant data on geothermal relations, structure of the crust, and the like, he concluded that fusion of earth material is possible in two zones. The first of these is in the crust, approximately midway between top and bottom. Fusion here is possible only under conditions of an especially rapid increase in temperature, and this leads to the formation of granitic magma. The second zone is at a depth of 100 km. Here the "normal" increase in temperature with depth proves to be sufficient to melt basalt at the existing pressures. The possibility of melting basalt with no special additional effects is also very important. The computed depth of this layer practically coincides with the depth of the low-velocity channel. From this follows the conclusion of Belousov, significant for the solution of our problem, that the Gutenberg low-velocity channel is a zone in which basaltic liquid is constantly produced.

It should be noted that at the start it is not important to us whether this liquid is actually present constantly in the low-velocity channel or whether it exists in such a state that the most insignificant additional effect is sufficient to cause the liquid to begin forming. Whatever the truth, basaltic liquids appear in some zones when tectonic movements occur [Sheinmann, 1962, 1963]. At my request, V. S. Safronov has computed the probable rate of collection of the smallest amount of this liquid in large bodies (at the same place). The results are extremely interesting: the time necessary for such liquid to collect in sufficient quantities to give rise to what might be called a magma is comparable to the geologic stage of the earth; in any case it is at least equivalent to the entire post-Precambrian history of the earth. From geologic evidence, however, we know that whole magmatic cycles, with their complex succession of magmas may be approximately one order shorter, but individual episodes, such as the entire history of postfolding basalts in any region, may even be two orders shorter. Without the aid of differential movements, shear displacements, and other manifestations of movements in fault zones that might serve, so to speak, as catalysts for a process of combining very small segregations into magmatic masses, the appearance of these magmatic masses would be impossible.

Tectonics thus permits drops and intergranular films of future basalt to combine and move upward. The initially segregated liquid undoubtedly forms both small drops (i.e., more or less equant bodies) and thin intergranular films (practically two-dimensional). The latter must become a distinctive lubricant and appreciably increase the creep properties of the material and, consequently, increase its reaction to tectonic movements and decrease the probability of earthquake foci developing in this medium. The simplicity of this solution is only apparent, however. The liquid, being segregated in small quantities, may move and collect in magma only under exceptional conditions. Supplementary effects are therefore necessary, either great decline in pressure over short distances or melting not of drops but of very appreciable amounts (see Chapter 14).

The rise of the newly formed magma that has collected may be represented as the simple floating of a light "bubble" in a very viscous liquid. However, the extraordinary rise of basalt where it has been possible to observe it, compels us to believe that the mechanism of magma rise is different, that the principal path of ascent is prepared not by the magma itself, as would be true if it is a matter of floating. The path is prepared by a different agency: the magma uses "weakened zones" along which tectonic movements take place, i.e., deep fractures, although the existence of long-lived fractures at depth is impossible.

Observations in Hawaii give us some idea concerning the rate at which magma rises in a tectonically calm region [see Powers, 1955, or Eaton and Murata, 1960]. These authors have shown that when there are no special impediments, the lava rises at a very appreciable rate (see Table 4).

Of greatest interest are the data for 1942, when observations were more complete. Such observations permit us to state that magma does not float up slowly in a very viscous medium;

TABLE 4

| Volcano | Year | Depth at which movement was noted, km | Rate, m/hr |
|---|---|---|---|
| Mauna Loa | 1933 | 55 | 7,5 |
| • | 1935 | 43—44 | 30,5 |
| • | 1940 | 58—66 | 10,75 |
| • | 1942 | 50 | 35 |
| Kilauea | 1934 | 43 | 14 |

it is forcibly extruded along a channel, against what appears to be very little resistance. The cause of this extrusion may be merely the pressure of the surrounding masses, their weight. The weight of the column of magma is considerably less than the weight of the surrounding rocks plus the water in the ocean.

Let us recall in this regard that there are other proofs of the active penetration of basalts, such as their injection into the crust by separating and crushing layers of rock and even thrusting up such ground material in the form of distinctive pipes about the ends of sills. The author observed these phenomena in the region of the Siberian traps [Sheinmann, 1956].

One of the main characteristics of the described process of the separating out of basaltic magma is the possible content of such material in continental matter. Two approaches are possible in this respect. Vinogradov [1959, 1961], in experimenting with the substance of stony meteorites, probably similar in composition to the substance of the mantle, showed that it is possible to separate out about 7% of light liquids, similar to basalt, by zone refining. It should be noted that repeated separation by zone refining, as done in experiments, is an exceptionally effective method. In the other approach, we started from a computation of the amount of basalt that might be obtained from the substance of the mantle (selected, of course, according to our "taste") in order that the residue be dunite as the extreme member of a possible differentiation series. Such computations were made in due course by the author, and they gave a maximum yield of basalt of 15% (by weight). If we change the composition of the initial substance and take, instead of meimechite,* "rich" in alkalies, other possible representatives of the primary mantle substance, the figure is lowered. It is also lowered if the separation does not proceed to its conclusion, and if the residual is not dunite but one of the peridotites. Thus, this 15% is hardly an achievable maximum, and we obtain very good agreement between calculation and experiment.

However, these computations and experiments cannot be considered proofs entirely. As noted in Chapter 14, the separation of such small quantities of liquid from this "tough paste" is very difficult, and our knowledge at present compels us to assume a much more extensive fusion, as a result of which very mobile "magmatic quicksand" is formed. This "quicksand," if the volume is rather large, should be an obstacle to the passage of transverse waves. However, the difficulty encountered in attempting to establish a magma chamber beneath Klyuchevskii Peak [Gorshkov] indicates that even the presence of a considerable volume of "magmatic quicksand" is not easy to detect. If this mobile material is confined to a rather narrow channel, it is likely that seismic methods may find it impossible to detect the liquid, or may find it extremely difficult.

The principal conclusion that may be drawn from the above discussion is the following, as it appears to us: the generation and movement of magma in extrageosynclinal deep-seated tectonic zones may be divided into three stages. The first is the separation of films and drops in the substance of the mantle. This fusion of basaltic liquid is apparently caused by the state of material in the low-velocity channel and is not associated with any addition of energy to produce melting. The effect of tectonics may consist in facilitating fusion as a result of

---

*Meimechite, named for Meimechi River, is an ultramafic rock containing phenocrysts of olivine, pseudomorphs of pale yellow serpentine after olivine, and black opaque glass, with small amygdules of serpentine or carbonate. — Transl.

reducing pressure and, correspondingly, in increasing the amount of liquid, but this possibility should not be exaggerated. The second stage involves the transfer of the films and drops that have formed and their collection into large masses, i.e., the filtering of the liquid. This requires special conditions (see Chapter 14). The separation of small quantities of liquid is most likely a rare phenomenon. It is easy to separate liquid from a solid mass only when the liquid component represents more than half the volume. Lastly, the third stage, the rise of the magma that has formed, is related only to the energy of the earth's attraction, although the liquid squeezed out in this way follows a path formed by tectonic movements, aided, perhaps, by heating of the earlier parts of the magma.

## Silicic Magmas

The question of extrageosynclinal silicic magmas commonly arises. It is necessary, however, to consider the ordinary confusion that appears. As a matter of fact, without taking into account the specific character of the "granitic stage" of geosynclinal magmatism, it is easy to separate its late manifestations from the development of the geosyncline. We shall return to this question somewhat later. We now note merely that the development of fold regions does not end suddenly and that the conditions permitting granitic magma to form arise at a very late stage in the overall process, at the very end. It is natural that these conditions may exist at any time after culmination of the folding. More than this, they are characteristic of this period. But it is impossible to separate such granitic magma from the entire development and only when we make a completely formal approach are we able to draw the boundary between intrusion of late genuine granites at the time of postfolding elevation of the region and the formation of more complex silicic intrusions chiefly in the epoch of folding.

There are exceptions, where silicic magmas do not appear to be related to the geosynclinal process. The first of these is represented by volcanic and volcanic-plutonic belts. In recent years we in the USSR have intensified our study of such belts (they were first distinguished by E. K. Ustiev, and we devote Chapter 12 to their consideration). We shall note here merely that volcanic belts are genetically related very closely to fold zones, and, therefore, we cannot consider silicic magmas in belts as proof of the existence of granites independent of geosynclines.

The other exception is clearly intrinsic in extrageosynclinal zones. The best known of these is the type of ring intrusion in Southwest Africa (we shall touch on these formations in more detail in Chapter 12 in connection with the problem of silicic crustal magmas as a whole).

## Alkalic and Ultrabasic Complexes

Within extrageosynclinal zones we commonly encounter complexes of igneous rocks that can appear during differentiation only from an alkalic-ultramafic magma. These complexes are well known, have been repeatedly described in the literature, and therefore need no supplementary discussion here. We shall note first that this variety of magma, judging from the geologic environment in which it appears and from its petrographic relations, may be of two kinds. To the first we should assign all, or almost all, complexes of this composition on the continents. Most, if not all, such complexes are directly associated with eroded volcanoes and are preserved both as volcanic rocks proper and in vent facies and subvolcanic bodies.

But, to what extent is this variety of complex typical of the ocean? A number of oceanic islands have volcanoes the lavas of which correspond rather well with lavas of this type on the continents. It is remarkable, however, that neither on the Marquesas Islands, nor on Tahiti, Tubuai, or other similar islands, despite the presence of nepheline basalts, basanites, and related rocks, has anyone found the extreme members of the series (nepheline and alkalic

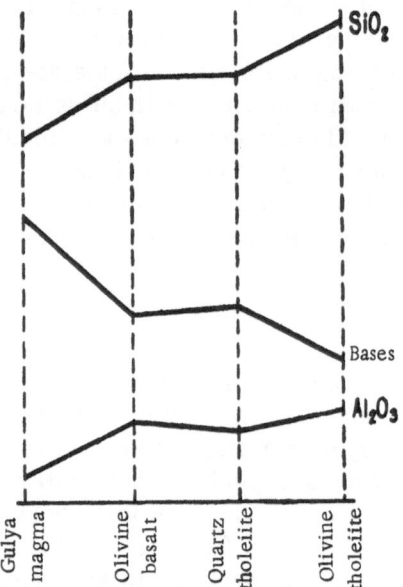

Fig. 30. Comparison of silica, alumina, and total principal bases in an alkalic ultrabasic complex, alkalic basalt, and tholeiites.

syenites) or, even more, hydrothermal formations such as carbonatites.* It is possible that this lack is found in the fact that no oceanic volcano has been destroyed to such an extent that the subvolcanic body has been exposed, but among the lavas and on continents there are generally no extreme products of differentiation. We should not forget, however, that carbonatite bodies are common in vent facies (East Africa), but up to the present time they are completely unknown in oceanic regions. We cannot now say, therefore, that alkalic–ultrabasic complexes of this type have developed both on continents and in the ocean (the author of the present work was still convinced but a short time ago of the presence of such typical complexes on oceanic islands). It now appears most likely that such complexes are not present in the oceans.

One of the cardinal questions is the problem of the relations between basaltic and alkalic ultrabasic rocks. There are adherents of both views: those who hold that the two magmas are independent and those who believe alkalic ultrabasic rocks are derived from basaltic magma. It is therefore important to ascertain to what degree the two rock series are interrelated in the geologic environment. If alkalic ultrabasic rocks are genetically related to basalts, we should be able to observe corresponding geologic connections. In this case, alkalic–ultrabasic complexes could not appear independent of basalts, and we could find them only together with the latter. This type of relationship is actually the normal one. Provinces of basaltic magma, trap fields (Tasmania, southeastern Australia, Colombia, the Appalachian foreland, the Karroo dolerites) are lacking alkalic ultrabasic rocks. The same applies to the olivine basalts of Ethiopia. In places, basalts are associated with approximately simultaneous alkalic ultrabasic complexes. For example, rocks of the latter type are found in the vicinities of Gujurat and Bombay (the Deccan trap field), in northern Siberia (the Gulya field within the trap region), and also in the region of the Parana traps (Jacupiranga). However, the view in question clearly contradicts the fact that alkalic ultrabasic complexes are developed in regions where no simultaneous basalts are present. This is the situation on the plains of North America, where several small complexes of this type (Magnet Cove intrusion and others), dated as Cretaceous, entirely lack "accompanying basalts." The same is true of our Kola province. It is difficult to find contemporaneous basalts also for the Sayan intrusions or on the Aldan shield. Even the bodies of the East African province, celebrated for their great extent and volume, contradict the hypothesis that there is a genetic relationship with basalts. The amount of basalt here is so small that it would be extremely difficult to derive from these small quantities of magma the numerous volcanic formations composed of alkalic ultrabasic rocks.

The second objection against the view under discussion, not so clear but still very significant, is that, in assuming a genetic connection between alkalic ultrabasic magmas and basalt, it is very difficult to imagine the separation of the first from tholeiitic magma. As we may readily see from the accompanying diagram (Fig. 30), olivine basalt lies between the two

*Sodic carbonatitic lavas of Ol Doinyo Lengai and other volcanoes of East Africa show that, under certain conditions, not only may hydrothermal carbonatites exist, but magmatic carbonatites as well.

indicated magmas. We should thus expect that tholeiitic magma should first yield olivine basalt and only after this produce an ultrabasic magma rich in alkalies. We shall see below that something like this does take place under certain conditions. But the direct transition from tholeiite to alkalic ultrabasite is unlikely at best, even if it is generally possible. Furthermore, the typical magma for the basin of the Parana, for Siberia, and for the Deccan was not tholeiitic. It is true that relatively small amounts of trachybasalts are known from the vicinity of the Gulya intrusive, but such rocks are apparently absent in the other fields.

Petrochemical proof is generally advanced in favor of a connection between alkalic ultrabasic complexes and basalts. In the final analysis, these reduce to the fact that present ultrabasic rocks (i.e., "Alpine type") are characterized by very low contents of titanium, calcium, aluminum, and alkalies, whereas basaltic magma is notably rich in these elements. Alkalic ultrabasic rocks, therefore, containing all these elements in notable quantities, must be related to basalts and cannot be correlated with ordinary ultrabasic rocks [Kuznetsov, 1964, p. 295]. This latter statement is not altogether accurate, however. If we approach the problem from the viewpoint of studying petrologic formations, we find no grounds for doubting that alkalic ultrabasic rocks differ appreciably from dunite-harzburgite rocks. However, the geologic conditions do not permit us to combine into a single group alkalic ultrabasic volcanic rocks with hypabyssal intrusive bodies (subvolcanic bodies), on the one hand, or olivine basalts with trachybasalt series, on the other. Kuznetsov is not entirely correct when he maintains that the first are not found without the second. In addition, having subdivided "alkalic basalts" into volcanic and hypabyssal facies according to mode of occurrence, he actually rejects from consideration their genetic unity, although it is easy to show in a series of examples that we are speaking of the products of a single magmatic center, at times following one after the other, at times appearing simultaneously, and, consequently, being inseparable, from the genetic point of view.

In contrast to the approach from the viewpoint of classifying formations, we attempted to reconstruct the relations giving rise to different complexes of magma. It was necessary to recognize the fact that it is impossible to obtain basaltic magma from rocks of the dunite-harzburgite series, since these rocks contain no, or almost no calcium or alkalies; and they are clearly low in aluminum. Dunite-harzburgite rocks are not the ancestors of basaltic magma, but represent the residual after it has been melted out. The question of the site of alkalic ultrabasic magmas under these conditions proves to be significant. As a matter of fact, the presence in them of some quantity of alkalies and calcium and of a sufficient content of aluminum leads us to assume that they have more or less preserved original features of the substance at depth from which the magma was generated. According to this view, this substance may give birth to basaltic liquid during differentiation, and the residual becomes ultrabasic material of the alkali-free series. The possibility of this approach appears geologically sound, but the petrochemical probability of it remains to be checked.

We have reservations at the start: in most cases it is not possible to observe alkalic ultrabasic complexes in any degree of completeness. Their relatively deep-lying parts have not been exposed. It is therefore desirable to take as a standard the possibility of a completely revealed body. A few volcanic rocks cannot be considered representative, since the most basic members of the series, poor in fluxing agents, do not occur at all in the volcanic phase, as a rule. It is for this reason, in particular, that we may explain why "alkalic basalts" are assigned only to the basaltic type. An apparently unique object of study appears to be the Gulya complex [Egorov, Gol'dburt, and Shikhorina, 1961; Épshtein, Anikeeva, and Mikhailova, 1961; and others]. The quantitative relations of different types of rock in this complex could not be determined with sufficient accuracy. But, as a first approximation, the relations were established according to the area occupied by each rock. Using a map prepared by Egorov, Gol'dburt, and Shikhorina [1961], we have obtained for the exposed part of the body the following:

TABLE 5

| Component | Gulya | Altai-Sayan | Vietnam | Sikhoté-Alin' | Olivine basalt |
|---|---|---|---|---|---|
| $SiO_2$ | 40.2 | 44.2 | 45.2 | 45.9 | 47.9 |
| $TiO_2$ | 2.0 | 0.01 | 0.06 | 0.32 | 3.4 |
| $Al_2O_3$ | 5.1 | 1.4 | 1.2 | 1.9 | 14.8 |
| $Fe_2O_3$ | 7.9 | 4.9 | 5.8 | 4.5 | 12.4 |
| FeO | 7.7 | 2.5 | 3.4 | 3.1 | |
| MgO | 27.5 | 46.9 | 42.8 | 42.3 | 7.8 |
| CaO | 7.2 | 0.6 | 0.4 | 0.9 | 8.9 |
| $Na_2O$ | 1.3 | 0.05 | 0.1 | 0.4 | 4.7 |
| $K_2O$ | 0.9 | 0.03 | 0.2 | 0.1 | |
| Totals | 99.8 | 100.6 | 99.3 | 100,0 | 99.9 |

Dunite . . . . . . . . . . . . . . . . . . . . . . . . . . . . . . . . . . . 58.5% of the area
Meimechite . . . . . . . . . . . . . . . . . . . . . . . . . . . 13.5
Ankaratrite, khatangite, etc. . . . . . . . . . . . . . . . . . . . 24.3
Peridotite . . . . . . . . . . . . . . . . . . . . . . . . . . . . . 1.7
Ore-bearing pyroxenite . . . . . . . . . . . . . . . . . . . . . 1.3
Ijolite-melteigite . . . . . . . . . . . . . . . . . . . . . . . 0.7
Nepheline and alkalic syenites . . . . . . . . . . . . . . . . 0.1

We should point out that within the unexposed parts, about two-thirds of the body, there are grounds for assuming only ultrabasic differentiates. Furthermore, it appears most likely that the thickness of the dunites is greater than that of the ijolites and syenites. The role of the latter in the intrusion as a whole must therefore be smaller than indicated by the figures in the list. On the other hand, the volcanic facies contain widespread rocks with relatively large contents of alkalies, calcium, and aluminum, and with insignificant magnesium. It is still difficult to compute volume, but, as a first approximation, it may be stated that all corrections compensate each other, since the first two are opposite to the third. Then, on the basis of the cited figures and the average compositions of the rocks, we obtain the composition for the magma that gave rise to this complex (Table 5).

According to the silica and magnesia contents, this magma is of ultrabasic composition. Unusual aspects are the somewhat elevated contents of titanium, calcium, alkalies, and, in lesser measure, aluminum and iron. For comparison we have furnished average compositions for the Altai-Sayan, Vietnam, and Sikhoté-Alin' ultrabasic rocks (the first after Kuznetsov [1964], the others after Izokh [1965], recomputed to water-free analyses) and for average olivine basalt (average for the basalts of New Zealand, Samoa, and St. Helena [Sheinmann, 1964₃]). From Table 5 we see that the Altai-Sayan ultrabasic rocks are exceptionally poor in alkalies and titanium, but the Sikhoté-Alin' rocks are notably rich in these elements.

A comparison shows that either the view that the Gulya magma is near a mixture of harzburgite-dunite and basalt will not withstand criticism, or that, if this view is nevertheless true, we have computed the composition of the Gulya magma improperly, and the quantity of its silicic derivatives is greater than was assumed (in particular, the greater number of volcanic rocks in the complex). Independent of this, one's eye is struck by the fact that almost all the principal elements of the Gulya magma suggest a mixture of basalts and dunite-harzburgites, but, since the Gulya rocks are not geologically related directly either to one or the other, it is most logical to assume that they represent a great similarity to the composition

TABLE 6

| | Altai-Sayan | Vietnam | Sikhoté-Alin' | Gulya | Olivine basalt |
|---|---|---|---|---|---|
| $TiO_2$ | 0.01 | 0.06 | 0.32 | 2.0 | 3.4 |
| Alkalies | 0.08 | 0.3 | 0.4 | 2.2 | 4.7 |
| $K_2O$ | 0.03 | 0.2 | 0.1 | 0.9 | |

of the initial substance of the mantle (with proper consideration, of course, of some changes during rise of the mantle, particuarly the apparent loss of silica and the enrichment by some secondary elements, such as potassium and titanium).*

Titanium is commonly mentioned as an element typical of "basaltic" magmas, so that its presence points directly to the impossibility of comparing it with ultrabasic rocks. However, it would be proper to associate it paragentically with the alkalies first, especially with potassium (Table 6).

No marked correlation with other elements has been noted. Thus, in alkalic ultrabasic complexes, derivatives of very specific magmas are developed, and these rise through the continental crust at the same time and, under certain geologic conditions, with basalt, but unrelated to the basalt nevertheless. This magma is ultrabasic, since its silica content does not exceed the content in typical ultrabasic rocks. But it differs from the latter by its higher contents of alkalies, titanium, calcium, aluminum, and iron and by lower magnesia. These features lead us to assume, with great confidence, that the magma of alkalic ultrabasic complexes is much nearer the initial material of the mantle in composition than the magma of dunite-harzburgite rocks. In other words, the alkalic ultrabasic magma may be (and should be) considered the product of direct melting of the mantle. We should keep in view the fact that differentiation of this melt under conditions of tectonic calm may readily lead to enrichment of the newly formed rocks in alkalies, a typical result under these conditions [Sheinmann, Apel'tsin, and Nechaeva, 1961]. These late changes, like others associated with the sojourn of the magma in the crust, may in some measure change the primary composition of the complexes. We must not forget this, of course, especially when there are secondary magma chambers.

Present-day outpouring of such magma is completely unknown to us, although very young rocks of this type are known along the African rift (the most recent lava flows and small cinder cones of the Bufumbira volcanoes at the northern end of Lake Kivu are still almost untouched by erosion [Holmes and Harwood, 1937]). The latter circumstance leads us to state that the tectonic conditions obtaining during the Bufumbira vulcanism did not differ substantially from present conditions.

A very characteristic feature of this kind of magmatism is the very small volume of magma. The largest bodies known to us are not to be compared with the basalt eruptions of trap provinces or of Ethiopia. The largest known mass of alkalic ultrabasic magma formed the Gulya intrusive, which has an area of 2000 km² and is surrounded by a lava field of about the same area. The total volume, including the huge volcano, now gone, can scarcely be more

---

*Experiments of Bultitude and Green [1967] have shown that such magmas appear under specific conditions as a result of melting of the mantle.

than 20,000 km$^3$. Generally, the amount of this kind of magma welling up from depths is much less. Bodies rarely have a cross-sectional area measured in the tens of square kilometers.

It is hard to state to what degree the insignificant amount of ascending magma corresponds to the small size of the chamber. It is fully possible that at the depth of the mantle rather large volumes of magma may be formed under proper conditions. But, in contrast to basalt, which is easily squeezed upward and probably ascends almost without residue to the upper parts of the crust and to the surface, heavy ultrabasic magmas may rise to the surface only when conditions are especially favorable. The weight of the overlying rocks proves to be insufficient to squeeze these magmas out; supplementary forces are required for this. Besides the rapid removal of basaltic magma from the chamber, conditions obtain for rapid collection of the magma and, consequently, for further ascent, whereas for ultrabasic magma, the operation of the chamber most likely ceases as the magma is squeezed out. Because of these differences, only a small part of this magma is expelled, nothing comparable with the volumes of basaltic magma.*  Still, the small size of the chambers represents a very probable factor.

All the above discussion concerns the first type of alkalic ultrabasic complex. The second type differs from the first not so much in the assemblage of rocks but in the conditions of its deep manifestation. Until recently it had not been found on the continents, and even now it has been established beyond doubt only in the Pacific Ocean. The typical representative of this type is found in the alkalic ultrabasic rocks of the volcanic series at Honolulu and Koloa in the Hawaiian Islands. As a result of the studies of Powers [1955], Winchell [1947], Macdonald [1949$_1$, 1949$_2$], and Macdonald and Katsura [1964], an indisputable connection has been established between the nepheline basalts and basanites of Oahu, Maui, and Kauai and basalts. These alkalic lavas are very young. They followed olivine basalts, which, in turn, came after tholeiites, and the relations among the components of both pair are alike. Just as the olivine-basalt magma of Oahu is found to be residual from tholeiitic lava after the first ceased pouring out, nepheline-basalt magma is the residue from olivine basalt. This conclusion appears all the more likely in view of the fact that the same relations hold between the alkalic lavas of the Lahaina Series and olivine basalts on Maui as between rocks of the same composition in the Koloa Series and olivine basalts on Kauai. The appearance of nepheline basalts and basanites after basalts is characteristic of the Hawaiian archipelago. As a result, we have noted three evolutionary stages of the primary magma of this region: 1) tholeiitic basalts with rare accompanying differentiates; 2) first residual magma — olivine basalts, with mugearites and trachytes; and 3) second residual magma — nepheline basalts and basanites.

Apart from Hawaii, a similar picture is observed in the East Carolina Islands, in the island groups of Truk and Ponape, and, possibly, on Kusaie.†  The islands of Ponape and Truk are residuals of two large shield volcanoes of Hawaiian type. In their visible parts, the volcanoes are composed of olivine basalt, typically "oceanic." The most basic rocks are very similar to the Hawaiian nepheline basalts and basanites [Stark and Hay, 1963; Yagi, 1960]. As in Hawaii, these rocks rest on basalts, with a break and erosion between. In particular, the alkalic lavas overlie eroded dikes that cut basalts. In all cases the amount of nepheline rocks is small, and these rocks constitute an insignificant part of the basalt mass. All these islands are characterized by very weak differentiation of the alkalic magma. Another interesting

---

*The absence of experimental data at the present time makes it impossible to cite figures. Experiments on the fusion of meimechites, performed by Yu. S. Genshaft and V. Nasedkinii, have shown that melting begins at 25 kbar and a temperature above 1500°C. In order that the liquid not crystallize, when a local increase in pressure occurs, squeezing the magma upward, it is necessary to assume the magma to be superheated.

†Tables of chemical analyses are given on pp. 139-166.

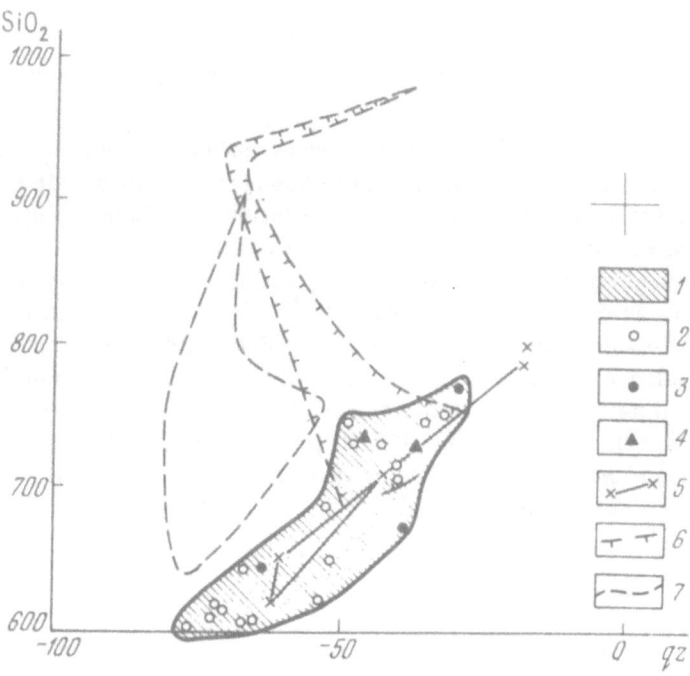

Fig. 31. qz diagram of nepheline basalts and associated rock types at Honolulu. 1) Total field of Hawaiian series of this type; 2) Honolulu Series; 3) Koloa Series; 4) Lahaina Series; 5) rocks of Truk and Ponape; 6) rocks of Tabuai; 7) rocks of the Cook Islands.

TABLE 7

| Component | Gulya magma | Nepheline basalt | Nepheline basalt |
|---|---|---|---|
| | | Honolulu series | |
| $SiO_2$ | 40.2 | 36.75 | 42.86 |
| $TiO_2$ | 2.0 | 2.41 | 2.94 |
| $Al_2O_3$ | 5.1 | 11.98 | 11.46 |
| $Fe_2O_3$ | 7.9 | 6.05 | 3.34 |
| FeO | 7.7 | 7.45 | 9.03 |
| MgO | 27.5 | 12.08 | 13.61 |
| CaO | 7.2 | 13.81 | 11.24 |
| $Na_2O$ | 1.3 | 4.75 | 3.02 |
| $K_2O$ | 0.9 | 0.91 | 0.93 |

feature is that on the qz diagram [Sheinmann, 1965] the nepheline basalts seem to continue, to the left, the basic tholeiitic sequence of rocks, with a typical tholeiitic value of K (23 to 25 in this case). Differentiation toward phonolites, ijolites, and such rocks is scarcely noted here in the basanites (see Fig. 31).

Thus, along with the above-considered primary alkalic ultrabasic magma, appearing independent of basaltic magma, there is a somewhat similar secondary magma. It appears as a distinctive residual of olivine-basalt magma, a residual strongly impoverished in silica and enriched in alkalies. From the example of the Hawaiian alkalic series we may determine the composition of this residual magma, which differs notably from the Gulya magma (Table 7).

These two types of magma are essentially different. It is sufficient, to see this, to compare the contents of aluminum, magnesium, calcium, and the alkalies. The similarity between them lies chiefly in a low silica content.

There is still another series of ultrabasic complexes in the ocean (such as on the Cook Islands, Tabuai, the Society Islands). For them the connection with basalt is vague. Their alkalic lavas are characterized by long-past differentiation, and their relationships differ markedly from the Hawaiian and Carolinian complexes. It is impossible now to settle the question of to which of the two types described above the magmas of these islands should be referred. Much depends on solution of this problem. If the complexes on the Cook Islands, Tabuai, and elsewhere, of this type, prove to be derivatives of basaltic magma, it may be stated that primary alkalic-ultrabasic magmas are not distinctive of regions with oceanic crust. The question then arises: what is the effect of the thick continental crust? If the magmas of the type in the Cook Islands and Tabuai prove to be unrelated to basaltic magma, differences between ocean basins and continents will have no effect on ultrabasic magmas. The first solution appears the more likely.

CHAPTER 9

# MAGMATIC FEATURES OF MIDOCEAN RIDGES

Above, in defining this type of deep-seated tectonic region, we noted that its significant differences are a somewhat greater depth of earthquake foci, as compared with extrageosynclinal regions, and a well-defined linearity of the zones.

The most significant tectonic feature of midocean ridges is the considerable and, judging from all we know, continuing uplift of its axial zone. This uplift is accompanied by volcanic phenomena, attesting to the rise of considerable masses of magma. This picture has become the accepted one for most geologists and geophysicists interested in midocean ridges. From the geophysical viewpoint, the most interesting phenomena are probably the following: the high heat flow and the presence beneath the ordinary oceanic crust of masses possessing seismic velocities intermediate between those for the crust and mantle (normally above 7 and below 8 km/sec). These data led to the view that deep vertical currents existed for the masses beneath the ridges and that the rising masses spread out laterally from the crest of the ridge. Most frequently this movement is considered a closed convection, the spreading mass sinking somewhere near the edge of the ocean and at depth moving back toward the roots of the midocean ridge.

No such sinking near the coasts of the oceans opposite the ridges has been observed, as we know, nor has anyone ascertained that material moves laterally from the ridge. Furthermore, present-day volcanic activity along the axes of midocean ridges is not nearly as intense as generally supposed. Commonly, active volcanoes are situated far from the zone of central rifts. There is no evidence of any intense rise of magma along the axis of a ridge. Near the central rift zone on the Mid-Atlantic Ridge we find the volcanoes of Jan Mayen and Iceland, the submarine mountains (inactive volcanoes) between Iceland and the Azores, and volcanoes in the western half of the Azores. Southward along the axis of the ridge are St. Paul's Rocks (these can scarcely be called a volcano), and Ascension and Bouvet Islands. But the volcanic structures of the greater part of the Azores are on an uplift transverse to the ridge and up to 400 km from the ridge. St. Helena is 600 km away from the central rifts of the ridge, Tristan de Cunha and Gough approximately 400 km away, and the Cape Verde Islands more than 1000 km away. The picture is the same for the Indian Ocean. Prince Edward, Amsterdam, and St. Paul Islands are on the ridge; but far from the axis, partly not even associated with the ridge, are the Crozet, Kerguelen, and Heard Islands, the entire Mascarene group, and the Ob and Lena Banks.

The question concerning in what measure the midocean ridges are volcanic features is completely regular. The continuations of such ridges upon the continents can in no way be considered due to the rise of magma. Masses of volcanic rocks are found here in individual, mostly rather small fields, generally rather widely separated from each other. More than this, extensive segments of such structures are found to be free of volcanoes or are accompanied by very insignificant volcanic phenomena (the Rhine and Baikal grabens, and, possibly, a great part of the Red Sea graben). It is true that in speaking of continental structures of the midocean-ridge type it is necessary to keep in mind that either it is the termination, the so-called "tail," of the ocean ridge (such is the zone of uplifts and grabens of the Red Sea and East Africa) or the structures are not generally continuous directly into the oceanic structures (such as the Rhine and Baikal features). It is therefore necessary to exercise some care in making our correlations. Still, after considering all that we have said above, we have some grounds for stating that volcanic activity associated with the formation of the midocean ridges is not very great, possibly no greater than the oceanic volcanic activity not associated with a ridge. We may express our doubt in considering the extent that oceanic basalts are related to midocean ridges.

A significant feature of midocean ridges is the "cushion" of rocks with low seismic velocities, immediately next to the crust. Only at some depth does the velocity of waves gradually increase till it reaches normal values. However, direct evidence concerning the existence of the low-velocity channel (waveguide) under a midocean ridge apparently does not exist. The possibility is not therefore excluded that the channel is here but vaguely expressed or that it disappears, such as under island arcs. But this is only an assumption. Most likely the appearance of this "cushion" must be explained by enrichment of the upper mantle in basaltic magma, which has risen from depth. A second aspect of the process is the consider-able warming of the material, due chiefly to the basalts rising from deeper zones. The nar-row zone of increased heat flow, strictly confined to the central rifts of the ridge, lead us to assume the presence of a zone of fractures extending to depths of at least 50-70 km (the depth from which magma rises under the island of Hawaii). The presence of this zone may be explained by the squeezing out of the basalt. The separation of basaltic liquid and its subse-quent ascension probably increase the mobility of the fracture zone.

The magma that forms under these conditions is generally of the same type as that form-ing in tectonically calm regions. Judging from existing data, only tholeiitic magma is generated in the central zone of ocean ridges. In its further development, toward the end of volcanic activity, this magma gives birth to small amounts of olivine basalt, which appears in the latest flows, forming the very crest of the volcanoes.

The volcanic fields associated with analogous structures on the continent have a different appearance. No tholeiitic basalts appear in East Africa or in the zone of Baikal grabens. Alkalic basalts form small fields as a rule; these might be considered derivatives of tholeiitic magma, if one wishes, that got trapped at some depth. However, the complete absence of tho-leiitic lavas makes this view difficult to accept. Furthermore, the huge fields of Ethiopian lava, where tholeiitic lavas are also lacking, would be very difficult to explain as a result of differentiation in a chamber of tholeiitic magma. We must therefore state that, within conti-nents, structures of the midocean-ridge type are characterized by basalts with high alkali con-tent. These either come from primary magma from the mantle or form by precipitation of orthopyroxenes at great depth (see Chapter 14). The restriction of this magma to continents indicates that the liquid has the ability to discard part of the relatively infusible minerals and to free itself from them; i.e., it indicates that the rise of magma is not so precipitate as it is under the ocean. It is possible that this is related to the presence of a great thickness of light rocks. When the magma chamber lies at a depth of 70-75 km beneath a continent, the weight of the column of basalt liquid barely counterbalances the weight of a corresponding column of crustal rocks (30 km) and the upper mantle (40-45 km), whereas under the ocean these columns are counterbalanced at a depth of approximately 60 km below sea level. As we have noted before, the rise of magma under the Hawaiian Islands begins at approximately this depth. Con-sequently, there is probably a significant difference in rates at which magma rises in oceanic and continental regions. With slower rise, the possibility of separation of precipitating crystals is greater. It is no less likely, however, that olivine basalts of the continents are the result of insufficient heating in the chamber, at the site of magma generation. Recently, large bodies of peridotites, most likely dunite-harzburgites, have been discovered in the central rift zones of midocean ridges. At the moment it is difficult to decide whether these are ancient rocks, having formed long before appearance of the ocean ridge, or whether they are young, the intrusion of which was related to formation of the ridge. If the rocks prove to be ancient, they will attest to differentiation of mantle material, similar to that originating in geosynclinal regions. However, we do not know the conditions that obtained at the time of formation of these bodies. If the ultrabasic rocks of the midocean ridges are young and their emplacement accompanies the formation of the ridge, the rocks of the dunite-harzburgite series prove to be phenomena not only of geosynclines but also of ocean ridges. And this becomes the most characteristic feature of magmatism in the ridges. We are finished with this suggestion of similarity with geosynclines, however. We have already seen that basaltic magma in midocean ridges is practically identical to the magma of other extrageosynclinal regions. Furthermore, we must not forget that peridotites of this type may be residual or they may have precipitated from magma during the formation of any basaltic magma. The exceptional character of midocean ridges then proves to be merely apparent, and such peridotites will be found in any basaltic province (at some depth, not at the surface).

CHAPTER 10

# GEOSYNCLINAL MAGMAS

Geosynclines represent the most complex magmatic zones. And, at the same time, they are best known, at least in the most prominent manifestations of the process. Probably the most distinctive feature of magmas in geosynclinal zones is their variety. This variety is clearly related to the energy of tectonic movements, but may be related in still greater measure to the great depths of the roots of these disturbances. The establishment of this feature becomes possible and, it seems to the author, indisputable when we compare regions of present-day deep-focus earthquakes with zones along continuations of the trend of these regions. We have already seen that series of young fold zones change along their trend to island arcs. In observing this we may trace the connection of individual zones into which such regions are divided. Thus, the deep-water trenches of island arcs are traced directly into foredeeps of fold zones; the zone of the island arc itself is traced into the uplifted fold zone. The separation of the arc into inner and outer, with a depressed zone between, may also be seen in fold zones. Backdeeps may also be clearly distinguished occasionally.

These structural relations are so obvious in a number of places that there can be no doubt about them. Such are the relations in Arakan Yoma; its foredeep and backdeep correspond to the double island arc of Indonesia with its deep-water foredeep and backdeep seas (see Fig. 1). Such is also the picture of the Lesser Antilles and their transition to the Andes of Venezuela (Fig. 32), the transition of the Kurile and Vityaz' ridges to the Kamchatka and Koryatskii fold zone, from the Patagonian Cordillera to the Scotia Ridge arc and on to the transition to Graham Land. At times it is suggested that we have to do with a more or less accidental coincidence along a single line of genetically different structures. The transitions are so systematic, however, and the connection between young fold zones and island arcs so obvious, that this kind of statement can hardly be given serious consideration. The unity of these structures is beyond doubt. We may merely suggest three possible combinations: 1) both zones are entirely identical; they once made up a single unit, part of which later subsided and was converted into an island arc; 2) the fold zone is a later stage of development, when general uplift led to transformation of the earlier geosyncline into a continent; and 3) the difference between a fold zone and an island arc is not one of stage but is caused by differences in the crust in which the process is at work.

There is little doubt that the first is the least likely solution. Nothing is proved by the idea that the Lesser Antilles are the result of "oceanization" of the Andes and that their site in the recent geologic past was a compressed and uplifted fold structure. Such are the relations of the Kamchatka—Koryatskii region and the Kuriles, southern Alaska, and the Aleutians, and so on. On the other hand, everything attests to the uplift of island ridges, although such uplift may be accompanied by relatively short-lived subsidence, of course. More than this, extensive subsidence of a regional character may be superposed, may be related entirely to other structures. But, even in this case, the arc will be distinguished against the general background as a region of relative, perhaps even absolute uplift. An example of such subsidence,

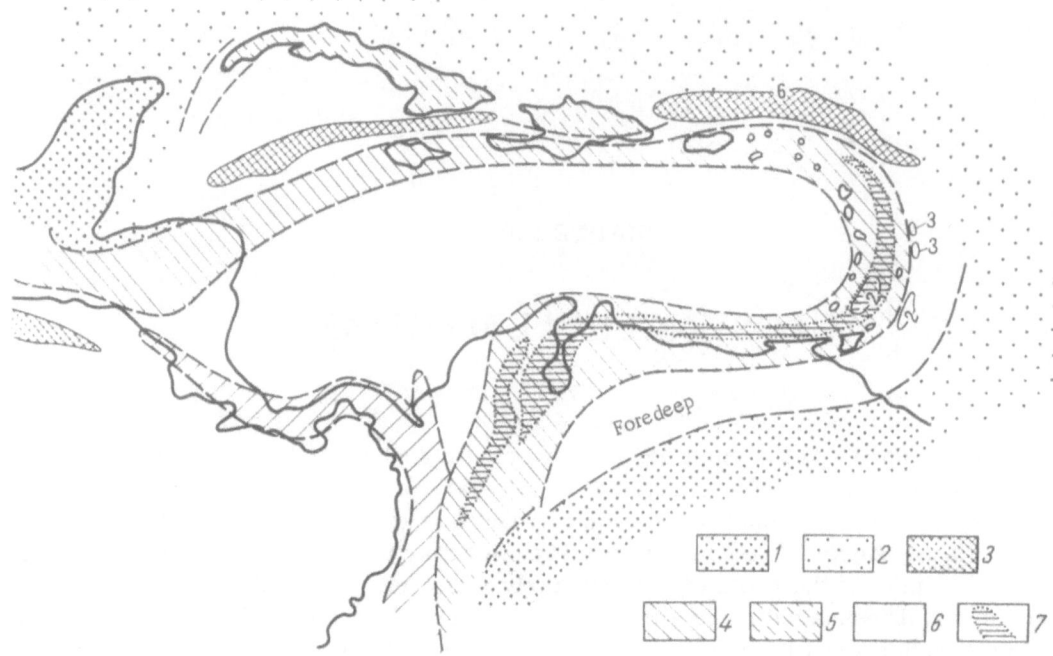

Fig. 32. Transition of the Antillean island arc into fold zones. 1) Platform; 2) frame of fold zones in the sea; 3) deep-water trench; 4) Venezuelan-Cordilleran fold zone; 5) Cuban arc, adjacent to the Venezuelan-Cordilleran fold zone; 6) Andean fold zone; 7) depression separating inner and outer ridges of island arc. Arabic numerals on map designate depth in km.

superposed on and alien to the structure of the arc is the region of the Aegean Sea with the drowned Cretan arc. An especially interesting and grandiose example is found in the huge region of the southwestern Pacific, where tremendous, perhaps ocean-wide, subsidence has been superposed on the complex geosynclinal zone of island arcs and, probably, previously existing fold zones.

There thus remain two possible solutions to the problem. We must state that the choice between them is not easy, since the history of the structure, especially when not superposed on an older fold zone or not preserved as a remnant of older massifs, is difficult to reconstruct. It is therefore difficult to distinguish whether we are dealing with the parallel development of structures on different frameworks (different types of crust) or with successive stages of a single structure. Apparently the most probable solution is the following. In the most general case the structure of an island arc precedes the structure of the fold zone. The rate of this process differs for different conditions, however. Depending on the conditions, the rate of transforming a geosyncline into an uplifted fold region varies for different parts of a single large region, even for geosynclines that begin at the same time. The relative durations of the different stages also vary.

As a result, where continental conditions existed long ago and were preserved, on the whole, during the formation of successive geosynclinal downwarps, the stage of deeply buried geosynclinal prisms (with oceanic or similar crust) may prove to be relatively short. The period of elevation of the island arc above the sea may also prove to be curtailed. Before this, complete conversion of the crust to continental type takes place. The succeeding stage of

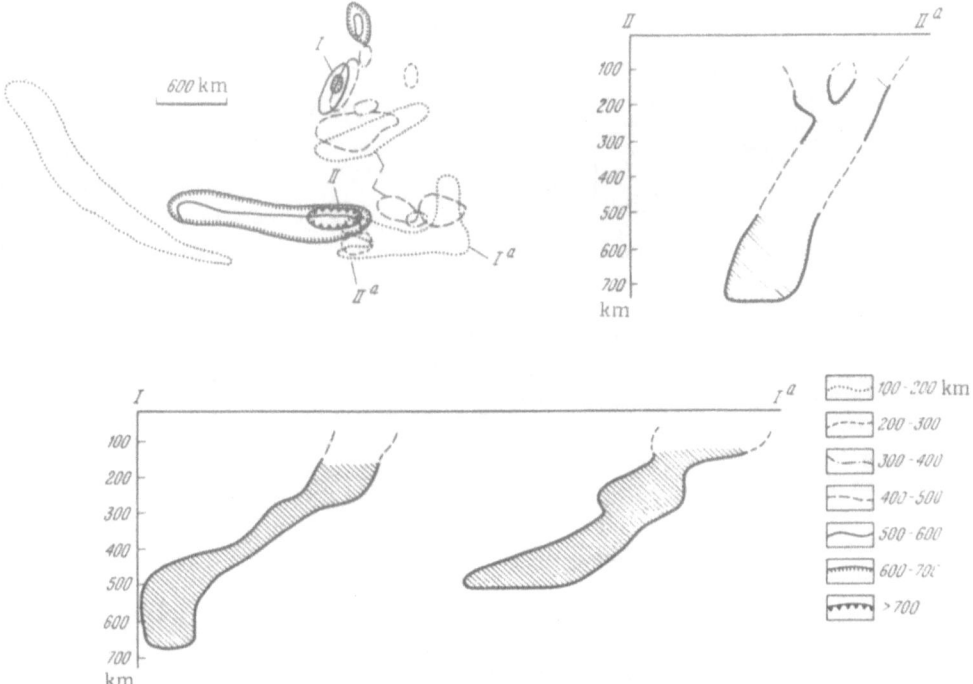

Fig. 33. Deep-focus earthquakes of Indonesia (contour interval, 100 km). I—I$^a$) Profile from Celebes arc to eastern edge of the Lesser Sunda Islands (double arc and two earthquake zones); II—II$^a$) profile in the region of Timor. Vertical scale the same as horizontal.

conversion probably takes place the more readily the smaller the segment of oceanic crust (it is unimportant whether these oceanic segments are formed anew during reactivation of geosynclinal conditions or whether they are inherited in some degree from earlier epochs).

On the other hand, in regions where there is no continent and the role of continental crust is small, the transformation of a geosyncline to a high uplifted land mass is difficult and may be extended over a long period of time. It is likely that under certain conditions the process is terminated by uplifts during which no continental mass is formed (perhaps the region of the southwestern Pacific). Thus, the effect of conditions under which the process takes place appears to be definite and appreciable, but it seems more likely that the basic fact is that the difference in type of region is merely a reflection of the difference in stage of a single process.

In our further discussion, the view of zones of island arcs as well as geosynclines of the present day will be the basis of our excursions into the zone of geosynclinal magmas.

Geosynclines and Deep-Focus Earthquakes

If we turn to what we said concerning the distribution of deep-focus earthquakes (Chapter 4), the connection between these earthquakes and geosynclines appears very convincing. It is well known that such earthquakes in each particular region lie within a more or less regular, comparatively narrow zone that is somewhat inclined at depth. The emergence of this zone at the surface is well marked by the position of a deep trench. There has been considerable controversy concerning the actual or imaginary existence of such zones of deep-focus earthquakes. We shall not dwell on this controversy, but will merely refer those interested to the literature [for example, Belousov and Rudich, 1960; Petrushevskii, 1964; Sheinmann, 1964]. Here it will be sufficient to furnish some profiles (Figs. 33, 34).

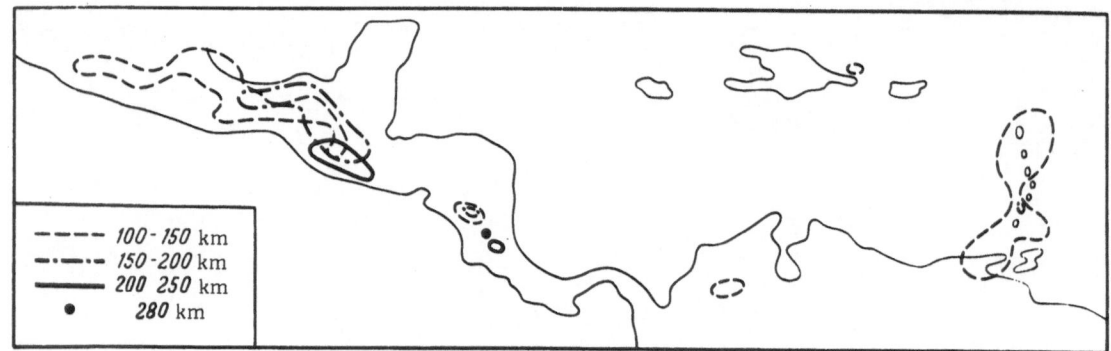

Fig. 34. Deep-focus earthquakes occurring in Central America (magnitudes no less than 6.5). After L. Koning.

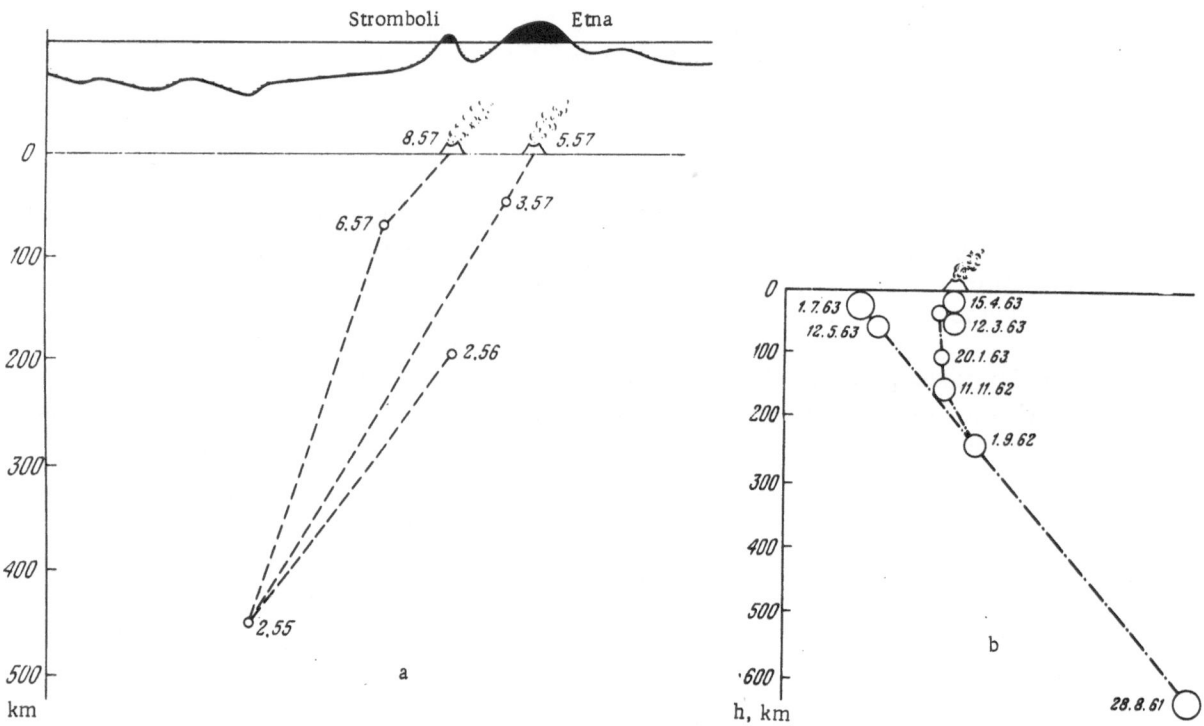

Fig. 35. Relation of deep-focus earthquakes to the rise of magma (after Blot [1964]). a) Earthquake of February 1955 and the eruption of Etna—Stromboli in May and August 1957; b) earthquake of August 1961 and eruption of Ambrym (New Hebrides) in April 1963. Arabic numerals at the circles indicate dates of earthquakes (day, month, year).

There is no question in the general outlines about the connection between the distribution of deep zones and geosynclines. The boundaries of the latter prove to be trenches in front of island arcs, corresponding in environment to foredeeps. Judging from the data both of geology (concerning foredeeps) and geophysics, the boundary may be placed even more precisely; the inner slope is within the geosyncline, the outer slope is the frame of the geosyncline. The backside boundary is less distinct, possibly being transitional, since the backdeep seas are still geosynclinal. The main part of modern geosynclines is thus above the upper part of the zone of deep-focus earthquakes. The zones of initial uplifts, corresponding to the central up-lifts of fold zones — the island arcs — are always situated approximately where the roof of the earthquake zone has been already buried to a depth of 100-150 km. This apparently corresponds to the depth at which deep fractures may be preserved for prolonged intervals (in the geologic sense).

## Magma from the Mantle

Mantle-generated magma of geosynclines is clearly restricted to earthquake zones. More than this, under other conditions it is possible to establish a direct connection between earthquakes and the movement of magma. Such information was long ago brought to light  The most interesting data may be found in the works of Blot [1964; Blot and Priam, 1963]. From Fig. 35 it may be seen that in all cases when it has been possible to establish such a connec-tion, the connection has been found similar to that between movement of magma and under-ground shocks observed on the island of Hawaii. In geosynclines, however, the shocks begin at considerable depths (100-250 km and even greater). The gradual rise of these foci toward volcanoes is seen in Fig. 35, and, when the shocks reach the vent of the volcano, eruption be-gins. This picture is occasionally supplemented by the resumption of shocks along the con-tinuation of the principal earthquake zone, after the outpouring of the lava, up to the emergence of this zone at the surface. Since all these earthquakes are tectonic and cause the movement of magma, the appearance of the indicated late earthquakes may be explained by the continuing unloading of stresses, but now outside the path of the rising magma. It is undoubtedly an im-portant fact that, independent of the rise of magma, a series of genetically related earthquakes is observed. The stresses thus appear at higher and higher levels. The shear displacements shift upward, and the mechanical energy pours outward, as it were.

It has not been possible to distinguish such series everywhere or to note the rise of magma from depth. The author's attempt to trace such series in the Kamchatka—Kurile arc showed it to be impossible in that region. The huge number of recorded shocks created a con-tinuous background at all depths, not permitting a series to be recognized. Still, we were successful in recognizing two or three, though they are very tentative and may prove to be imaginary. Apart from the background of individual shocks, another feature of this region should be noted. A connection between the rise of magma and earthquakes can be determined only when the magma rises from great depth to a vent. Explosions may be proved to be unre-lated to the rise of any appreciable amounts of lava. Kamchatka eruptions are now chiefly explosive, and, what is especially important, secondary magma chambers are present, in the Klyuchevskii group, lying at shallow depths [Gorshkov, 1958]. In general, there are almost no present-day outpourings of lava on the Kurile ridge, except for the northernmost end.

Deep magmas of modern geosynclines are well known, and they correspond to "fossil" magmas of fold zones. As we know, the geosynclinal series are typically calc-alkalic. On the qz diagram these rocks form a rather straight belt from olivine basalt and more basic rock such as picrite through tholeiite, quartz basalt, and andesite-basalt to andesite and even dacite. This, the so-called "principal sequence," is characterized by a more or less well-defined se-quence in time: as a rule, the more basic representatives of this series rise first. The "center of gravity" lies at olivine basalt. This is only a rule, of course. The first part of the magma to rise may be either basic or silicic, but always more basic than that which follows.

In future discussions, the rising magma has the composition of andesite-basalt. Conse-quently, the role of basic varieties in the series decreases, and the main bulk will be andesite-basalt. Ordinarily, this stage in the history of magmas in geosynclines corresponds to the time when an island arc is already well defined. This is an arc of the second type according to the classification of Belousov and Budich, i.e., an arc not yet accompanied, in our view, by any broad development of continental crust. Correspondingly, the epoch of more basic magma (basalt, spilite) in the geosyncline belongs to the time of greatest subsidence.

The later magma may become andesitic. At this time the rock series generally contains no basic basalts, although tholeiite and quartz basalt are rather widespread. Andesite-basalt is abundant. But the principal role is played by andesite proper (pyroxene andesite most frequently). Continuous sequences of andesite are common. More silicic dacite (and rhyolite) may occur with the andesite in subordinate amounts, generally very minor amounts. These rocks are associated with great uplifts of the region, the beginning of general inversion.

In view of the fact that andesitic vulcanism is commonly considered typical only of the late postfolding stage of geosynclinal development, we must pause briefly on this problem. That basaltic (spilitic) sequences are typical of the time of subsidence in geosynclinal development there appears to be little doubt. But ordinarily a long break in eruptive history is assumed, batholiths of complex composition being formed during the time of folding. And only after the region becomes mountainous is lava poured out again, now andesitic in composition. These are the relations outlined in the scheme of Stille. The picture is actually more complex, however. Kuznetsov [1964, pp. 104-106] has remarked that andesitic rocks appear "under very different geologic conditions and, particularly, at different stages in the life of a mobile zone." Andesite forms both before and after the emplacement of batholiths of gabbro-diorite-granodioritic composition, and its magma corresponds in composition to the magma of these batholiths. A characteristic feature for this andesite is its restriction to the zones of uplift. It may appear also during the formation of spilites, but, in contrast to the latter, it appears only on incipient uplifts within the geosyncline (the Silurian of the Urals). Possibly, and at times actually, andesite may directly, without break, succeed basalt (spilite) after the more basic lava has ceased pouring out, with the beginning of uplift within a downwarp. This is the picture observed in the Rudnyi Altai for the Upper Devonian—Lower Carboniferous. Lastly, andesite is exceptionally widespread after the termination of folding (the Pacific coast of the Americas, the Caucasus (Cenozoic)). The restriction of andesite to the zone of uplift means that this rock is readily destroyed by erosion and may not always be observed for this reason.

It should be noted that the composition of these formations is everywhere the same: predominant rocks are andesite-basalt and pyroxene andesite; basalt and silicic varieties occur in lesser quantities. A typical example of extensive development of andesite-basalt and pyroxene andesite following directly after outpouring of more basic lavas is the Kurile—Kamchatka arc (the eastern Kamchatka zone according to S. I. Naboko).

Thus, we should not delimit the basalt-spilite and andesite stages, considering their magmas to arive from different chambers. This is a single process in the development of vulcanism in geosynclines. There is no evidence for believing that andesitic magma must be derived from fusion of the crust. The same series of gradually ascending seismic shocks noted for the Hawaiian volcanoes has been observed for volcanoes with andesitic lavas (Paricutin, for example). The rise of andesitic magma from depths well below the base of the crust has thus been directly observed.

On the other hand, there appears to be no doubt that andesite has formed by melting out of the crust. Such andesite is apparently associated with the formation of batholiths. The chief criterion for distinguishing these two groups of very similar rocks is the geologic environment. It appears likely that in the future it will be possible to find some chemical differences as well, which may permit us to determine the site where the magmas are generated.

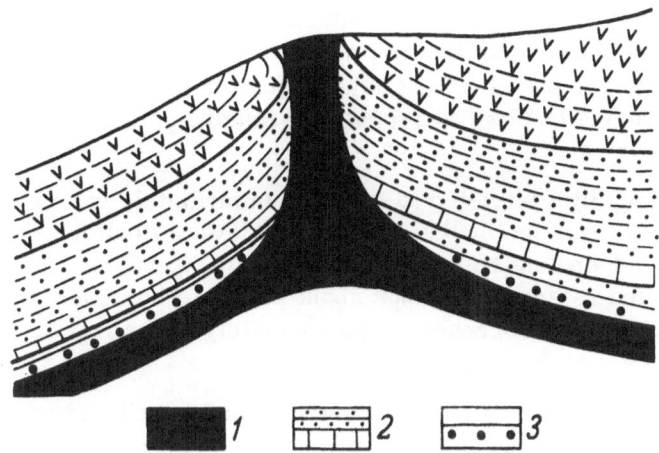

Fig. 36. Piercement of Kalychlinskii serpentinite body in sediments containing cobbles and fragments of this same serpentinite (after Knipper [1965], somewhat modified). 1) Serpentinite; 2) Cenomanian sandstone, pebble conglomerate, and limestone, with pebbles and cobbles of serpentinite (overlain by Lower Senonian volcanic rocks); 3) Cenomanian conglomerate with numerous pebbles and cobbles of serpentinite.

## Ultrabasic Rocks

It is well known that ultrabasic rocks are intruded in geosynclines at approximately the time the series of basic lavas are poured out. Since dunite-pyroxenite-gabbro rocks are generated from basic basaltic magma [Kuznetsov, 1964] and represent an example of deep-seated differentiation, we are here interested only in the dunite-harzburgite variety of ultrabasic rocks. It is thought that it is not very significant what actual near-surface structures the belt of these intrusions may be associated with. Certainly the connection between these belts and the downwarped zones of geosynclines has been considered significant. Later the view that gained the upper hand was that these ultrabasic rocks are restricted to deep fractures, and in this respect they came to be set in opposition to geosynclines. In fact, the controversy has led to a secondary problem. Dunite-harzburgite rocks are actually everywhere confined to deep fractures, but this in no way leads to denial of a direct connection with geosynclines and the period of subsidence of the geosynclines. This relationship is just as close as the first.

The question of how intrusions of this type take place is much more difficult. The question reduces to this: when do the intrusions occur and what is the state of the material at the time of intrusion? Observed intrusions of ultrabasic rocks into sedimentary rocks by no means always allow us to define the time of intrusion. The long-known "cold" contacts led many to believe that the material was already crystalline when it was injected, and was not a magma at all. Furthermore, Knipper [1965] described the formation in sedimentary rocks of several stages of detrital aprons, beds with older aprons being ruptured by the discordant body during the interval between formation of the older and younger aprons (Fig. 36). In other words, the intruded body was exposed to erosion and was then again covered by water and by younger sediments. As a result of later renewed rise, the body punched through these layers also, thrusting above the water, again being subjected to erosion. A new apron was formed, and so on and on.

This protrusion of solid rocks (it becomes possible after the appearance of a lubricant: serpentine) takes place much later than the primary intrusion and may prove to be entirely separated from it in time. There are grounds for assuming such renewal of movement even during the following epoch of folding.

It has been suggested that intrusions in such situations don't actually take place, that the ultrabasic rocks were in the solid state from the very first. We have data, however, that point to indisputable "hot" contacts, and these deter us from accepting the above view. Such "hot" contacts in New Zealand [Challis, 1965], are manifested by 800 m of hornfels. A difficulty in the detection of such contacts is the superposition of hydrothermal alteration, associated with serpentinization of the mass. Such "hot" contacts have been observed in Cornwall

[Green, 1964] and on Cyprus. "Hot" contacts are not frequently encountered, because the shifting of solid serpentinized bodies is apparently normal. And this fact strongly complicates the problem of determining the time of intrusion. In any case the time may be proved to be older than the age of the host rock containing the intrusive bodies (we repeat this conclusion following Knipper).

The intrusion of ultrabasic bodies is commonly restricted to the time immediately before folding and uplift. The author of the present work was inclined some time ago even to believe that intrusion took place at the very end of subsidence and before the principal folding and was the general rule. However, the information just cited raises misgivings about such late intrusion of ultrabasic rocks. One should expect the intrusion to correspond approximately to the rise of the main basaltic magmas.

The problem of the form in which the dunite-harzburgite rocks are intruded is still far from clear. We do not yet know how the melting conditions of olivine-rich or pure olivine rocks change with depth. We consequently do not know whether we have to do with complete fusion of material at depth or only partial, the residual olivine sinking in the liquid that has formed. It appears very probable that both situations may arise. Furthermore, when the magma moves upward it must come into an environment in which olivine will be precipitated from the melt. The newly formed crystals can hardly be distinguished from residual crystals if we take into account the continuous change in conditions under which the magma finds itself. It may be assumed that more or less pure melts as well as mixtures of crystals and liquids will be injected into the crust, the latter in some cases being so full of crystals as to be a mush. The appearance of extrusives of meimechite as well as intrusives demonstrates the possibility of this assumption. Whether ultrabasic intrusions are the result of ascent of magma or of magmatic mush, the rise of material requires special conditions. At the mantle-crust boundary, a column of such material would be counterbalanced by a column of the surrounding rocks. Therefore, the formation of intrusions near the surface without the influence of supplementary forces is very unlikely, if not impossible. Consequently, the rise of large masses becomes unlikely. The magma is squeezed out tectonically, and this process is easier if the mass of squeezed material is small.

## The Tectonofer

The phenomena described above are concentrated in a narrow zone (relatively, of course), extending to depth, to which deep-focus earthquakes are confined. Its shape only confirms the close connection between magma and movements. This zone has a distinctive but fully defined structure. It cannot be identified with a fracture or a fracture zone, although this is formally permissible from the structural point of view. One should in general be very careful about using the term deep fracture for phenomena under conditions of rapid healing of the ruptures. This use of the term leads to distortion of its basic meaning, since there is no breaking of material at depth. Regardless of the remarks the zone possesses features by no means characteristic of deep fractures. It forms not because it is the weakest place within some zone subjected to external forces or some restricted region between two blocks along which oppositely directed forces act.

As analysis of present-day movements at depth show, and also tectonic analysis of near-surface regions, there is no basis for seeking some special effects in the regions on the two sides of such a zone. It has been repeatedly asserted, and is asserted, that island arcs, and this means geosynclines also, appear only at the boundary between ocean and continent (consequently, the boundary between oppositely moving blocks is also noted). Strictly, this is the opinion almost universally held. On the basis of it, a scheme has been prepared in which account is taken of the difference between oceans and continents, convection currents moving

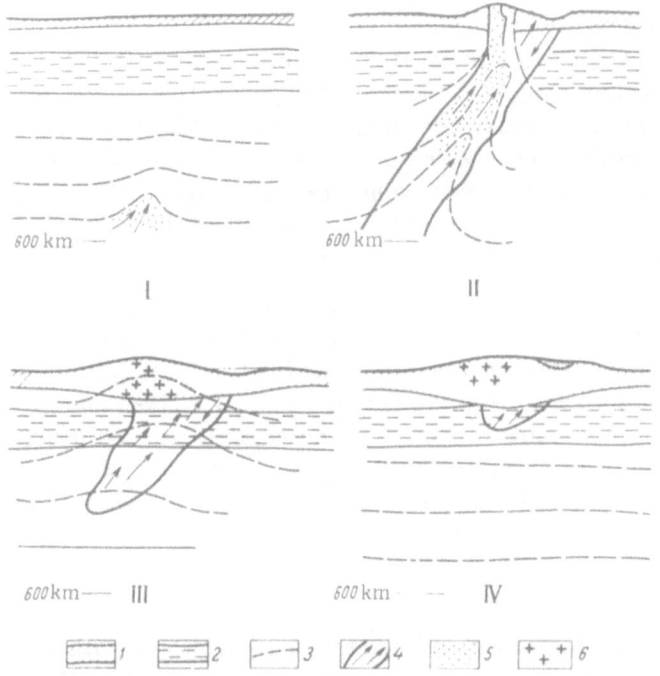

I

II

III

IV

Fig. 37. Formation and development of a tectonofer.
I) First stage of energy concentration, beginning of
growth of the tectonofer; II) full development (strong
heat flow, and mechanical energy moved toward the
surface); III) dying out begins (deep source of energy
that supplied tectonofer is exhausted; only the upper
part of the system is preserved); IV) disappearance of
tectonofer (entire store of energy subject to removal is
delivered to the surface). 1) Crust; 2) low-velocity
channel of Gutenberg; 3) isogeotherm; 4) tectonofer;
5) basic magma; 6) granite.

out somewhere under the continents, parts of the continent caving into the ocean, and so forth. This generalization is not supported by the facts, however. A very impressive fact is that modern structures of this kind, most of them active, outline the coasts of the Pacific ocean. It is forgotten that this distribution may not be repeated in other epochs or that among modern island arcs some are not associated with the ocean— continent boundary. We shall not yet speak of the probability of the wide distribution of such "exceptions" in the past, if geosynclinal zones of the past may be considered the same as island-arc zones. The author earlier remarked that the arc, more correctly the two arcs, of the Lesser Sunda Islands lies between two continental massifs; i.e., this represents an intracontinental geosyncline. The two arcs of the Mediterranean, still preserved, fail entirely to correspond to the view that arcs are marginal. The Scotia arc (South Sandwich), despite its direct connection with the continents of America and Antarctica, is bounded on both sides by ocean. Nor should we forget that the complex of Melanesian arcs and the Kermadec—Tonga arc are also bounded by

the ocean. Thus, there is no basis for restricting arcs to the boundary between ocean and continent.

It may be stated, consequently, that zones of deep-focus earthquakes and regions of modern geosynclines found above them cannot be associated with boundaries between ocean and continent, that such structural complexes must be considered independent of such boundaries, although many of them are found near the boundaries of modern oceans. The principal feature of modern complexes of this kind is the existence of a deep mobile zone, undoubtedly exceeding in depth, to an appreciable extent, tectonic phenomena beneath stable regions of any type. Such deep zones are characterized by the appearance of basic magmas, and the probable depth of formation of these magmas, judging from seismic data, is very considerable.

It may be stated, therefore, that we have to do here with appreciable flow of energy to the surface. This is indicated by the high values of heat flow in regions of island arcs. If we recall that in geosynclines the appearance of basic deep-seated magmas precedes batholiths and other silicic intrusions, we are forced to the conclusion that the zone of heating rises gradually from the depths of the mantle into the crust (this will be discussed in detail below). On the other hand, the comparison of depths of intense tectonic movements discussed above shows that, on the whole, the more mature stage is characterized by decrease in this depth. In other words, the process dies out upward, from the deep zone to the surface (Fig. 37).

The ascertainment of these facts is very important, since it determines the inadmissibility of believing that the process arises through the effects of exogenetic or near-surface factors. Beginning somewhere at great depth, the process, leading to the generation of deep-seated magmas and to earthquakes (and geosynclines above), gradually involves higher zones, ultimately reaching the upper layers of the earth's crust. At some stage the deep source begins to become exhausted, and the lower boundary of the active zone gradually rises. We should note that the zone embracing these phenomena is clearly limited, and it is the zone of deep-focus earthquakes.

The physical meaning of this kind of process has but a single interpretation. The formation of magmas shows that at depth heating takes place within the zone. Thermal energy is transferred upward. This is effected both as heat flow and as heat transfer by the ascending material. Thermal energy thus moves along this zone from the depths outward. The same is true of the mechanical energy; the removal of this energy outward is indicated by the sequence of earthquakes, the foci rising upward successively from depth (see Fig. 34). On the whole, consequently, the zone is a region along which a notably larger amount of energy moves to the earth's surface than through an equivalent cross-section in other regions of the earth. This ultimately is the cause for the development of geosynclines and for their transformation to fold regions.

The characteristics we have just discussed lead us to believe that we have to do with a new type of structural element, something that cannot be reduced to any known type. Its difference from any type of fracture we have already mentioned. Deep fractures have considerably less depth and are narrower, even the largest. Neither in the deep-seated occurrence nor in the nature of the disturbance can this structural feature be compared with zones of the described type. Some deep fractures are partial structural elements in the upper parts of such zones (such as fractures along which magma rises to volcanoes, for example). Other fractures, not associated with the zone, may cut across it, and so forth.

We think a possible term for this type of structure is tectonofer,* since it is with these deep structures that what we call the geosynclinal process is associated, the most powerful manifestation of tectonics.

In Chapter 14 we shall consider some features of the process in a tectonofer. Here we note only that the greatest heating apparently takes place at great depths. As the zone of heating rises, some of the store of heat is dissipated, and, correspondingly, the degree of heating declines as the zone rises. Apart from this, the width of the zone of heating may increase in the process. This may point in some degree to the appearance of more easily fusible magmas than those at greater depth. But the principal mechanism of this change is different, and we shall speak of this below (see Chapter 14).

It is hard to state just when a tectonofer begins to die. It might be thought that this takes place when activity declines at depth, resulting in the disappearance of deep-focus earthquakes. We might note, however, that this weakening of activity begins rather late, as witnessed by the Indonesian arc. This region is undoubtedly in a rather late stage (true, it is probably still prebatholithic), and it is marked by the maximum depth of earthquakes. Even the Calabrian arc, which is in transition to the fold-zone stage, is characterized by earthquakes at depths below 400 km.

———————

*H. H. Hess introduced the similar term, "tectogene" for the designation of a hypothetical fold of the crust sucked down into the mantle, resulting in the formation of a geosyncline. From the above discussion, it is clear that there is considerable difference in the meanings of the two terms.

We are therefore not on dangerous ground when we state that activity begins to die in the lower part of the tectonofer when general uplift in the geosynclinal zone begins. It is most likely that this dying is associated at first with diminution in the number and force of earthquakes at great depths, and only later does it disappear entirely. The rise of the zone of heating, as a result, outstrips the dying from below, so to speak, since, at the time batholiths are formed, i.e., when the zone of heating reaches the upper layer of the crust, the root of the tectonofer is just beginning to die. In any case there still exist (or might exist) very deep earthquakes. We should note that some decline in temperature at depth, leading to the disappearance of deep magmas, is not accompanied by any considerable diminution in mechanical activity. At this time the seismicity of the deep zones may be high.

Where the tectonofer intersects the low-velocity channel of Gutenberg, the latter apparently disappears. This disappearance is due to erosion, as it were, of the upper boundary of the low-velocity channel, because the upper part of the tectonofer when impregnated with magmatic material is identical to the channel in elastic properties. As a result, the region of low velocities rises to the upper boundary of the mantle.

The very melting of material in the tectonofer differs from melting in the low-velocity channel, not only in depth (the range of depths in the tectonofer is many times that in the channel) but in so-called genetic relations. We have seen that matter in the low-velocity channel but outside the tectonofer is near the point of partial fusion, and a comparatively small disturbance of the regime is sufficient to have melting begin. Penetration of deep fractures into the zone of the low-velocity channel is such a disturbance. As a result, the formation of magma and, even more, its ascent are stimulated by the formation of fractures in the surface shell of the earth. This, of course, is not an exogenetic process, but neither is it a deep-seated process. In contrast, fusion in the tectonofer is caused by increased arrival of energy from depth. Deep-focus earthquakes, whether their role is large or small makes no difference, are caused by the upward stream of energy. This difference in the conditions of fusion has a very significant meaning for understanding the energy conditions of the earth shell.

CHAPTER 11

# SILICIC MAGMAS (MAGMA FORMATION IN THE CRUST)

In deeply eroded ancient fold regions, levels are exposed at which silicic magma was once generated. From the field of abyssal geology proper, we now move over to conditions of investigation more customary to us, and it may be comparatively easy to draw some conclusions. The most important zone is that in which batholiths are generated, i.e., the sub-batholithic zone, which is not deep. Determination of the depths of denudation, which has exposed this zone, such as the ancient gneisses and gneissic granites of the Kola Peninsula and Karelia, gives values no greater than 15-20 km. It is possible that the depth may be appreciably less. Consequently, granitic magma is confined to the upper half of the continental crust: the "granitic" layer.

We know that essentially granodioritic batholithic complexes exist. Any transitions between these and granitic masses can hardly be sought in representatives of these melts from the depths of the crust (from the "basaltic" layer). The present cumulative views concerning the composition of this layer emphasize the low probability that this layer is predominantly basalt. Many data are to be found respecting this: metamorphic rocks of more or less basic composition [Rezanov, 1962; and others]; residual material after "distillation" of granitic melt from the metamorphic rocks being of basaltic composition, the deep-seated igneous rocks [Belousov, 1966₁], and, lastly, the view we have just discussed, that this layer is chiefly of andesitic composition [Green and Ringwood, 1966₂]. There should be no doubt that, at least in the upper part of the "basaltic" layer, a great role, if not the dominant role, must be played by metamorphic rocks.

We are faced with a curious phenomenon, which cannot yet be considered ultimately proved but which appears very probable. An earth shell has been marked as being dead, so to speak, in regard to the formation of magma. There are no clear indications of magma being generated in the lower part of the continental crust. It would seem that there can be no doubt about the sterility of the uppermost part of the mantle above the low-velocity channel of Gutenberg. As a result, the entire region from the upper boundary of the low-velocity channel up to the "granitic" layer is free from magma formation, more accurately from massive fusion. The probability exists, properly only for the "basaltic" layer, that magma may be generated under certain conditions here, but the resulting magma is of unusual type. If within the "basaltic" layer the granitic component has already been melted out of the rocks, and it is this that causes us to separate the layer from the "granitic" layer [Belousov, 1966], then, under ordinary conditions, even under geosynclines, temperatures can hardly be sufficient to produce fusion. Local (and rather rare) magma chambers may appear either because of the presence of readily fusible rocks, preserved at the site, though this is unlikely, or because of an especially great local rise in temperature. Charnockite or anorthite may be evidence of such formation of magma.

In connection with the problem of composition of the lower part of the crust and the upper part of the mantle, let us pause to consider the possibility of some of the basic magma becoming trapped during its ascent. It is a widely held view that most of the magma does not reach the surface and that the volcanic facies constitutes but a small part of the total fused material [Belousov, 1966₁]. The grounds for such a conclusion, however, appear rather improbable.

Basic magma in the upper stage, i.e., when it abandons the primary chamber, finds itself under considerable excess hydrostatic pressure (computations of this have been made many times). Ordinarily the magma moves very quickly, like water in a pipe. This is not the slow passage of magma making a way for itself, as must be the case of asthenoliths, but is nearly the free rise of a liquid. There are few channels that obstruct the passage of the magma, that fundamentally change the picture of rising magma, and the view of the asthenolith must be replaced by one of a narrow stream under high pressure. Under these conditions, considerable cooling of the magma (below the melting point at the corresponding pressure) might be expected only near the surface, where the excess of pressure under which the stream exists is much less than below. Here a considerable part of the magma, if emergence to the surface is difficult, penetrates weakened zones (cross-cutting fractures, bedding planes), actively separating the rocks. However, even taking into account the existence of lopoliths, one can hardly compute the ratio of trapped or captured magma to that poured out to be greater than one to one.

Of course, it is possible that large quantities of magma may be trapped at great depths, an example being the secondary chamber under the Klyuchevskii Volcano. The appearance of such chambers is possible, however, only under conditions of marked obstruction to upward passage of the magma. These conditions are not common, and basalts generally pour out freely. On the whole, we must admit that the earth's crust, although it is enriched to some degree with basic rocks, must not be assigned an exaggerated role in this process. A great deal of the magma, if not most, pours out on the surface or is trapped near the surface.

At present it is not always possible to make a reliable distinction between two principal groups of andesites. The problem applies, for example, to the andesites of Sumatra, for which, in analogy with the andesites and andesite-basalts of the rest of the Indonesian arc, appear to be probably of deep-seated origin. But, we should not forget the rather thick crust under Sumatra, allowing the melting out of silicic magma and the presence of associated granodiorites.

There is little doubt that the melting out of silicic magma is facilitated to a great degree by a higher silica content of the crustal rocks. The continental crust, at least the upper half, if not the greater part, apparently consists of rocks that, as a rule, formed at the surface (see above). Great significance therefore applies to differentiation through exogenetic factors [Frolova, 1951; Sheinmann, 1963₂]. This effect reduces chiefly to solution in sea water of the principal bases (particularly iron, magnesium, and calcium), and, as a result, to enrichment of the clastic sediments in silica. After the sediments are involved in tectonic subsidence, this cannot lead to silicification of the crust. Thus, melting out of silicic magma is facilitated.

The act of melting in the crust is caused by temperatures abnormally high for such shallow depths. As we know, at a "normal" temperature gradient, we expect temperatures, even at the base of the crust, that are far from the values needed to melt even quartz-feldspar eutectics. At the depth from which we derive batholiths and silicic lavas, i.e., at 15-20 km, under ordinary conditions such temperatures are entirely improbable without special effects from below. Thus, the principal condition for the generation of silicic magma, and granite in particular, is a strong influx of endogenetic heat, permitting the temperature to rise several hundred degrees (see, for example, the upper part of Belousov's diagram, Fig. 29). This rise in temperature cannot be expected from some mass in the crust unusually rich in radioactive elements; the normal temperature gradient in the continental crust is the result of the large concentration of these elements in it.

Furthermore, current data [Belousov, 1966₁, p. 19] show that, for the metamorphism observed at the base of geosynclinal zones and for melting of silicic magma, it would be necessary to increase the "normal" geothermal gradient by a factor of 3-5 over the value observed. This points directly to a greatly increased introduction of heat into these zones from depth. The principal cause of the generation of large masses of silicic magma can thus be only heating from below, heat supplied from the deep interior.

The restriction of silicic magmas to the late stages of geosynclinal development, particularly to the period of folding or to the following period, was noted as early as the beginning of the twentieth century, if not earlier. This connection became widely accepted and, at last, came to be assigned a causal character: the greatest movements expressed in folding lead in some manner to the appearance of magma. The interrelations are undoubtedly much more complex than this, however, since granitic intrusions take place after the folding also.

A most significant circumstance is that folding and silicic magmas precede the gradual ascent of activity in the tectonofer. This activity leads, as we have seen, to the appearance of deep-seated andesites. There is nothing convincing in the belief that with further rise of the tectonofer the zone of activity embraces the upper mantle and, finally, the crust. It is known that activity in the tectonofer corresponds both to marked and contrasted movements and to appreciable heating. Apparently, it is to this rise of the tectonofer into the crust that we assign the beginning of uplift and folding in the geosyncline above the tectonofer and the possibility of fusion in the crust.

Recently, Kuznetsov [1966] investigated possible mechanisms by which granites are generated. The greater part of his work is devoted to the distribution of intrusion in the upper structural stage. The part of his work important for our purposes is that which treats of the appearance of magma. A substantial confirmation of the author's views is found in Kuznetsov's conclusion concerning the very shallow generation of magma (the formation of the largest bodies at the site of their fusion), which attests to a very appreciable elevation of temperature in the zone of granite formation. Important also is confirmation of the great role of water that comes into range of the magma chamber and the role of solution of alkalies (especially potassium). The introduction of such material considerably lowers the melting point and, consequently, fosters the generation of silicic magma. However, even if we accept its great significance, we must as usual consider the very great introduction of heat to the region of the magma chamber from the deep zones of the earth, and we must thus search for the fundamental dependence of the melting out of silicic magma on the zone of maximum energy emission, i.e., on the tectonofer.

It would seem that, with gradual heating beneath a geosyncline, the most easily fusible mixture should melt out first; i.e., the first silicic intrusions should be alaskitic granites. This is not so, however; the change in silicic magmas takes place in the same way as in basic magmas, from silica-poor to silica-rich varieties. The principal epoch for batholith emplacement, nearly coinciding in time with the period of uplift (partial inversion, more rarely total) and folding, is always characterized by intermediate magmas: granodiorite, soda granite. And only later, at the very end of folding, or even after folding, granite magma proper is injected, at times approaching alaskite in composition at the very end of the process.

When extensive folding does not terminate the process, and uplifts accompanying the folding are repeated in geosynclinal downwarps remaining from the first phase (this took place, for example, in some regions of the Asiatic Caledonides), the second phase is accompanied by new batholiths, again of intermediate composition, and only after this comes the intrusion of granitic magma proper. There can be no doubt of the fact that this order of intrusion attests to progressive cooling of the region of the magma chamber. We are forced to believe that, during the period of its heating, the readily fusible magma did not rise for some reason. Obviously this can be explained in only one way: the zone of heating rises so rapidly that the melting out of eutectic mixtures, undoubtedly taking place at the beginning of the process, is overlapped by a more general melting of matter before the first melt has opportunity to be injected. The fact that this phenomenon is repeated with the renewal of extensive folding merely confirms the view we have expressed. This overlapping of primary silicic magma and intermediate magma may be explained in the following manner. The amount of eutectic melt capable of being extracted is undoubtedly small, and its extraction from the remaining solid mass must therefore be difficult (this is clearly seen in regions where migmatites are developed). With more complete melting of the material, the amount of liquid phase increases, and, even if it does not exceed the solid phase in volume (then the solid matter would simply settle out of the liquid), it may be squeezed out more easily (and, consequently, more quickly), and therefore may overtake the eutectic liquid.

After the injection and differentiation of magma of batholiths, the temperature in the area of the chamber gradually declines and a period begins during which only eutectics or mixtures near eutectic composition can be melted out. Another cause of this phenomenon, undoubtedly facilitating the first, may be the gradual increase in introduction of fluxing agents and, primarily, alkalies (potassium in particular). A considerable introduction of these constituents on a par with silica-bearing solutions leads to the relative impoverishment of the magma in calcium and, hence, to the appearance of granites proper.

It should be noted that the picture described by Kuznetsov (see above) refers basically just to this last stage, the granitic stage, and that during the epoch of batholith formation in the first stage, the penetration of fluxing agents from below to the magma chamber is small. It is likely that the second stage comes a long time later, so that the liquid may be squeezed out and may accumulate in amounts sufficient for injection. This circumstance determines the considerable break in time before the intrusion of later granites proper. The widely held opinion concerning this is that later granites are not related to the folding, but are completely independent manifestations of magmatism. This view is supported by the appearance of granitic batholith-like stocks and tabular bodies, both beyond the limits of the fold zone proper, as is common, and where geosynclinal development has ceased.

It appears that in all these cases a particular error occurs in the analysis of geological data. This reduces to an artificial restriction of the view concerning the range of the principal tectonic process in the life of the earth for the past billion years, a process that we conveniently call the geosynclinal process. This restriction follows different, but in general similar, paths. One of these leads to the conclusion that after evolution of the so-called orogenic stage, this stage is removed entirely from the geosynclinal stage and is actually set in opposition to it just as the platform opposes the geosyncline. Concerning the fundamental difference in character of the last two structures and concerning the fact that one cannot consider the platform state a result of simple dying of activity in the geosynclinal zone after termination of folding, the author has already felt forced to speak earlier. This is no place to repeat what was said there [Sheinmann, 1955].

We must clearly keep in mind two circumstances when seeking a solution to the problem of extrageosynclinal granites.

The first involves the fact that the development of geosynclinal belts, i.e., the steepest and longest-lived geosynclinal structures, in which all the others enter as component parts, is by no means always terminated abruptly, so to speak: up to a certain epoch geosynclines exist in the belt, but after folding there are none. Actually the picture is generally just the reverse. At the end of development of the belt, the nature of the geosynclinal process in it changes more or less distinctly. In the literature this is most frequently indicated only as a curtailment in the area in which geosynclinal downwarps exist at this time and in which folding develops. This, for example, is the picture we draw for the transition from the Baikal structures to the Alpine structures in Europe.

There is a different type of change in the process, however, which may be observed at the end of life of the belt or of parts of it. In this case, after formation of the fold zone and formal (we emphasize this) termination of the geosynclinal process, relatively shallow and, apparently, always short-lived downwarps are preserved or are formed anew, and are then filled with sediments and replaced by zones of uplift and folding. The area occupied by these zones is considerably less than that occupied by the geosynclines prior to folding. These features may be disconnected, since in the downwarping process individual patches within the belt are involved. Furthermore, the existence of such small downwarped segments in the belt support the fact that, on the whole, the belt has already become a more or less stable region. The gradual involvement, in the course of development of the belt, of ever more parts in prolonged uplift is directly related to (in general progressive) transformation of the crust within it to continental, even with increase in thickness.

Under these conditions, the appearance or survival within the belt of the just-mentioned final geosynclinal downwarps may be accompanied by relatively weakened downwarping, as if they had become involved in a general uplift. However, none of this allows us to reject the geosynclinal nature of these downwarps or the presence (let it be weakened, dying out) of the geosynclinal process. If ordinarily the extinction of this process appears only, or almost only, in curtailment of the area involved in geosynclinal subsidence, then, in the discussed example, it is possible to observe the actual extinction with complete loss of the geosyncline.

As we noted previously [Sheinmann, 1956], examples of such late appearances of the process may be found in the Hercynian downwarps in the eastern Appalachians, the residual Mesozoic geosynclines of Transbaikalia, or the later Caledonian and early Hercynian of the Sayan—Altai region. We may add that it is very probably true also of the "residual-geosynclinal" nature of such basins as the southern Caspian, the Black Sea, and, at least part of, the Mediterranean. According to Belousov, such late geosynclinal downwarps belong to the complex makeup of the group of parageosynclines. There are no real grounds for tearing this stage away from the preceding stages of development of the belt. It is not necessary to propose some kind of specific resurgence of the belt (probably best to transfer here the term "revival" applied by Nagibina [1958, 1960]), since development of the belt did not break off; it continued. If we approach these phenomena in this way, then the exceptional character of the "silicic intrusions of Transbaikalia" or of the "late intrusions of Tuva—Altai" disappears. And in this case there exists a direct connection between granites and folding. The entire process, as in the case of a fully developed geosyncline, is surely due to deep-seated phenomena. This is apparently the way in which Kheraskov [1963] approached this question. But a weak development, or even absence, of subcrustal magmas (they may be only andesites, obviously; basalts do not appear) indicates a relatively shallow depth of the tectonic process. Perhaps we should imagine the following to be true: that there is no new bundle of deep-seated energy, as in the case of a normal geosyncline, and everything is due to heating of the residues of the tectonofer previously in existence here. It is more probable, however, that a new store of energy is created, though a much smaller store. It is this lessened heating that accounts for the absence of deep-seated basaltic magma. The comparatively low temperatures produce melting only of andesites (the "andesitic temperature trough"). Weakness of movement determines stunted geosynclinal basins and comparatively early extinction of the entire process. The melting of granitic material is meager, and this distinguishes this phenomenon from that observed in normal geosynclines. It should be noted that typical granodioritic magmas with their complex batholiths are not formed, or are but rarely formed. The heating is insufficient for melting of this magma.

The other circumstance that we should mention here is the appearance of granites outside the zone of folding. This type of granitic intrusion is considered a special indicator from the viewpoint of extrageosynclinal granites. The central and northern parts of Transbaikalia and the Stanovoi Range are especially typical in this respect. Here, as is well known, Mesozoic (Jurassic) granites are widely distributed. Moreover, their abundance is greater than in the Transbaikalian Mesozoic fold zone. In order to investigate the question of whether these granites are in any way dependent on the geosynclinal process, decisive significance is found in their spatial distribution and their time of intrusion. Their indisputable restriction (Transbaikalia–Stanovik) to a broad belt along the margin of the active zone has long been known. And only failure to recognize the connection between this zone and geosynclinal development has permitted workers to consider the Stanovik granite as an example of extrageosynclinal, more correctly nongeosynclinal granite. We should note here also the similarity in time between the Mesozoic Transbaikalian folding and these granites. Essentially it is not necessary to prove the absence of some connection, but to explain why melting in the upper half of the crust was not concentrated in the geosynclinal zone, but embraced neighboring regions about the framework.

There is no doubt of great supplementary heating of the crust, which cannot be explained by enrichment in radioactive elements. If the cause of melting should really be this, then the granites themselves should be enriched in radioactive elements, and extrageosynclinal intrusions could be readily distinguished from geosynclinal granites. This is not so, however, and the only cause of melting proves to be heating from below. On the other hand, the characteristic of extrageosynclinal granites proves to be the absence of traces of magma formation in the mantle, i.e., the much weaker process at depth than that observed in the tectonofer.

Apparently the most likely way to solve the problem of origin of silicic magma in the region framing the fold zone is to consider the zone of heating to expand markedly in its upper stages. Such expansion might be primarily the result of heating not only the "crown" of the zone but the sides as well. This type of heating is **natural**. In some rare circumstances, deep, steep fractures, facilitating heating, may prove to be restricted chiefly to the margin of the region. This leads to the principal development of magma outside the fold zone proper and on one side of it (the Mesozoic granites of Transbaikalia–Stanovik).

If the paths of heating are distributed over such an extrageosynclinal area more or less uniformly, a field of granites appears. If these paths prove to be a narrow "bundle" of fractures, a volcanic-plutonic belt appears.

The view that granitic magma migrates horizontally from the tectonofer for tens and even a few hundreds of kilometers does not appear likely. There are difficulties with it: 1) it is inconceivable how forces might drive magma horizontally for such distances, and 2) it is impossible to imagine how magma, with such displacement, can depart a zone of heating and at great distances pass into a cold region. The great viscosity of silicic magma does not permit it to move under these conditions with sufficient speed to avoid crystallization on the way. As a result, we must admit that the chamber of extrageosynclinal intrusions is beneath them, and that the clear connection in space and time with granites of fold belts is due to the unity of the deep-seated process, to the dependence of heating the broad zone of the crust on deep-seated phenomena in the tectonofer. A study of these phenomena, based on the recognition of these connections, appears to us the most promising direction for investigation in this field.

# CHAPTER 12

# VOLCANIC (VOLCANIC-PLUTONIC) BELTS

In recent years, after the works of Ustiev and his co-workers [Ustiev, 1959, $1963_1$, $1963_2$, 1965; Speranskaya, 1961] on the study of the Okhotsk volcanic belt, this type of structure has gained considerable attention. The distinctive character of vulcanism, its superposition on older structural plans and its apparent independence of geosynclines, has given rise to extreme views concerning the complete independence of volcanic belts both of platforms and of geosynclines [Komarov, 1960; Komarov and Khrenov, 1963; Komarov, Odintsov, and Khrenov, 1964; Khrenov and others, 1965]. Since such belts are primarily belts of magma formation and only the ascent of magma defines the designated structure itself, we must touch on them here, even in the briefest way.

Volcanic belts have received the most study in our country of any of these structural belts. There is no reason to give descriptions, therefore. Such material is widely known. We may note that volcanic belts appear on top of very different older structures, and they can in no way be thought to be confined to any one type of these older structures. The appearance of such belts is well known along the borders of developing geosynclinal zones, such as the northern half of the Okhotsk—Chukotskii belt, with which the study of this relationship has begun. This relationship is by no means necessary, however, and it may not even be the most common one. In the Okhotsk—Chukotskii belt, two parts may be distinguished: 1) the northeastern, bordering the region of the Kamchatka—Koryatskii folding and superposed on Mesozoic structures, and 2) the southwestern, also superposed on Mesozoic structures over much of it, but far removed from the zone of Tertiary—Quaternary folding. It is difficult now to determine the age of the structures covered by the Sea of Okhotsk more precisely.

The Sikhote—Alin' belt is bounded on the west by the Tertiary—Quaternary geosynclinal zone and, like the northeastern part of the Okhotsk—Chukotskii region, is superposed on Mesozoic structures. However, along with these marginal belts, in the terminology of Ustiev, there are so-called "intraframework" belts of the Selenga—Vitim type, situated at a marked distance from the fold belt. There are no appreciable differences between these belts and the marginal belts, except for their positions relative to the fold belt. This fact is significant to us in attempting to relate magmatism and tectonics.

The principal difficulty in these investigations is the complete uncertainty concerning the nature of the movements at depth during the formation of the belt, since there are no presently forming structures of this type. All our assumptions must therefore be built merely on comparisons with near-surface structures approximately simultaneous with the vulcanism and on comparisons with phenomena in geosynclinal zones. Such comparisons have led to some preliminary conclusions:

1. Volcanic phenomena are accompanied by considerable tectonic activity. The latter indicates vertical movements and not very strong folding. The range of the first may be determined, for example, by the accumulation of 6-9 km of tuffs and lavas with interbeds of

coarse sediments (the Tamir Series in the Selenga—Vitim belt). There also we may see large open folds and monoclines with dips up to 40-50°.

2. Volcanic activity is clearly separated into two·independent stages. It begins, as a rule, chiefly with silicic magmas. The average composition of rocks of this stage were computed for western Transbaikalia by Saltykovskii [1966] as liparite-dacite. Basic and intermediate rocks play a subordinate role at this time; representatives of such rocks are both basalts and andesites. The accumulations and strong tectonic activity mentioned in (1) are associated with this stage. The second stage is characterized by weak tectonic activity (there is no moderate subsidence of comparative small areas of folding) and basic magma. Intrusive bodies, normal for the first stage, are not common or are lacking entirely. The Jurassic basalts and Mesozoic basins of the Selenga part of Dauriya correspond to this stage.

3. It may be stated that the marginal belts are characterized by rocks of the calcic series with normal granites (only the very latest fractions of the magma acquire a more alkalic character and give rise to alkali granites and syenites). In contrast to this, belts some distance from the edge of fold zones are characterized by magmas with more alkalic tendencies, giving rise to alkali granites, syenites, trachyandesites, and trachybasalts. Rocks of the "principal sequence" are absent or are relatively rare. This picture may be observed in the Okhotsk and Selenga—Vitim belts.

4. In length of time required for formation, volcanic belts apparently always somewhat outstrip development in the neighboring geosynclines. The long epoch of subsidence in geosynclines essentially has no analog in the volcanic belt. The only manifestation of this tendency is subsidence of individual segments against a background of general uplift. Thus, the silicic magmas of the Okhotsk belt (Cretaceous) correspond to the epoch of subsidence in the neighboring geosyncline on the east. In western Transbaikalia, silicic magmas ascended in the Triassic—Jurassic, i.e., at the time of subsidence in eastern Transbaikalia. If we wish to make a comparison with a geosynclinal belt, development of a volcanic belt begins with an epoch of welling of silicic magma and then goes quickly into the second stage, when nongeosynclinal type of basic magma appears, which is most easily correlated with trap magmas or olivine-basalt provinces. Repetition of the magmatic process appears to be possible in rare circumstances. In any case, we may thus treat the sequence in the Selenga part of the Selenga—Vitim belt, described by Kiselev and Saltykovskii [1967]. Tremendous quantities of silicic lavas were poured out here along early northwest-trending fractures of the Malkhanskii zone at the very end of the Permian (?) and Early Triassic. After extensive dislocations in Late Triassic time, the chiefly trachybasaltic Chernoyarovskii Series, essentially indistinguishable from the basaltic series of the second stage, was formed. In Late Triassic time, the orientation of the defining structures changed to northeasterly, and the process began anew, so to speak. Then, simultaneously with the trachybasalts of the Chernoyarovskii Series, the substantially silicic Borgoi Series was formed in the basin of the Dzhida, and subsidence occurred anew. After new dislocations, as early as the Middle-Upper Jurassic, trachybasalts of the Ichetuiskii and Khilok Series were poured out.

To these features of volcanic belts we must add a few words concerning the possible structure at depth. The absence of the first magmatic stage of geosynclines (ophiolites) is not grounds for assuming the existence of a tectonofer at depth below the volcanic belt. The indicated broad connection between volcanic belts and geosynclines, in both space and time, shows that we should seek for even a deeper relationship between them, and, since a causal connection between geosynclines and volcanic belts is hardly possible, it is better to consider them both consequences of a single cause. To us this cause is the tectonofer. If this is true, we must then believe that occasionally, in the zone where long-lived fractures reach down to the tectonofer or nearly reach it, heating of the crust by deep-seated heat takes place not only along the zone where a geosyncline is formed but also along branching lateral zones, which do not serve as channels for magma and which extend beyond the limits of the geosyncline.

5. Meanwhile, the complete absence of information concerning tectonic conditions at depth under volcanic-plutonic belts allows us only to identify a few concrete facts concerning the relationship. We may add that we should expect to find a volcanic belt on the same side of the geosyncline (or young fold zone) on which the tectonofer is buried. The emergence of associated zones, forming a volcanic belt at the surface, is most readily imagined as an upward branch of the tectonofer and not as a horizontal branch. In case of an almost vertical tectonofer, one must expect the volcanic belt to be as likely to form on one side as the other.

Of course, the mechanism we have just suggested is very primitive, and can hardly serve as an adequate description of the phenomenon. But we present it as a first and very crude attempt. It may be noted that there are some data in favor of it. One fact is the position of the Okhotsk–Chukotskii belt with, so to speak, the "back" side of the Kamchatka tectonofer. Another is found in the volcanic belts of the Rocky Mountains in their relation to the Cordilleran zone. Still another is the Hungarian volcanoes in back of the Carpathians. For most cases, however, it is impossible to establish the probable inclination of a tectonofer. And, in general, this method of checking cannot be considered strict.

CHAPTER 13

# MAGMA AND THE SUBSIDENCE OF OCEAN BASINS (CONCERNING "OCEANIZATION")

In our day the idea of subsidence of segments of the earth near elevated zones and the formation of these into oceans on the spot has become rather widespread. By the term "oceanization" or any substitute for it, it would appear that a variety of phenomena are referred to. Before anything, we should probably specify precisely what we mean by this term.

Without doubt, typical oceanic crust has formed in the place of segments of continental crust or of transitional crust [Belousov, 1955 and other papers; Muratov, 1957; Petrushevskii, 1964, Sheinmann, 1958]. It is true that this process meets with very serious objections on the side of physics [Lyustikh, 1959, 1961]. Absolutely indisputable evidence of geology and biology, however (areal distribution of animals and plants), leads us to believe we have proved that land recently existed where now we find a deeply subsided zone with typical oceanic crust. It hardly seems necessary to repeat proofs already presented or to cite new data that are essentially superfluous. We are obliged therefore to ask physicists to analyze this problem once more, and to try to find a satisfactory solution. We need hardly doubt that, however improbable the described phenomenon appears to the physicist, an explanation must be found if the phenomenon actually exists. The indicated subsidence is clearly not uniform, but is associated with very different kinds of geological environment. And, although we have no clear data concerning the geology, particularly the tectonics, of deep-seated regions, the environment near the surface changes so much that these changes can only reflect great differences at depth. Furthermore, these differences conform in great measure to types of deep-seated tectonics (see Chapter 4).

Two types of subsidence apparently occur to create basins with oceanic crust. Examples of the first are the Sea of Japan, the southern margin of the Bering Sea, the seas of the Indonesian archipelago and Melanesia, the Caribbean Sea, and the deep basins of the Mediterranean (or some of them). All these seas are characterized by a direct connection with zones of modern geosynclines or geosynclines recently transformed to fold belts. It would be more accurate to say that these seas occur within still-existing geosynclinal belts (in the sense of this term as used by the author [Sheinmann, 1955]). The subsidence itself is here associated with the geosynclinal process, particularly with the generation or the process of regeneration of geosynclinal downwarps. The first type of newly forming oceanic segments should therefore be called geosynclinal. Considering the possible mechanism, we must keep in mind the conditions existing above the upper part of the tectonofer.

The second type of newly formed oceanic district differs strongly from the first in its geologic features. Examples of it are all the coastal zones of the Indian Ocean (except the region of the Indonesian arc) and all the Atlantic Ocean, except for the Caribbean and Scotian (South-Sandwich) arcs. Subsidence, geologically recent beyond doubt, may be verified here at a number of places. Furthermore, as is well known, there are strong grounds for our believing these oceanic zones to be entirely youthful, since the subsidence producing them and the corresponding changes in type of crust could not have begun before the Jurassic; and it continues up to the present day. A peculiarity of this type of subsidence — and this has also been described — is its complete independence of geosynclines. The subsidence has been superposed on existing structure as something alien to it, not continuing the development of that structure. Correspondingly, the appearance of new regions with oceanic crust cannot be associated with activity of tectonofers. No recent activity at depth during this type of subsidence has been verified, and even in the upper levels seismic activity may be slight. This difference between the two types has long been known, but it has been applied only to the nature of coasts: "Atlantic" and "Pacific" types of coasts.

It would probably be proper to give names to the different types of subsidence leading to the appearance of new zones with oceanic crust. Since the first type of subsidence has to do with one of the manifestations of geosynclinal life (geosynclinal zone), we can hardly speak of the formation of ocean in the proper sense of the word, even if such zones join the modern ocean.

We should note one other feature, deriving from geosynclinal aspects: the formation of the first type of zone with newly produced oceanic crust always occurs next to zones of uplift, commonly alternating with them. During subsidence of the second type, the new segments are organically connected with the ocean, and prove to be part of it. It becomes difficult to draw a line between such newly formed zones and the previously existing oceanic region. It would be logical to use the term "ocean-formation" for this type of process.

The term oceanization, introduced by Belousov, might be applied in this case, but, unfortunately, the word for Belousov already has a fully defined meaning: the conversion of continental or any intermediate-type crust to oceanic crust. The appearance of a segment with oceanic crust does not necessarily mean the appearance of a new segment of ocean, and the term therefore fails to conform entirely to its etymological significance (an oceanic crust may appear also in a geosyncline).

The author finds it difficult to select a term for the appearance of new, strictly oceanic areas. We might perhaps settle on "ocean-formation" or "thalassogenesis." It assumes that the above-described difference between the two types is very significant, and that only subsidence of the second type corresponds to the formation of oceanic segments. We should note that the above discussion is fundamentally related to the study of the transition zone between continent and ocean. Since geosynclinal zones appear above specific deep-seated structures (tectonofers) in no way related to the separation between ocean and continent and may grade from continent to ocean and the reverse, study of the junction between ocean and continent should be made primarily where there are no geosynclines, i.e., in regions with Atlantic-type coasts. Study of the transition from continent to ocean in the region of island arcs leads to an understanding of the transition from geosynclines to their oceanic or continental framework, but it is difficult to use in interpreting the transition from continent to ocean. This, in particular, is the situation in the studies of the region of the Sea of Okhotsk and the Kurile—Kamchatka arc.

Special interest is found in the western coasts of both Americas. It is well known that we have difficulty assigning these coasts to one of the two types distinguished by Suess. The boundary between ocean and continent is here parallel to the structures arising from geosynclines, and, it should be said, young geosynclines. On the other hand, there are no typical present-day geosynclines with their island arcs and seas.

Apparently, conditions are simplest along the coast of North America, where there is nothing to point to the existence of a tectonofer. There are no deep-focus earthquakes here, nor ocean trenches, nor geosynclinal type of vulcanism. Probably, geosynclines were not formed after termination of Mesozoic folding, and the final dying of activity at depth took place sometime in the Tertiary. Recent subsidence that led to capture by the ocean of the margins of fold zones has been superposed on fully formed young structure in such a way that the new boundary between continent and ocean coincides with the trend of the folding. It is this latter circumstance that distinguishes the west coast of North America from coasts of Atlantic type. The difference is not merely one of geometry, but is clearly related to deep-seated peculiarities of the region. In particular, despite the similarity to Atlantic type of coast, the Pacific ocean opposite North America has nothing resembling midocean ridges.

Conditions change substantially south of Lower California. The appearance of an ocean trench, deep-focus earthquakes, volcanoes with andesite-basaltic lava — all uniformly point to the existence of a still very active tectonofer at depth. We should thus search for island arcs here. But there are none. In Central America, the role is played by the isthmus itself. The position of the deep trench corresponds to the slope of the tectonofer toward the east, so that the Caribbean is the backdeep, similar to the Seas of Okhotsk and Japan (it may be noted that the Caribbean plays the same role relative to the Antillean arc, and we have to do with what is called a two-sided orogen). There are no data on strictly oceanic subsidence (ocean-formation or thalassogenesis) along the west coast of Central America.

Farther south, both the deep-water trench and the zone of deep-focus earthquakes and andesitic magma are preserved. The Antillean structures of southern Venezuela are connected to the western (Andean) branch, but this does not exist farther south. The Andes here, and down to Patagonia, are associated with a still "living" deep-seated zone descending to the east. This tectonofer (or the tectonofer replacing it along the strike) is a direct continuation of the Central American tectonofer, and the Andes take the place of the island arc, elevated to great height, as it were.

There is no doubt that very young subsidence has occurred along the coast, with the formation of ocean deeps. And here, as along the North American coast, the fractures along which subsidence occurs are parallel to the fold structures. It appears most likely that here too we have to do with ocean-formation, superposed on the young folding. But, in contrast to North America, this superposition took place before the tectonofer died out. If this is so, the region would be a positive illustration of the fact that the processes of geosynclinal arrangement (i.e., the "superstructure" on the tectonofer) and subsidence leading to the appearance of new oceanic segments are to a considerable degree independent of each other, to such an extent that in rare cases one may be superposed on the other. We could not find actual confirmation, however, that such superposition has actually taken place along the west coast of South America, and this discussion must be considered merely as preliminary.

There is considerable interest in the question concerning the degree to which thalassogenesis is related to the formation of magma. There is little doubt that magma most likely ascends during the described processes along deep fractures along which the subsiding occurs. Consequently, volcanoes should most probably appear along the boundary between continent and ocean. It is well known, however, that they do not appear here. A number of volcanoes occur along the coasts of South and Central America of course (on the Pacific side), but they are associated with the still-active young fold belt and its lavas: andesite and andesite-basalt. Nowhere along the coast of the Atlantic Ocean or the Indian Ocean, nor along the extrageosynclinal part of Antarctica, are there volcanoes. The only exception is the region of the Mozambique Channel. But even here the vulcanism is not confined to the African coast, i.e., to the boundary between ocean and continent. Volcanic islands (Comoro Islands) are out in the ocean, in the middle of the channel, and the volcanic zone is continued to the northern part of Madagascar. Thus, the principal fractures accompanying thalassogenesis do not conduct magma to the surface.

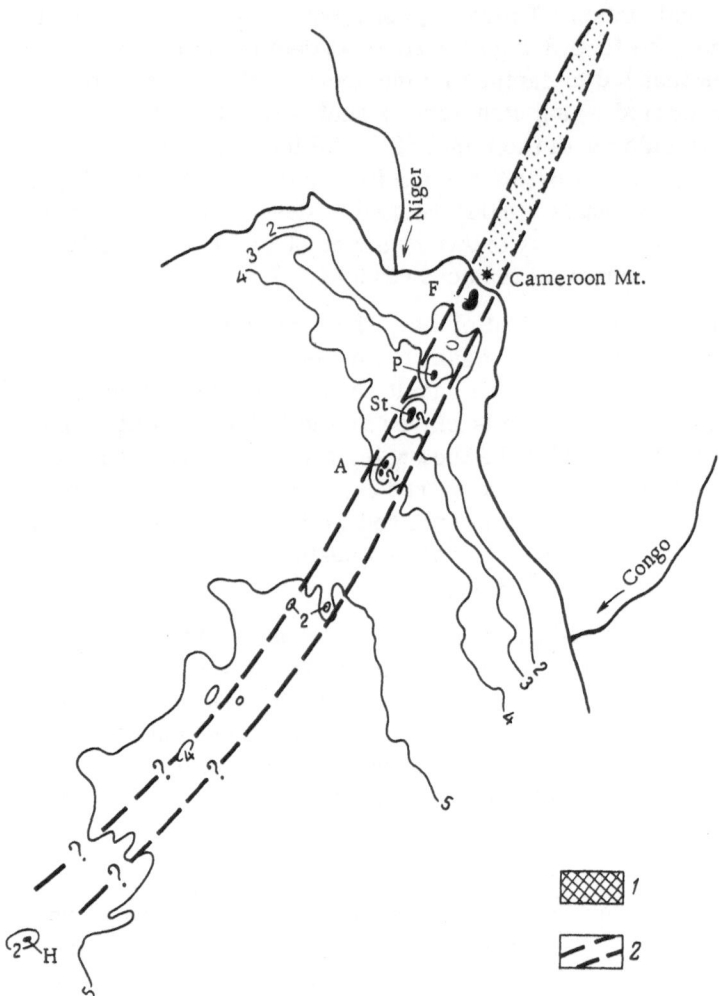

Perhaps, because it appears especially instructive in this respect, we should consider the series of Quaternary volcanoes on the islands of Principe and Fernando Poo through Cameroon* toward Lake Chad (Fig. 38). Alkalic basalt lavas are here confined to a remarkably straight fracture of great length, passing across the coast and being clearly independent of subsidence in the Gulf of Guinea (a very young part of the Atlantic Ocean).† Of most interest is the fact that in the region where this fracture crosses the margin of the continent there is no hint of any distribution of volcanic phenomena along this boundary. Even in this situation the ocean-continent boundary remains magmatically sterile. It is true that the greatest activity of the volcanoes occurs at the intersection of fracture and boundary. Here we find the huge Cameroon Volcano, and not far away in the sea are the volcanoes of Principe, Fernando Poo, and other islands. But these lie on the same fracture, and the volcanic phenomena are not spread along the coast. The increase in activity is explained by the fact that near the intersection of the indicated fracture with bounding fractures of the continent the transmitting capacity of the main fracture for magma proves to be high.

We see a somewhat different picture when we turn to the volcanic phenomena within the sunken or subsiding segments. Within these we occasionally find extensive vulcanism. Madeira, the Canary Islands, the Cape Verde Islands, Atol das Rocas,

Fig. 38. Relations of the deep fracture and the volcanic line Annobón—Cameroon—Chad with the zone of subsidence in the Gulf of Guinea (isobaths in thousand-meter intervals). F) Fernando Poo; P) Principe; ST) São Tome; A) Annobón; H) St. Helena. 1) Belt of volcanoes and lava fields (from end of Cretaceous to present); 2) continuation of this zone into the ocean (volcanic islands and volcanoes). The slight bend in the zone is the result of incompletely successful projection (in orthogonal projection this zone is almost ideally straight).

---

*Tables of chemical analyses are given on pp. 139-166.

†Subsidence here began in the Cretaceous, but the sediments indicate that the depth of formation was rather insignificant. Therefore, despite the fact that the tectonic line determining the boundary between the present continent and ocean was marked out in the second half of the Mesozoic, the appearance of ocean depths and the annexation of the Gulf of Guinea by the ocean dates most likely from later time.

Fig. 39. Relations of Madeira, Canary, and Cape Verde Islands to the continental zone (isobaths at thousand-meter intervals).

Arquipelago Fernando de Noronha, and Trinidad, against the Atlantic coast of South America, suggesting the vulcanism of the Mozambique Channel, are examples. To these we may add numerous submarine volcanic peaks. It is still not clear, however, whether these conditions are characteristic only for recently subsided zones or whether we have to do here with a general picture for all oceans. Apart from this, we must discover to what extent vulcanism of this type is associated with ocean formation.

The Madeira — Canary Islands complexes and, in great measure, perhaps, the Cape Verde Islands cannot be considered with any great probability to represent residual magmas from primary tholeiitic basalts.

Canary Islands and Madeira. A complete stratigraphic section of the Canary archipelago includes the following [Furon, 1963]. The oldest exposed rocks are dolerites (Cretaceous ?). The islands rose later, most likely at the very end of the Cretaceous. The Paleogene is represented by volcanic rocks (Furon mentioned acgerine rhyolites and trachytes, but it has been impossible to check these descriptions, since no one has confirmed them petrographically). The rocks rest on an eroded surface of basic rocks, which have been folded. Above, unconformably, rest white tuffs, trachytes, and phonolites belonging to the first half of the Miocene. Vindobonian fossils are found higher in the tuff sequence.

The Lower Miocene is represented by underwater flows, which later gave way to eruptions on land, the so-called "basalt plateau" with horizons of terra rossa. The second half of the Pliocene was characterized by intense volcanic activity, accompanied by faulting and other deformation. Layers with Tyrrhenian fossils are found on the "basalt plateau," and these are overlain by basic lavas with layers of marine sediments and terra rossa.

If we consider all existing rock analyses from the Canary Islands as a whole, we get the impression of chaotic distribution of points on the diagram. It is therefore necessary to consider each island separately.

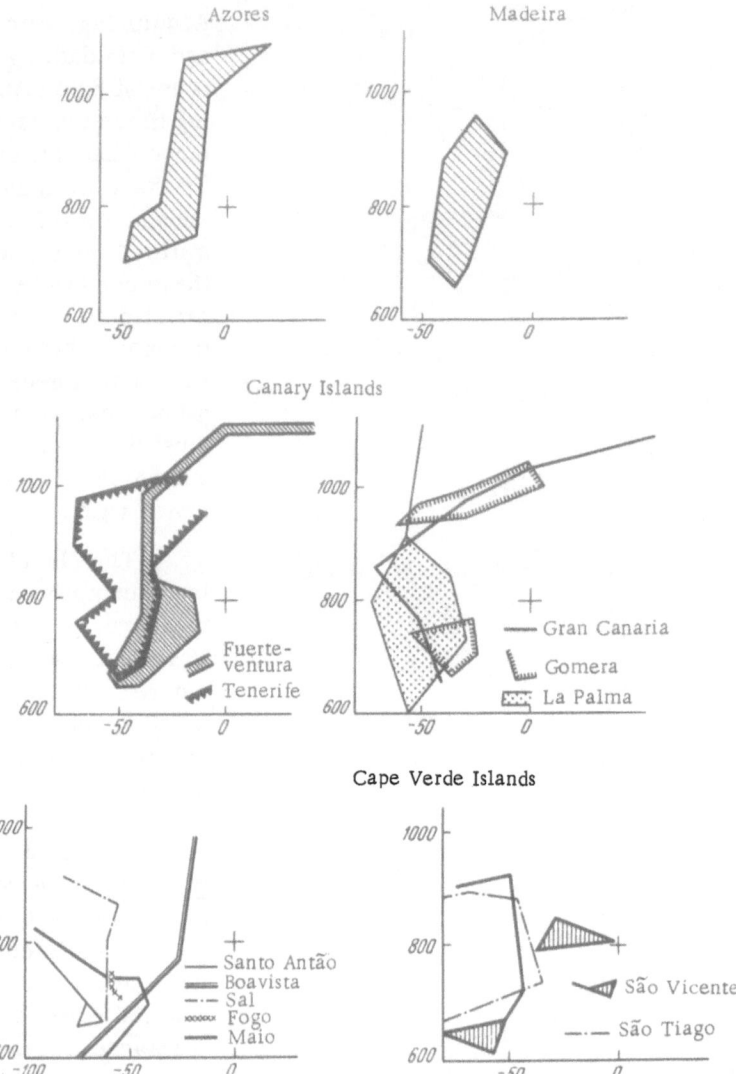

Fig. 40. qz diagrams for rocks of Madeira, Azores, Canary
Islands, and Cape Verde Islands. Typical olivine-basalt
series: on the Azores, Madeira, Fuerteventura and La Palma
(Canary Islands), and Boavista (Cape Verde Islands). Series
with the character of alkalic-ultrabasic complexes are found
on Tenerife and Gran Canaria (Canary Islands), Santo Antão,
Fogo, São Tiago, and Maio (Cape Verde Islands). Both lines
are present on São Vicente (Cape Verde Islands). The lavas
on Gomera (Canary Islands) give no clear picture.

We have at our disposal analyses for the rocks of Fuerteventura (26 analyses), Tenerife
(35 analyses), La Palma (11 analyses), Gomera (17 analyses), and Gran Canaria (8 analyses).
See Fig. 40 for results.

On Fuerteventura, the nearest island to Africa (Fig. 39), occurs a widespread ancient
series of Tertiary age (Miocene and post-Miocene according to Bourcart and Jeremine [1938]).
It includes both intrusives and extrusives. The intrusives include peridotite, pyroxenite, feld-
spar-bearing pyroxenite, olivine gabbro, essexite, diorite with olivine and hornblende, and

alkalic syenite. The lavas include picrite, olivine basalt, labradorite basalt, basanite, spilite (?), quartz keratophyre, comendite, and rhyolite (alkali rich). Young lavas are picrite and olivine basalt. It is easy to see from the diagram (Fig. 40) that the rocks of the island fit well in the lineage of oceanic basalts. In the basic and intermediate varieties, K = 0; i.e., we have to do with a typical olivine-basalt association. The most silicic varieties have been clearly enriched in silica, and the amount of alkalies in them has been reduced.

From Gran Canaria, farther to the west, we have analyses of limburgite, analcime basanite, trachytite (more basic than trachyte), silica-rich "phonolite," and pantellerite. The most basic rocks of the series (limburgite, analcime basanite) are rich in alkalies and are probably related to rocks of the alkalic-ultrabasic group. The most basic of the trachytites (46.7% $SiO_2$) is characterized by twice as much alkalies (about 8%) as basalts with the same silica content. The more silicic "phonolites" indicate a lower content of alkalies and an increased quantity of silica. The most silicic differentiate, pantellerite, indicates that at the end of the process the accumulation of alkalies ceased and the residual magma acquired the character of an alkalic rhyolite.

On the whole, the qz diagram is most suggestive of the curves of alkalic-ultrabasic complexes with long-past differentiation (negative K). We note some hint of this in the diagram for Fuerteventura with its extension in the field of basic rocks.

The island of Tenerife, the next to the west, is characterized by a complex group of rocks. It includes olivine basanite, tephrite and trachyte-tephrite, trachyte-basanite, olivine basalt, basalt, trachydolerite, trachyandesite, foyaite, and a series of phonolites (with nosean, analcime, haüynite, obsidian, vitrophyre). The lavas are represented by two series. The older forms the foundation of the island (basic lavas, in lesser degree pyroclastic rocks, and last, trachyte and phonolite). Rocks of the second series form parasitic cones at the base. Lavas of eruptions recorded by Europeans are trachydolerites for the most part. A distinctive feature of this magmatic center and of the rocks derived from it is that the rocks are generally similar to the alkalic-ultrabasic series but include some features of the olivine-basalt series. There is practically no petrochemical difference between the young and older series of the island. It is thus indisputable that the observed features are not random: the magma preserved its distinctive features at least through the Neogene and the Quaternary. Any proposed hypothesis must take this into account, must recognize the inapplicability of the view that the magma is the residual of an alkalic-basalt melt at depth (such as the Honolulu Series on Oahu), and must seek the development of the lava in a short interval of time and subsequent prolonged existence of the magma chamber or in repeated melting out in a single region from masses identical in composition and having the same temperature and pressure conditions.

We have obtained analyses from Gomera for augite-labradorite basalt (including very basic varieties with 40% $SiO_2$), dolerite-basalt trachydolerite, essexite-diabase, phonolite, phonolite-trachyte, and trachyte. The points for most of these rocks lie fairly well on the curve for olivine basalt. Apparently this island is covered with the same complex as that found on Fuerteventura.

The westernmost of the islands, La Palma, is characterized by rocks generally similar to those described above: limburgite, haüynite tephrite, essexite and essexite-diabase, trachydolerite, and sodalite trachyandesite. On the qz diagram, these rocks form a rather broad field with a narrow vertical extension toward more silicic varieties. K is approximately 0. The oldest rock on the island, chloritized and epidotized essexitic diabase, makes up the framework (ancient dome). A younger dome lies on top, also composed of rocks with high alkali content. The youngest lavas are limburgitic trachybasalt. The most significant geologic feature of the island is the possibility of dating its formation [Zavaristkii, 1944]: the older dome was formed no earlier than the middle (end ?) of the Cretaceous and no later than

the end of the Paleogene; the upper dome is Lower Miocene and has continued to develop up to
the present time. Thus, data on La Palma permit us to date the formation of the island precisely. It is true that we should consider the possibility that some of the rocks poured out before the island was formed, while it was still part of the continent.

These brief descriptions lead us to state that, on the whole, the volcanic rocks of the
Canary Islands combine the features of the olivine-basalt and alkalic-ultrabasic associations,
creating a distinctive mixed type of complex.

Madeira and Porto Santo are very similar in their volcanic associations. Madeira, the best studied of the group, consists chiefly of trachybasaltic rocks. The age of the island and of the vulcanism is comparatively old. According to Hartung and Wolff the volcanic
rocks rest on Middle Miocene rocks. According to Gagel, Middle Miocene sediments were deposited in a bay that penetrated an already existing volcanic structure. At the base of the section on the island occur underwater lavas and white tuffs, assigned to the Lower Miocene
[Furon, 1963]. The Pliocene is represented chiefly by volcanic rocks containing a layer with
lignite. Lacustrine sediments were formed in the Quaternary, and volcanic rocks represented
by basalt are present. Furon thus appears to agree with Gagel. Volcanic activity on the island
ceased sufficiently long ago to permit erosion to destroy the volcano and to expose subvolcanic
bodies. Judging from the presence of lavas covering a layer with Quaternary flora, the end of
volcanic activity cannot be placed earlier than the Lower Quaternary.

The main bulk of rocks on Madeira* consists of trachydolerite. Under this term are
combined both relatively silicic alkalic basalts (52-55% $SiO_2$) and very basic basalts (42-46%
$SiO_2$). The latter include varieties containing nepheline, and, consequently, they should be given a
different name. Erosion has exposed hypabyssal correlatives of the basic lavas: essexite and
essexite-diabase. More basic differentiates are also found in some quantity, probably the result of settling out of dark minerals: these rocks have been given the name madeirites. The
content of silica drops to 40%, making the rocks very basic essexite-dolerites. Limburgites
are also present.

Trachyandesites and trachytes are of secondary importance. Plutonic correlatives of
these rocks are also known (akerite, sodalite syenite).

On the whole the entire rock series is typically olivine-basaltic and is characterized
only by a somewhat diffuse outline on the diagram (see Fig. 40).

Cape Verde Islands. These islands are assumed to lie at a site where systems of
deep fractures intersect at almost right angles. The principal system trends to the west-
northwest. Along it occur the volcanoes that form the islands of Santo Antão, São Vicente,
Santa Luzia, Branco, Raso, São Nicolau, and Boavista (Fig. 41). This line is cut by two almost
north-trending fractures. At the intersection of the western of these with the main fracture
lies São Nicolau. Southward on this same fracture lies Fogo (on which is found one of the
active volcanoes of the archipelago). The eastern fracture passes through Sal and along the
western coast of Boavista to Maio. To what extent this treatment of the tectonics [Bacelar,
1932] can be considered proved on the basis of data from the literature is a difficult matter.

What appears to be indisputable is the still incompletely lost connection between the
African continent and the islands. It is true that this is morphologically marked only by a
salient of the sea floor at depths between 3000 and 4000 m, i.e., abyssal depths of oceanic
order, so that there are no grounds for expecting continental crust to be preserved under this
salient. The base of the islands, rising above this salient approximately 2000 m and completely
separated from the zone of continental slope, has the shape of a horseshoe, open to the west,
and all the islands sit on this single submarine ridge.

_____
*Tables of chemical analyses are given on pp. 139-166.

Fig. 41. Cape Verde Islands (isobaths at thousand-meter intervals).

The geologic structure of the islands very clearly indicates a direct connection with Africa in the past. According to Furon [1963], the oldest known rocks belong to the second half of the Upper Jurassic and are cherty limestones, by no means deep-water formations. On Maio are found Valanginian-Hauterivian somewhat metamorphosed limestones, covered with basalts, which are also Lower Cretaceous. Above these rocks occur Barremian bituminous marls, variegated Aptian clays, and Senonian (?) metalimestones.

A similar section has been measured on São Nicolau (Upper Jurassic, Lower Cretaceous, Senonian limestones). Here a faunal correlation has been established for the eastern half of the archipelago with Morocco and Portugal, and for the western half with Central America. During the Eocene, sedimentary rocks continued to accumulate (Lutetian marls on São Nicolau). From the Miocene (Lower ?) to the Vindobonian transgression, limestones with foraminifers, starfish, and calcareous algae were deposited on São Nicolau, Maio, and São Tiago; and volcanic rocks were widely formed: "basalt," trachyte, and phonolite. Volcanic rocks are again encountered above the Vindobonian limestones.

The character of the sediments from Jurassic to Miocene time points to a shallow sea and to a close connection with Africa. A pre-Tertiary connection with America is also indicated, not nearly so sharply expressed. Apparently, individual islands on the base of Africa can be assumed to have existed only since the Tertiary. As will be seen below, this change did not lead to any fundamental change in the region of magmatism. On Maio, intrusive rocks cut Mesozoic limestones that were deposited before the advent of oceanic conditions. It is thought that the oldest rocks are Cretaceous volcanic rocks (to a considerable degree, apparently, by analogy with Madeira and the Canary Islands). Apart from Maio, deep-seated rocks of the same age (?) have been described from São Vicente and São Tiago. Alkalic granites, syenites, foyaites, essexites, theralites, and olivine gabbro from these islands have been described. The pre-Barremian lavas of Maio apparently also belong to this stage.

The vulcanism of each island is somewhat distinctive.

On Santo Antão, made up of a number of volcanic cones, rocks of the alkalic-ultrabasic association are developed: limburgite, nepheline basalt and basanite, trachybasalt, nepheline tephrite, trachyte, phonolite, nephelinite, leucitite, and lavas with haüynite. Of these we had analyses only for limburgite, tephrite, and leucitite (in which sodium was much more abundant than potassium, however).

The island of São Vicente is distinguished by being made up of rocks of two series. One of these consists of basalt, basaltic trachydolerite, and gabbro-dolerite. We should probably also assign the trachyte of Monte Fateixa, which has a somewhat high alkali content, to this series. The other series is nearly akin to those on Santo Antão: sodalite phonolite, nepheline syenite, basanite, ankaratrite, melilite ankaratrite. On the diagram these two groups of rocks diverge from each other appreciably. The island is an oval caldera. The floor of the caldera is covered by phonolites; younger rocks on the north may correspond to the group of oceanic basalts.

From the island of Sal we have analyses of ankaratrite rich in olivine, ankaratrite, nepheline monzonite, and phonolite. All analyses agreee well with the association of alkalic-ultrabasic rocks.

Very basic basalt, dolerite, and syenite-monzonite from Boavista are undoubtedly of the olivine-basalt association.

On Maio, apart from the intrusive rocks we have already mentioned, younger volcanic rocks occur. At the base lie phonolites; and these are covered by basic rocks (ankaratrite, ankaramite, basalt), which, in their high alkali content, correspond to the basic representatives of the alkalic-ultrabasic association.

Intrusive rocks ("rocks of the basement") are exposed in a valley on São Tiago. In the tables of chemical analyses (see pp. 139-166) these are represented as nepheline monzonite, nepheline syenite, and essexite. Possibly we may place teschenite here also. Of volcanic rocks, we have an analysis of phonolite. Apart from these, trachyte, basalt, and limburgite are also known from the island. All five available analyses refer to rocks oversaturated with alkalies and belong to the group of alkalic-ultrabasic rocks. Apparently the remaining rocks belong to the same group, since rocks described as trachytes on Santo Antão and basalts on Maio are rich in alkalies and are related to the alkalic-ultrabasic association.

On Fogo, as we have pointed out, occurs one of the active volcanoes of the archipelago. The last weak eruption of this volcano took place at the beginning of the twentieth century. The lavas of the volcano (it is composed chiefly of tuffs and other pyroclastic rocks) consist of basanite, teschenite, and porphyritic (plagioclase) basalt. The last rock is the youngest, very basic (42-43% $SiO_2$) and rich in alkalies (total of 6-6.5%), so that we can hardly consider it basalt in the strict sense of the word (any more than a number of other basalts on these islands). The somewhat more silicic teschenite (45% $SiO_2$) is also oversaturated with alkalies (about 7% alkalies). There is no analysis of the basanite. On the whole, all these rocks, present in the central peak (within the somma), are typical of the ultrabasic group.

It is by no means easy to correlate the stratigraphic and volcanic characteristics of the archipelago, but from all we have noted above, we suggest a probable young age (Miocene ?) for all the alkalic-ultrabasic series of the Cape Verde Islands. However, these same rocks might possibly be represented in the Lower Cretaceous of Maio.

As we have noted, we have information on rocks only of the alkalic-ultrabasic series from Santo Antão, Sal, Maio, São Tiago, and Fogo. This simple picture is destroyed on São Vicente, on which we find both alkalic-ultrabasic members and rocks of the olivine-basalt group. Unfortunately, we have no data on the geologic relations of these rocks, even on their geographic distribution (for the series of analyses, indications are given only for the island of São Vicente without specifying the locality). We may therefore merely assume that here also, as on some of the Pacific Islands, and in a number of continental provinces, channels existed simultaneously to permit basaltic and ultrabasic magma to ascend. Lastly, the island of Boavista is composed entirely of the olivine-basalt series. C. T. de Assunçao and others [1958] recently described remains of a ring complex on Fogo and Brava, with carbonatites. The intrusions are of continental type.

Fernando de Noronha.* This island (off the Brazilian coast) is represented by only two analyses: ankaratrite and basanite. There is no doubt concerning the presence in some measure of basic representatives of the alkalic-ultrabasic association.

Comoro Islands. The rocks of the Comoro Islands and of the northern end of Madagascar were described by Lacroix [1923] and, more recently, by de Saint Ours [1960]. On the whole this province strongly suggests the province of the Canary Islands and Cape Verde Islands with their alkalic basalts, basalts, and alkali-rich rocks, up to phonolites and related varieties.

From this brief summary of rocks from islands lying in recently subsided segments of the ocean, some conclusions may be drawn concerning the high alkali content of the magmas. All these islands, or almost all, were part of a platform before subsidence began. Of course, the indicated type of volcanic series may appear even where there is no evidence of the presence of recent continental character (such as at Tahiti), but such regional predominance of magmas oversaturated with alkalies is not observed there. However, this type of extrusive and intrusive series is not necessarily an accompaniment of a region of young ocean formation. Nothing like this is to be found in the North Atlantic, in the Arabian Sea, along the western shore of the North Atlantic, or other similar regions. Furthermore, the provinces we are considering are not yet so closely related to ocean formation. It would be more nearly correct to say that the age of vulcanism in these provinces is such that we cannot state beyond doubt that it is association with subsidence down to the ocean floor. We have seen that the beginning of volcanic activity on Madeira, Canary Islands, and Cape Verde Islands might belong to the time when a land mass still existed in this part of the Atlantic, i.e., prior to the formation of ocean here.

Thus, we have practically no proof of the appearance of any considerable quantity of magma in connection with the settling of segments and the transition of crust to oceanic type. The single type of vulcanism to be found in most young oceanic regions and over most of the oceans leads us to the principal conclusion that the formation of magma here coincides chiefly with a period when the ocean was already in existence. We do not know if subsidence of the ocean floor continues under these conditions. It seems more probable that vulcanism is connected not with bounding fractures but with some other type of fractures within the ocean basin itself. In any case, volcanoes are not associated with fractures that border sunken blocks. It appears most likely that we have to do in this case with common vulcanism of extrageosynclinal type (in the sense we use the term in the first chapters of this book).

In conclusion, we must note the following:

Fractures bounding zones of ocean formation have proved to be insufficiently deep to penetrate the zone of potential magma formation. It might be assumed, of course, that the specific conditions arising during such subsidence, which we do not know, lead to the disappearance or very deep burial of the zone of potential fusion. However, without regard to the possibility itself that such conditions, unknown to us, might arise, the assumption is superfluous,

---

*Tables of chemical analyses are on pp. 139-166.

since the low-velocity channel of Gutenburg lies at a depth under the oceans, including the youngest segments of ocean, less than under the continents. The conclusion that the fractures accompanying thalassogenesis are relatively shallow therefore remains in force and is one of the possibilities. As a result of thalassogenesis, questions arise, outside the framework of our considerations thus far, concerning the connection between tectonics and the melting of magma.

The very fact of a relatively shallow occurrence of such fractures is very important, since it is confirmation of the view that the activity of processes of elevation of the continents and formation of oceans, as well as the tectonics of platform regions, may be detected by us at comparatively shallow depths, much shallower than the geosynclinal process. If elevation of the continents and subsidence of oceans is in any way related to great depths, then movements at these depths are so slow and have such negligible contrast that our instruments cannot register them.

# CHAPTER 14

# CONDITIONS FOR THE APPEARANCE OF MAGMA

Until now we have tried to examine the ways that the tectonic life of a region might be connected with the formation of magmas. The question concerning the conditions that cause magmas to appear and that lead to subsequent evolution of these magmas has been left to one side. In this chapter we shall try to fill in this gap.

## Basaltic Magmas

In almost all cases, magma is the product of **partial fusion of some substance.** Such fusion creates drops or films of liquid within the solid mass. Drops might be formed if any entire minerals are melted in the host rock. Films develop when readily fusible parts are separated out so that part of the grain remains behind (relatively infusible compounds), and the residual grain is coated by the liquid film. It is most likely that during fusion of any rock liquid will be separated in both ways, and, as a result, both drop-like inclusions and very thin films of liquid are present. If the amount of this segregated liquid is sufficient, the rock is converted to a more or less liquid mush. Intergranular films play a very large part in changing the viscosity of the rock. Such a mass is much more mobile than solid rocks; its density is less than that of the solid rock, and it may be assumed that it is slowly floating upward, leaving the more solid material behind [Belousov, $1966_1$, $1966_2$]. Since the liquid that has separated off occupies much less volume that the remaining solid matter, there is no reason for assuming the appearance of "antiasthenoliths." In fact, in the region where the liquid has been melted out, there remains, after the liquid has moved upward, an almost continuous layer of material that has lost its basaltic component. As a result, in the zone of melting there remains an almost continuous layer of ultrabasic rock, containing little easily fusible material. Ringwood [1962; Clark and Ringwood, 1964] and the present author [Sheinmann, 1961] arrived at this opinion almost simultaneously. The restoration of equilibrium, which has been disturbed by the isolation within the mantle of relatively light masses, is very difficult under these conditions, and it can hardly take place in a short time.

Let us look back to the time when the mush-like mass was moving upward with difficulty. Whatever the composition of the segregated liquid, as it rose, exchange reactions must have inevitably taken place between it and the solid rock, since, as noted by Green and Ringwood [$1966_3$], a mass always tends toward that composition of liquid phase that corresponds to the temperature and pressure conditions at each given level. At any moment, probably, not too far from the level at which melting has begun, perhaps at the very place of its segregation, the liquid separates from the solid phase and continues its rise, freed from a large part of the solid inclusions.

The separation of crystals may take place in only two ways: by the settling of heavier minerals or as a result of filter pressing, squeezing out the liquid much as water is squeezed out of a sponge. These, which at first glance appear to be rather simple processes, encounter serious obstacles when we approach them from the viewpoint of physics.

Crystal settling. Crystals settle readily if the density difference is sufficiently large and the number of crystals small. There is no doubt about attaining the first condition, since the difference in density produced by the melting of basalt amounts approximately to 0.6-0.7.

The number of solid phases is another matter. The settling of heavy minerals will take place readily so long as they are surrounded by liquid, and it will cease when the grains come in contact. The latter situation arises when the volumes of liquid and solid phase are approximately equal. Thus, as a first approximation, we might state that, as a result of settling, the amount of freed liquid will be equal to the total volume of liquid minus the volume of the solid phase. In other words, in order for magma to appear, it is necessary for more than half the substance to be melted.

This situation is necessary in order for such separation of magma to take place in a rather short time. The rate of settling of crystals may be determined, although crudely, by using Stokes's law. If we assume the viscosity of basaltic liquid at a depth of about 100 km to be $10^5$-$10^7$ poise, the size of grain to be ~1 mm, and the density difference to be 0.6-0.7, then the settling rate amounts to 0.1-10 cm/yr; i.e., in order for the crystals to sink 1 km, $10^5$-$10^6$ years would be required. In fact, however, the rate must be greater, since sinking grains will stick together when they come into contact. If we assume that such adherence takes place on each second meeting of grains, then settling of 1 km occupies in all but 10 to 100 years (the settling rate of aggregates with a cross-section diameter > 6 cm amounts to about 100-150 m per year). There is thus no reason for doubting the actual validity of the settling process.

After the grains come in contact with each other and the liquid is preserved only in the interstices, the situation changes markedly. There is now a solid framework, and the floating of the liquid becomes impossible, although it may circulate through existing channels. In order to squeeze the liquid out, a great drop in pressure is necessary, since the rising liquid must disrupt the grains; i.e., the limiting determinative possibility of movement is the viscosity of the solid phase and not of the liquid. Under ordinary conditions, such drop in pressure is impossible. However, we should search for the possibility of great drops in pressure under some special conditions. Such an attempt was made by Gzovskii [1963].

It may thus be said that the separation of magma from continental matter takes place very quickly when the liquid content is much greater than the crystal content. The question of the conditions under which small amounts of liquid may be squeezed out still requires investigation, although the possibility of such squeezing out under certain conditions appears very probable. Geological observations on crystalline rocks also point to this possibility. If it were impossible for such liquid to be squeezed out, it would not be possible for small migmatitic veins to appear in crystalline schists or for granophyre veins and zones to appear in bodies of basic rocks.

Still, we should not overestimate the easy appearance of conditions for filter pressing. This phenomena is more likely accidental, and we cannot base the separation of large magmatic masses on such accidents.

It should be remembered that there is still another mechanism for separating small amounts of liquid. If the solid grains in the mush are not equidimensional, but have a large flat face, they may become systematically packed (like bricks) if they are agitated for a long time. With such packing, very little liquid remains between the grains, and the melt may be effectively removed even if there is not much of it. The necessary agitation may be effected by seismic shocks.

As soon as the liquid becomes magma and separates from the solid phase, evolution of the magma begins. Change in its composition is associated directly with its rise, when both pressure and temperature change. Various changes may occur. If the magma rises through rocks with temperatures much lower than its own, a layer of congealed rock forms along the walls of the channel, isolating the liquid. This idea has been repeatedly expressed by the author, and was recently repeated by Green and Ringwood [1966$_3$]. In this case we have to do with a closed system. It may be assumed that the temperature of the magma will be approximately the same as that of the country rock. Exchange reactions are then inevitable, and the magma will become somewhat enriched in low-melting components [Green and Ringwood, 1966$_3$]. However, we can hardly expect any considerable assimilation, since the walls of the channel are made up of rocks that have in great measure already endured melting by rocks [Sheinmann, 1961]. Ringwood and Green have noted that such "insubordinate" elements, i.e., elements the contents of which are practically independent of the contents of the "principal" elements, may be K, Ti, P, U, Th, and some others. Lastly, change in the composition of magma may take place by mixing with other rocks or by picking up blocks of rock of non-basaltic composition. We shall not consider these possibilities, since what interests us is change in the magma itself during its ascent, not possible contamination in the upper levels.

Apparently the most powerful mechanism of magma evolution is settling. We should keep in mind, however, that 1) sufficient liquid is necessary, and 2) it is necessary to account for the effect of the rate of rise of the magma on the possibility of settling.

We have already seen that the rate of ascent reaches values of tens of meters per hour. Since the channel cannot be a straight tube with smooth walls, the rise of liquid along it must cause considerable complications in flow, which make the settling of crystals difficult, if not altogether impossible. The settling rate of grains at a viscosity of the magma of $10^6$ poise (i.e., in the zone where the magma is generated) amounts to $14/10^7$ cm/sec for crystals 1 mm across and $14/10^5$ cm/sec for aggregates 6 cm across. Even at the surface of the earth (viscosity of $10^2$ poise), the corresponding settling rates are 0.014 and 1.4 cm/sec. The rate at which the magma rises, observed beneath actual volcanoes, is 10-20 cm/sec. Thus, basic magma that rises rather quickly can hardly change composition as a result of crystal settling, but the possibility of assimilating fragments of the walls during rapid ascent is also small, since the effect of the walls is slight within the mantle, and upward a chilled crust on the relatively cold walls appears.

The most complete information concerning the behavior of magmas of different composition at great pressures appears in the recent works of Ringwood and Green [Green and Ringwood, 1966$_1$, 1966$_2$, 1966$_3$; Ringwood and Green, 1966]. Their data show that at great depths one should expect melting of a very basic liquid, with a composition between picrite and olivine basalt. This view is apparently based on the fact that the temperature at depths below 100 km produces fusion of just such liquid, and that a sufficient amount of it may be formed in this way to allow easy separation of it. This view appears highly probable, but thus far we cannot consider it proved.

Under conditions of tectonic calm (extrageosynclinal regions), it is possible to determine the temperature at depth by using the data of Lyubimova [1959] and to compare this with the melting temperatures (after Ringwood and Green [1966]). However, comparisons of this kind should be made with great care, the more so when we recall that notable heating at depth occurs beneath zones of recent vulcanism. It is therefore better to base our conclusions on geologic and geophysical data that permit us to determine the depth from which magmas of olivine tholeiite and similar rocks rise. As already noted, earthquakes are shallow in this zone, and one should not expect any rapid movement of magma at great depths. The aseismic character of these depths most likely attests to slow ascent of the magmatic "mush," and this corresponds to movement of asthenoliths [Belousov, 1966$_1$, 1966$_2$]. The deepest seismic shocks recorded in these regions are from depths of 50-70 km, and are directly related to the beginning of a rapid ascent of magma. We may therefore assume that the solid and liquid phases separate at about this depth; i.e., this is the depth at which basaltic magma appears. Its rapid ascent to the surface undoubtedly prevents settling of crystals, and the magma of Mauna Loa and Kilauea, for example, reaches the surface without having cooled sufficiently to drop below the melting point of olivine before spreading out. This mechanism may be assumed for the zones beneath trap fields at the time of extrusion.

Knowling the melting point at different pressures, and assuming that at the time liquid separates from the solid phase both phases are in an equilibrium state, we may determine the temperature at a depth of about 60 km (18 kbar) to be approximately 1350°C. This value seems to be based on firmer ground than computed temperatures obtained from general considerations. In both cases tholeiitic magmas (olivine tholeiite most probably) reach the surface. The appearance of more silicic magmas (quartz tholeiite) is possible if differentiation takes place at depth, by settling of either garnet (at a depth of about 100 km) or olivine (at a depth of 30-35 km, perhaps less) (see Fig. 42). It may be seen from Fig. 42 that olivine basalt and olivine tholeiite separate at a depth of about 100 km by the congelation of garnets at first. With further drop in temperature, clinopyroxenes join the garnet. It should be recalled that sufficiently rapid and effective settling of crystals is possible only when the volume of crystals is much less than that of the liquid. However, pyroxenes begin to precipitate out under these conditions, when the solid phase represents more than 50%. We should note further that increasing silica content, as a result of the settling out of garnet, suggests the separation of magma at depths much greater than the depths of recorded earthquakes, and it is therefore assumed to be relatively improbable.

If magmatic movements remain at shallow depths (less than 30-35 km), there is every reason to expect separation of olivine from the magma. And, in this case, the most advantageous conditions for rapid settling of crystals coincide with the temperatures and pressures existing when olivine alone precipitates out. The results of computation of the corresponding changes in magmatic composition are shown in Table 8.

For simplifying the computation, it was assumed that olivine crystals with the ratio Mg:Fe = 2:1 contain no admixture of other elements. In truth, such admixtures are present, and in the liquid separating out it has been found that there is somewhat less $Al_2O_3$ and CaO and somewhat more MgO and FeO. After settling of the crystals, the liquid corresponds to olivine-free and quartz tholeiites probably, but not necessarily with high alumina content.

Thus, in order for tholeiitic basalts to arise, forming trap fields, the following conditions must be met:

1. Separation of basaltic drops and films in the mantle material at depths of about 100 km (the material in the low-velocity channel is apparently near this state at the start);
2. Slow rise of this crystal-liquid mush, unaccompanied by earthquakes;
3. The appearance of fractures at about 60 km (or somewhat deeper), greatly facilitating passage to the surface;
4. Rapid rise of the liquid that has separated from crystals;
5. But, magma does not form secondary chambers at shallow depth; we have to do with homogeneous lavas (olivine tholeiite);
6. If such a secondary chamber forms no deeper than 30 km from the surface, rocks rich in $SiO_2$ may form because of precipitation and settling of olivine: olivine-free and quartz tholeiites (but andesite does not appear). Tasmania is possibly an example of the last type.

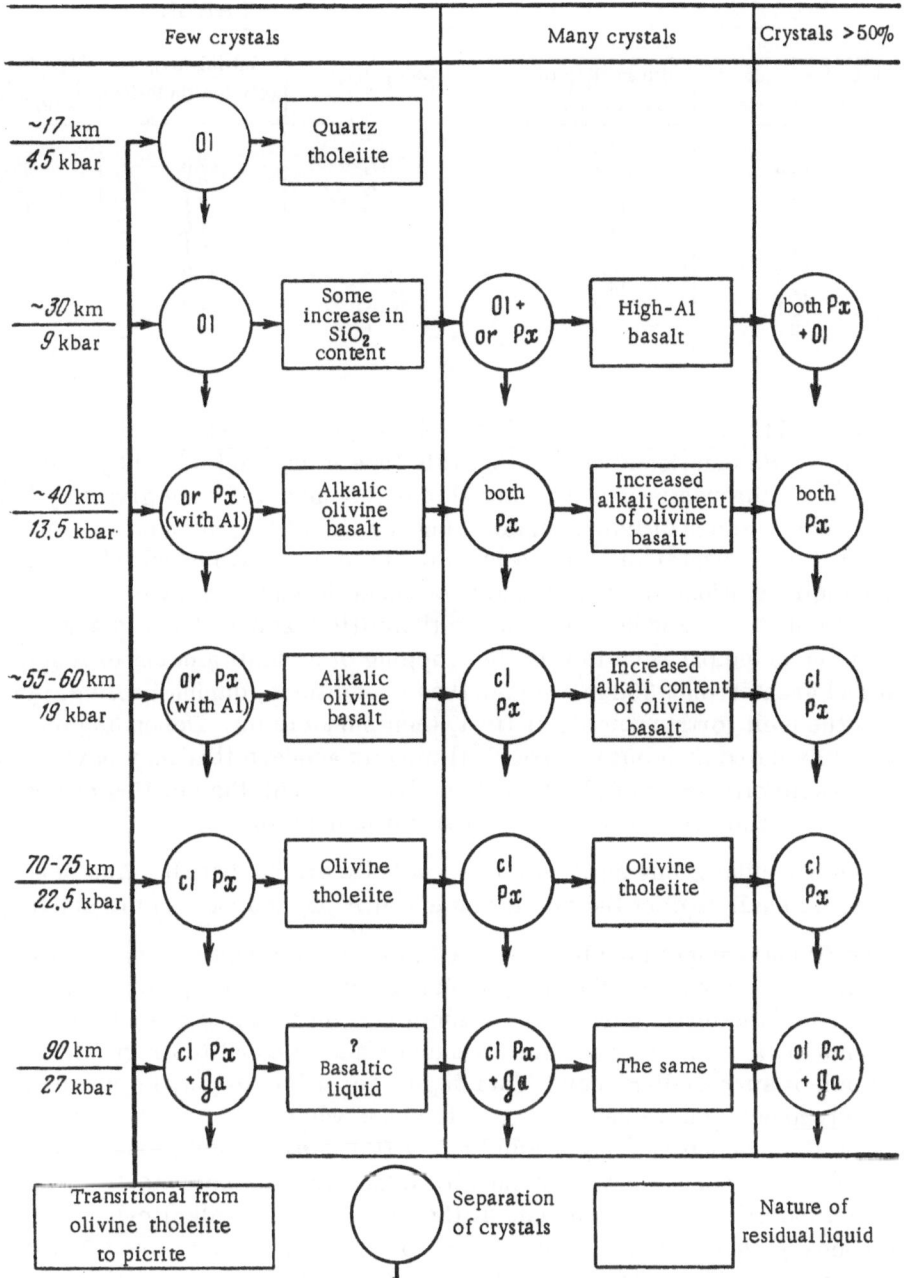

Fig. 42. Changes in composition of magmatic liquid as a result of crystals precipitating out at different temperatures and pressures. The primary liquid corresponds in composition to a rock intermediate between olivine tholeiite and picrite (after data from Ringwood and Green [1966]. Ol) olivine; Px) pyroxene; or) ortho-; cl) clino-; ga) garnet.

## Alkalic Basalts

The conditions for the appearance of basalts somewhat richer in alkalies (from olivine basalt to alkalic basalt) are somewhat different. Two lineages may be distinguished here.

TABLE 8

| Initial composition | | Composition of liquid | |
|---|---|---|---|
| | | with settling of 20% olivine | with settling of 30% olivine |
| SiO$_2$ | 47.0 | 48.3 | 49.1 |
| TiO$_2$ | 2.0 | 2.5 | 2.9 |
| Al$_2$O$_3$ | 13.1 | 16.4 | 18.7 |
| Fe$_2$O$_3$ | 1.0 | 1.3 | 1.4 |
| FeO | 10.1 | 7.4 | 5.4 |
| MgO | 14.6 | 7.8 | 5.0 |
| CaO | 10.2 | 12.7 | 14.4 |
| Na$_2$O | 1.7 | 2.1 | 2.4 |
| K$_2$O | 0.1 | 0.1 | 0.1 |

TABLE 9

| Component | Initial magma (olivine tholeiite) | After settling out of 10% rhombic pyroxene |
|---|---|---|
| SiO$_2$ | 47.0 | 45.5 |
| TiO$_2$ | 2.0 | 2.2 |
| Al$_2$O$_3$ | 13.1 | 14.7 |
| Fe$_2$O$_3$ | 1.0 | 1.1 |
| FeO | 10.1 | 6.4 |
| MgO | 14.6 | 11.9 |
| CaO | 10.2 | 11.3 |
| Na$_2$O | 1.7 | 2.0 |
| K$_2$O | 0.1 | 0.1 |

Olivine Basalt, Basanite, and Nepheline Basalt of the Hawaiian – Carolinian Type. We mentioned before that this type is characterized by tholeiitic lavas experiencing almost no differentiation, forming shield volcanoes. These give way to small masses of olivine basalt, and still later, terminating the life of the volcano, basanites and nepheline basalts appear (Chapter 8). There is no doubt of a genetic connection relating all three groups, but their relations in volume and time make it appear very probable that the alkalic rocks are distinctive residuals of primary tholeiitic magma. This is a very real possibility. Apparently the principal condition is the trapping of a small amount of magma at depths less than 60 km and greater than 35 km. Under these conditions, during cooling, pyroxene precipitates from the melt (orthorhombic at first, then monoclinic). Depending on the amount (it may be appreciably more than 50%), a residual magma appears that may have the composition of basalt with some olivine or of alkalic olivine basalt. With the settling of about 10% rhombic pyroxene, the residual liquid is near alkalic basalt (Table 9).

The resulting secondary magma is very near alkalic olivine basalt. And here, for the sake of simplicity, we shall neglect the small amount of Al$_2$O$_3$ and CaO in the settling pyroxenes.

Such holding or entrapment may be represented in the following manner: if the ascent of basaltic liquid from depth ceases, the fundamental condition for its pouring out on the surface is destroyed, since the lower part of the channel becomes closed at that depth where the weight of the column of liquid from that point to the surface is equal to or greater than the weight of the column of surrounding rock. This depth, according to petrologic and physico-chemical data, corresponds, as we have seen, to the zone where pyroxenes have settled out of the basaltic magma (i.e., approximately 30-50 km). After these crystals have settled out, the liquid may be again squeezed upward, even with weak tectonic movements, creating a cap of alkali basalt on a tholeiitic shield volcano. If the residue of this alkalic-basalt liquid is preserved in the secondary chamber after some period of volcanic activity, then further separation of orthopyroxenes from it at the same pressures (10-18 kbar) leads to the formation of a new residual magma, this one very near nepheline basalt or basanite. The latter might form as a result of pyroxenes settling out of an olivine-bearing basalt magma. The settling of pyroxenes from alkalic-basalt melt (20-30% crystals) creates a nephelinitic magma, oversaturated in aluminum.

Basalt Complexes with High Alkali Content (Ethiopian Type). We still do not know if it is possible for alkalic olivine basalt to melt directly from the mantle or, if this is possible, at what temperatures and pressures it may occur. We must assume that proper experiments will be set up in the near future to check on this. But, direct melting of olivine-basalt magma is apparently possible, creating provinces of the Ethiopian type. This may now be considered proved. Another way for alkalic-basalt magma to form is differentia-

TABLE 10

| Component | Olivine-bearing basalt | Liquid after settling out of 10% enstatite | Liquid after settling out of 20% enstatite | Alkalic olivine basalt (for comparison) |
|---|---|---|---|---|
| $SiO_2$ | 47.1 | 45.9 | 44.4 | 45.4 |
| $TiO_2$ | 2.3 | 2.6 | 2.9 | 2.5 |
| $Al_2O_3$ | 14.2 | 15.8 | 17.8 | 14.7 |
| $Fe_2O_3$ | 0.4 | 0.4 | 0.5 | 1.9 |
| FeO | 10.6 | 9.3 | 7.8 | 12.4 |
| MgO | 12.7 | 12.0 | 11.1 | 10.4 |
| CaO | 9.9 | 11.0 | 12.4 | 9.1 |
| $Na_2O$ | 2.2 | 2.4 | 2.8 | 2.6 |
| $K_2O$ | 0.4 | 0.4 | 0.5 | 0.8 |

tion of olivine-bearing basalt magma or magma transitional between olivine and picrite. If only enstatite, containing no impurities, precipitates out of olivine-basalt magma, the result will be as shown in Table 10.

When making a comparison with alkalic olivine basalt, we should keep in mind that, in nature, enstatite contains some quantity of $Al_2O_3$ and CaO. To this fact we should add that at decreasing pressures, from 18 to 13 kbar, monoclinic pyroxene joins enstatite. Thus, to increase our accuracy, we must state that the initial liquid after crystals have settled out will be poorer in aluminum and calcium, which replace part of the precipitating FeO and MgO. As a result, after settling of pyroxenes from olivine-bearing basalt, alkalic olivine basalt is formed in both cases. This process may be expected at a depth of about 50 km at a temperature of 1350-1400° or at 60 km at 1370-1430°. At both greater and lesser depths, rhombic pyroxene does not precipitate first from the magma, yielding locally to clinopyroxene in some places, olivine in others.

We should note here that the limiting depth for the appearance of alkalic basalt is that depth from which the first seismic shocks are recorded before eruptions on the Hawaiian Islands. We suggest that at this depth magmas separate from mixtures of liquid and the solid phase. It may be thought that, when the further path upward is free and the magma rushes to the surface, crystal settling is ineffective. Tholeiite is poured out, and only a small part of the magma changes to the composition of alkalic basalt. If at these depths (60 km) the magma is trapped and the temperature falls to the liquidus boundary, then crystals settle, and it is an alkalic-basalt magma that reaches the surface. The same result is probably obtained by primary melting out at this depth (if, in general, this is possible under the temperature conditions).

Solution to the problem of how the alkalic-basalt magma forms is very desirable, since, without this information, it is difficult to understand why we encounter only alkalic olivine basalts in regions of great rifts on the land and only, or almost only, tholeiite in the rifts of midocean ridges. It is also essential to discover the conditions of primary melting of tholeiite and alkalic basalt in order to establish the cause of extraordinary predominance of the first in trap fields. The most probable explanation is that the mantle changes so in composition from region to region that, under any particular conditions, tholeiitic magma melts out at one locality, alkalic-basalt magma at another. As a matter of fact, it appears most likely that it all reduces to a change in the melting conditions.

The mechanism of formation and further evolution of magma of typical calc-alkalic type in island arcs is somewhat different. As is well known, this type is characterized by a series of silica-saturated rocks from basic basalt to tholeiite—andesite-basalt—andesite, and, finally, to dacite (the last, if it is present, is nowhere abundant).

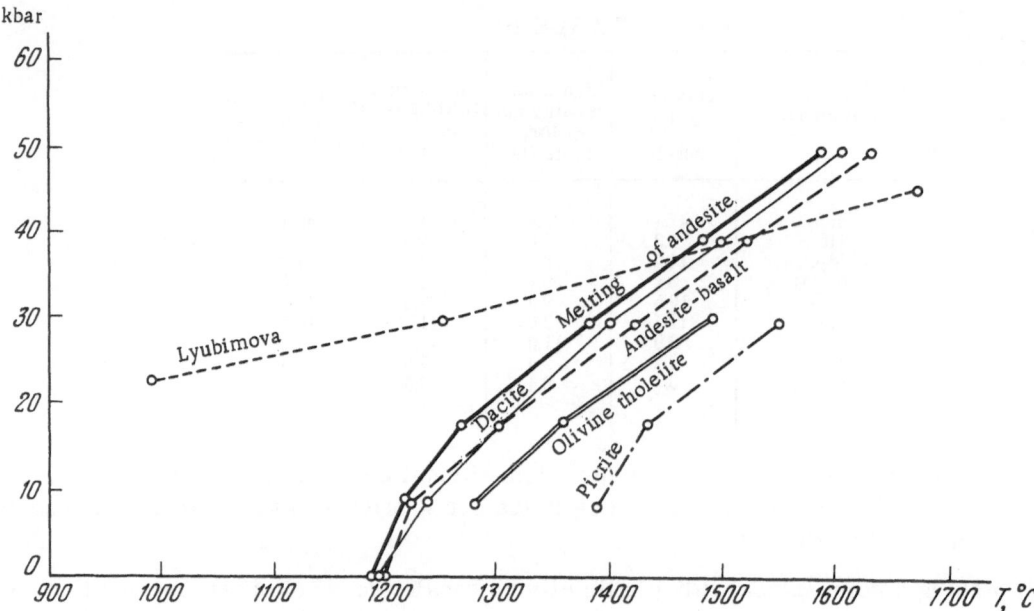

Fig. 43. Comparison of temperature increase within the mantle (after Lyubimova) with melting data for high pressures. Lyubimova's curve should be bent upward on its right end in order to satisfy experimental data.

Thus, in reconstructing the assumed conditions for melting out of this magma, we must satisfy this entire sequence, and must show that, in contrast to the case already analyzed, not only do some basalts appear (occasionally accompanied by a small volume of rhyolite), but the continuous basalt-andesite series is present. Another feature of this series, already mentioned in the preceding chapters, is the very considerable depth at which the process takes place. The magma separates out at depths down to 200 km, possibly even **at greater depths**, i.e., at pressures of 50-60 kbar or more. For such conditions experimental data are rather scarce, and it is **therefore impossible to reconstruct the environment without a great deal** more uncertainty than for shallower depths.

A comparison of assumed temperatures at depth with data on rock melting at high pressure shows considerable disagreement. Thus, the temperature curve as given by E. A. Lyubimova intersects the melting curves not only for tholeiite but also for picrite (Fig. 43). If Lyubimova's curve were accurate, all basalts, including picrites, would be in a liquid state at a depth of about 150 km, and in the low-velocity channel of Gutenberg there would be a mobile "mush," in which the liquid phase would constitute at least 50% of the total volume. This is not what we observe, however, and we are obliged to believe that, beginning at a depth of approximately 100 km, the curve of normal temperatures bends strongly upward, so that the gradient here must not exceed 2.5° per kilometer. This applies to both oceanic and continental regions. Only under these conditions are the data on melting, even for dry rocks, satisfied. If any appreciable amount of water vapor is present, the melting point should be depressed and there would have to be a strong deflection of the curve of "normal" temperatures toward the pressure axis.

We know that basaltic lavas rise from below in geosynclines during the initial stage of their development. Consequently, at least for depths of 150-200 km, the temperature and pressure favor the melting of basalt (most probably even picrite basalt).

Minor extrapolation of the melting curve for basalt shows that at 50-60 kbar we should expect its melting point to be at least 1600°, perhaps even 1700° (according to Magnitskii [1965,

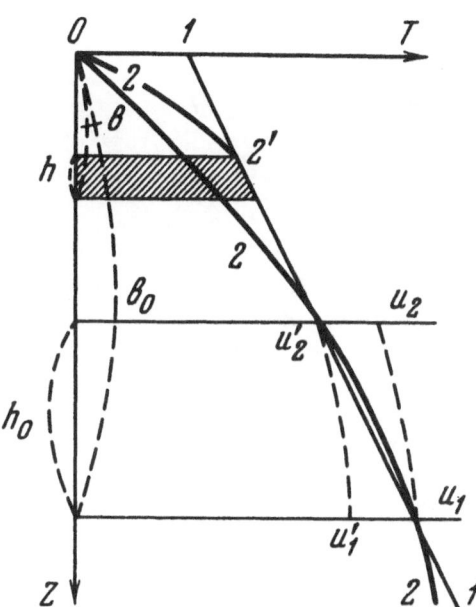

Fig. 44. The mechanism replacing zone melting (zone refining) at depth (after Magnitskii [1965]). 1 — 1) Melting temperature; 2 — 2) temperature at depth; $u_1 - u_2$) adiabatic curve for rise of heated liquid to roof of zone of melting; $u_2 - u_2'$) difference in temperature of rising liquid; $u_2' - u_1'$) adiabatic curve of descending substance, yielding excess heat; $u_1' - u_1$) difference in temperature of descending substance and temperature at the level $B_0$; $h_0$) zone of melting.

p. 20] the minimum temperature, adiabatic, is 1300°; according to electrical-conductivity data, it is 1200-1500°C). Thus, at these depths in the analyzed example, we must seek a temperature higher than "normal" by 200°, perhaps even more. Magma appearing here, or having arrived from still greater depths, is subjected to strong lateral effects of solid moving masses (deep-focus earthquakes), and these effects may aid in separating the melt from solid particles (the possibility of considerable drops in pressure during high-contrast movements).

Up to the time when development of zones of low viscosity (tentatively we call them "fractures") leads to a rapid ascent of magma, the conditions described by Magnitskii are created, replacing zone refining in its pure form. According to Magnitskii [1965, pp. 330-335], if a zone of melting appears, the following takes place (see Fig. 44): the zone is limited from above by the depth A, at which the temperature curve at depth intersects the pressure curve for the substance; from below the pressure is limited by the depth B, where these curves again intersect, as a result of which T at depth again proves to be lower than T of the melting point. Within these limits melting occurs (so long as the conditions are preserved). It should be noted that melting is not necessarily complete. It is sufficient that the liquid be free to move through the interstices between crystals, and this is possible with 40-50% liquid, possibly even with a lower

amount. Since the adiabatic temperature gradient (about 0.5°C/km) is undoubtedly lower than the gradient in the melt ($\geq$ 1°C/km), the rising and adiabatically cooled magma is displaced from level B to level A, is overheated, and will yield its "excess" temperature, heating the crown of the magma chamber and melting it, i.e., raising the zone of melting (curve 1). Having given up its excess heat, this part of the magma begins to sink, but is again adiabatically heated. At level B it becomes supercooled, and crystals with the highest melting point precipitate out, settling to the floor of the chamber and somewhat raising the level B. The heat of crystallization goes to warm up the remaining liquid. When the temperature of the liquid is comparable to the temperature of the surrounding mass, the process begins to repeat. As a result there will be a slow rise in the zone of melting, and the liquid will be enriched in the easily fusible components under the given conditions.

Let us pause to consider the conditions in the tectonofer. The influx of deep-seated heat into its lower part creates an initial zone of melting at depth. The conditions under which the process described by Magnitskii becomes possible may be stated simply: the liquid must make up a sufficiently large part of the total mass, and this is possible, of course, if its composition is very close to that of the mantle. More accurately, it will always be either of picritic composition or intermediate between picrite and olivine basalt. As soon as conditions develop for sufficiently easy displacement of the liquid within the chamber, the process just described

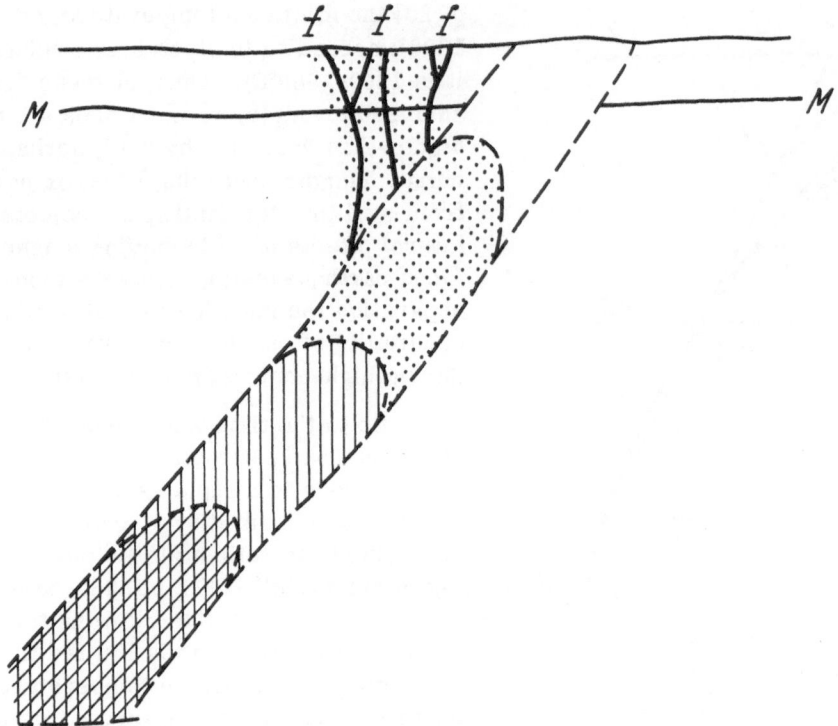

Fig. 45. Sequential rise of the zone of heating in a tectonofer and the
site of emerging magma (and excess heat) along fractures. The
"floor" of the zone of heating before the tectonofer begins to die practically fails to rise during a very long time interval.

begins, and the roof of the chamber will rise, and the liquid becomes less basic (Fig. 45). We
should take into account the fact that the influx of heat from depth will increase in the young
tectonofer. The chamber will be heated in greater measure, and, as a result, the upward
movement of the chamber will become more rapid than it would from a single heating (according to Magnitskii). On the other hand, this heating makes the "floor" of the chamber practically immobile, or, more accurately, in its upward movement it will lag noticeably behind the
rise of the "roof." As a result, the size of the zone of melting increases as the zone moves
upward, and, at the same time, the liquid in it gradually becomes somewhat more siliceous.
Movements within the tectonofer will cause the liquid to be squeezed out (within the zone of
movement) and will cause this liquid to rise more rapidly. This phenomenon will in some measure destroy normal circulation of the melt, of course, but only partially, near the zone of im-
pending shear displacement (from observations above deep-focus earthquakes we know that
such shear displacements are repeated for a prolonged interval from a single focus: i.e.,
there is a tendency for zones of shear displacement to operate for a long time). Outside this
zone of shear displacement the size of the magma chamber should increase and the liquid
should continue to grow more silicic. Since all this takes place at great depth, we have every
right to assume that the change in composition of the melt conforms to the thermal "andesite
trough" [Green and Ringwood, $1966_2$]. As experiments on melting at high pressures have
shown, at 18 kbar and more, andesite has the lowest melting point (Fig. 46, prepared from
data of experiments by Green and Ringwood). This andesite trough is not destroyed even at
pressures of 40-50 kbar, and it probably exists at even higher pressures; i.e., at a depth of
40-30 km and less, the decline in melting point of the more silicic rocks (particularly dacite)
gradually outstrips the drop in melting point T of andesite, and in the continental crust, even

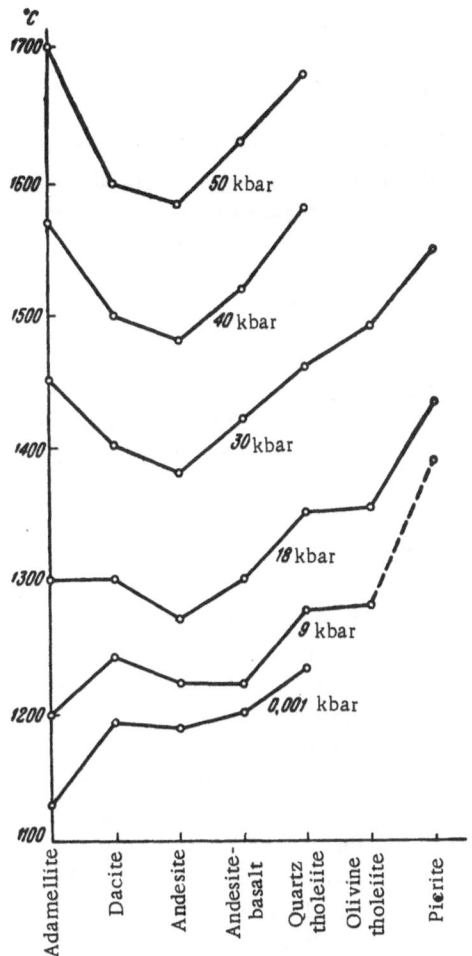

Fig. 46. The andesite thermal trough (each curve corresponds to a particular pressure) (after data of Ringwood and Green [1966]).

more at the earth's surface, the more silicic rocks have the lowest melting points.

Taking into account the andesite trough at great depths, we should expect the composition of the liquid in the upper part of the magma chamber to tend toward andesite even though the rising chamber in the tectonofer is undoubtedly being heated. This tendency will become stronger when the influx of supplementary heat from depth ceases. Then the chamber begins to develop fully according to Magnitskii's scheme: its ascent becomes slower, and gradually the loss of excess heat leads to more rapid approach of the magma to andesitic composition. As we know, the appearance of andesitic magma proper generally coincides with, or is near in time to, the principal folding and inversion; i.e., it belongs to the late stages in development of the tectonofer. At this time the deep chamber terminates its activity, and the heat rising through the tectonofer begins to act on the crust (melting out of silicic magmas).

It is entirely possible that separation of liquid from the solid phase and further migration of the magma chamber take place at rather early stages: filter pressing in the tectonofer under these conditions is marked, but it is not the general mechanism. The liquid in this chamber has most likely been superheated, but this cannot make it become more basic. The described convection will hinder this in the chamber, and when the chamber begins to rise along zones of weakness, it will affect the rate of this rise.

Apparently the direct melting of andesite from the mantle is impossible, although no experiments on this have been described in print. Even should it be possible for liquid of this composition to be melted out, the amount would be very small. Thus, andesite appears only after some form of basalt has been removed from the mantle.

Ringwood and Green [1966] suggest that basaltic liquid must be converted first to eclogite, and only after this is partial melting out of andesite possible. It is hard to understand this statement. Basaltic (eclogitic) melt, occurring under these conditions, may discard in the form of crystals any part of itself, and it will tend toward andesite (if, of course, the temperature drops sufficiently and approaches the melting point of andesite). Consequently, the change from basalt to andesite-basalt, and then to andesite, indicates merely a gradual drop in temperature in the active zone, possibly in connection with uplift near the surface. Such a drop in temperature may be associated with the general lowering of the energy level in the tectonofer and the shift of principal activity in the tectonofer to the uppermost levels of the mantle, even to the crust. The nature of the rising magma, typical calc-alkalic magma, indicates that, in rising, it is not trapped at intermediate depths and therefore cannot be changed to alkalic basalt (as a result of precipitation of orthopyroxene). This corresponds with seismic data, attesting to the rapid rise of magma from depths of 100-200 km. The existence of the andesitic thermal trough at great depths indicates that more silicic andesite cannot form there by differentiation. Such differentiation becomes possible only at low pressures near the surface, when the melting point of dacite is less than that of andesite.

All the preceding discussion will be considerably altered if it is assumed that a more or less dry basaltic magma, without substantial change in composition, reaches high levels that contain abundant water. Then, if crystals first settle out, giving rise to a magma of the composition of quartz tholeiite, the formation of andesite-basalt and then andesite becomes much easier. This way of forming andesitic magma should not be overlooked therefore.

## Ultrabasic Complexes

The question concerning to what extent ultrabasic rocks may represent the substance of the mantle has not been finally answered. However, there are no grounds for including dunite-harzburgite as a possibility. This combination is most probably residual after melting out of basalt and andesite. Its very composition is contrary to the idea that it represents original mantle material. From such material (dunite-harzburgite) it is impossible to separate any large amount of basaltic liquid. Furthermore, such rocks are found chiefly in geosynclines (and, apparently, on the crests of midocean ridges; possibly also on the ocean floor). The huge basalt fields on the continents are completely lacking such rocks. The question of possible fusion of ultrabasic rocks of this type is still vague. Proper experiments have not been performed. Geologic data have led workers to propose both intrusion in the solid state and regular intrusion. Existence of the latter are demonstrated by the presence of "hot" contacts, which may be observed if the serpentinite mass does not move after emplacement (such, for example, as the Red Hills massif on South Island of New Zealand, and the intrusion on Cyprus). Judging from these data, it may be assumed that ultrabasic magma, possibly a mixture of liquid and olivine crystals, may appear at the time of maximum heating at depth, at a time comparable to that of the melting out of basalt under a geosyncline. Such conditions probably existed (or do exist) under ocean ridges. The possibility of intrusion and even extrusion of such "olivine mush" is proved by the meimechites in the Gulya field of Siberia.

It is very hard to answer the question concerning the nature of magma that yields alkalic-ultrabasic complexes. We have already had occasion to remark that until now it has been impossible, with complete assurance, to recognize this kind of complex in the oceans. In this connection it is not clear what role the continental crust plays in their formation. There appears no doubt that magmas of this type cannot be generated within the crust. Geologic conditions do not permit their formation from basalt.

The principal features of such magma are: very low silica content, high contents of titanium, alkalies, aluminum, and calcium (and correspondingly low content of magnesium). These features lead us to refer alkalic-ultrabasic complexes to the gabbroidal clan, since ultrabasic rocks are characterized by low contents of aluminum and calcium [Kuznetsov, 1964]. However, such a definition means that dunite-harzburgite is the only representative of ultrabasic complexes. Such restriction can hardly be successful. The first and decisive feature of ultrabasic rocks should be the silica content, responsible for the formation of olivine and accompanying pyroxene. Two principal groups may be distinguished among rocks with such low silica content. One of these is characterized by predominance of MgO and a low iron content. Aluminum, calcium, and alkalies are practically absent. This represents the dunite-harzburgite association. The second group is characterized by the same low silica content, but among the bases, along with MgO, which is dominant even here, iron also appears in considerable quantities and also appreciable calcium and aluminum (we should not forget that we ought to expect an increase in the role of iron at the expense of magnesium at very great depths). Alkalies also are more abundant. Whereas dunite-harzburgite cannot be the parent material for

TABLE 11

| Analysis No. Component | 1 | 2 | 3 | 4 | 5 | 6 | 7 |
|---|---|---|---|---|---|---|---|
| $SiO_2$ | 40.2 | 41.9 | 45.2 | 42.0 | 45.5 | 44.5 | 45.8 |
| $TiO_2$ | 2.0 | 2.5 | 0.5 | 0.6 | 0.6 | 0.6 | 0.6 |
| $Al_2O_3$ | 5.1 | 7.3 | 4.3 | 5.4 | 2.0 | 5.4 | 5.4 |
| $Fe_2O_3$ | 7.9 | 7.8 | 4.3 | 5.4 | 4.8 | 5.4 | 5.4 |
| FeO | 7.7 | 7.2 | 4.6 | 0.3 | 3.2 | 1.0 | 5.8 |
| MgO | 27.5 | 22.2 | 37.8 | 42.5 | 40.3 | 39.0 | 33.0 |
| CaO | 7.2 | 8.1 | 2.6 | 3.3 | 2.9 | 3.3 | 3.3 |
| $Na_2O$ | 1.3 | 1.9 | 0.5 | 0.6 | 0.6 | 0.6 | 0.6 |
| $K_2O$ | 0.9 | 1.3 | 0.1 | 0.1 | 0.1 | 0.1 | 0.1 |

Note: 1) Weighted average composition of rocks of the Gulya field on the assumption that trachybasalt, trachyandesite, trachyte, and the like are not related to the Gulya complex; 2) the same, considering the trachybasalts, etc.; 3) "pyrolite," with a composition dunite-harzburgite: olivine basalt = 4:1; 4) residual liquid after settling out of 20% enstatite from the pyrolite; 5) the same after settling out of 10% pyrope-almandine; 6) the same after 20% olivine with MgO:FeO = 2:1; 7) the same after 20% forsterite.

basalt, but is most likely the residual material after basalt is melted out, the second group of rocks may evolve basalt [Sheinmann, 1964₃].

Thus, the question arises: what is the relationship between magma of this "alkalic-ultrabasic type" and the primary substance of the mantle? It is commonly assumed that this primary substance is some mixture of dunite-harzburgite and basalt. It would appear that it would then be necessary to find the composition of what should prove to be parental material for the alkalic-ultrabasic complexes also. However, attempts to obtain from "pyrolite" (mixtures of dunite-harzburgite and basalt) a magma of the type responsible for the Gulya rocks, for example, have been unsuccessful. If we, in imagination, precipitate **garnet, enstatite, olivine,** or a mixture of these from the assumed "pyrolite," we obtain a residue in no way similar to alkalic-ultrabasic magma. As an example, we have given some results, omitting the details of computation (see Table 11).

It is easy to see that it is impossible in this way to obtain the Gulya magma, since it is not possible to obtain the necessary changes in the relations of $SiO_2$, $Al_2O_3$, FeO, MgO, and CaO.

We might be somewhat more successful in obtaining the relations if we assumed that magma of the Gulya type breaks up into a solid phase (dunite + harzburgite) and liquid, near picrite in composition. Then, with the ratio dunite:picrite = 2:1, we obtain the following (which may be considered column 8 in Table 11.

$$(8) \begin{cases} SiO_2 \dots\dots\dots 40.6 \\ TiO_2 \dots\dots\dots 1.0 \\ Al_2O_3 \dots\dots\dots 4.6 \\ Fe_2O_3 \dots\dots\dots 5.3 \\ FeO \dots\dots\dots 8.4 \\ MgO \dots\dots\dots 32.5 \\ CaO \dots\dots\dots 4.4 \\ Na_2O \dots\dots\dots 0.7 \\ K_2O \dots\dots\dots 0.2 \end{cases}$$

Which is similar to column 1

TABLE 12

| Component \ Analysis No. | 9 | 10 | 11 | 12 | 13 | 14 | 15 |
|---|---|---|---|---|---|---|---|
| $SiO_2$ | 45.5 | 44.1 | 46.0 | 44.4 | 45.4 | 42.2 | 44.2 |
| $TiO_2$ | 1.9 | 2.1 | 2.4 | 2.4 | 2.5 | 3.1 | 3.1 |
| $Al_2O_3$ | 12.4 | 13.8 | 9.2 | 6.1 | 14.7 | 18.4 | 15.2 |
| $Fe_2O_3$ | 0.9 | 1.0 | 1.1 | 1.1 | 1.9 | 2.4 | 2.4 |
| FeO | 8.7 | 7.2 | 6.6 | 1.8 | 12.4 | 10.0 | 10.6 |
| MgO | 18.8 | 18.8 | 20.0 | 15.5 | 10.4 | 8.2 | 8.8 |
| CaO | 9.7 | 10.8 | 12.1 | 12.1 | 9.1 | 11.4 | 11.4 |
| $Na_2O$ | 1.6 | 1.8 | 2.0 | 2.0 | 2.6 | 3.2 | 3.2 |
| $K_2O$ | 0.1 | 0.1 | 0.2 | 0.2 | 0.8 | 1.0 | 1.0 |

All these computations are but illustrations to show the possibilities. They are in no way proofs of any actual process. We should not forget that on its way and at the time of congealing in a subvolcanic body, magma might extract from the countray rock some of the more mobile compounds, chiefly $TiO_2$ and $K_2O$.

We have made computations on the assumption that the Gulya magma is derived from a basaltic basic liquid having the composition of picrite (column 9 in Table 12).

With reference to Table 12: Column 10 is the result of settling of 10% enstatite; the lower content of MgO and very high $Al_2O_3$ show that it is impossible to arrive at the Gulya magma on this path. Column 11 is the result of settling of 20% garnet; the $SiO_2$ content increases, not decreases. Column 12 represents the settling of 10% enstatite and 10% garnet; the silica content declines, but along with it, the MgO content and, very markedly, the iron content also declines. It might be assumed that the parental material is alkalic olivine basalt (column 13). However, neither the settling of enstatite (20%, results shown in column 14), or of garnet, or of a mixture of the two (10% + 10%, shown in column 15) leads close to the composition of the Gulya magma, since in all these examples the MgO content declines and the $Al_2O_3$ content increases during mineral settling. If we assume the precipitation of olivine, the contents of $SiO_2$ and $Al_2O_3$ increase and the content of MgO declines.

We are thus forced to the conclusion that the magma of alkalic-ultrabasic complexes cannot be obtained by crystal settling at depth, nor from a possible pyrolite, nor from basic or alkalic basalts, especially from the more silicic varieties. It would appear that the only possible source of this magma must be direct fusion of mantle material.

Some success in this regard was achieved by Bultitude and Green [1967] in Canberra. They showed that liquid of the composition of olivine nephelinite could be obtained during the melting of rock of pyrolite type at high pressures (27 kbar and more), in the presence of large quantities of water, and at relatively low temperatures. We thus have data now to show that alkalic-ultrabasic magmas may be obtained under special conditions: at depths of at least 70 km, at low temperatures (~1200°C), with a notable content of water, and also with low amounts, which requires special conditions also for separation of the liquid. We have thus obtained an explanation of the rarity of these magmas, their consistently small volumes, and their total or almost total absence in the ocean, where deep zones are strongly heated.

It is very important to investigate the problem of whether such complexes are restricted to continents or whether nepheline lavas on oceanic islands indicate the rise of such magmas under the oceans as well. In the first case, it is necessary to investigate the effect of continental crust, which most likely points not to contaminated primary magma but to the creation of special conditions at depth. The capture of mobile components by the magma during its rise is apparently an undoubted fact, and in some cases it may lead to creation of very distinctive magmas (Bufumbira, Vesuvius, and elsewhere).

CHAPTER 15

# CONCLUSIONS

In order to solve the problem of the connection between the occurrence of magma and tectonics, we cannot examine already created structures and already solidified magmatic rocks. It is necessary to penetrate to the deep-seated processes that accompany the formation of magma. In order to accomplish this there is but one course: try to perceive modern tectonic movements at depth and the present-day formation of magma there. It is true that the insufficiency of data does not allow us to exclude data on earlier structures (near the surface, the others we do not observe) and earlier magmas. But such data must be compared with present-day data and explained on the basis of the latter.

Study of the distribution of earthquakes at depth has permitted us to distinguish three types of regions with deep-seated movements: extrageosynclinal, midocean ridges, and geosynclinal. The first two correspond to zones of relatively slow and low-contrast movements at great depths. Earthquakes in these zones take place at depths of a few tens of kilometers. The geosynclinal type is characterized by earthquakes at depths down to several hundred kilometers. This type of zone includes modern island arcs (modern geosynclines) as well as young fold belts, which have already completed their development but in the recent geologic past went through the stage of island arc. This comparison is based on the transition from arc along the strike to fold regions and the presence of deep-focus seismic activity, not quite extinct, beneath some segments of these fold belts. The comparison has been greatly aided by the study of isostatic gravity anomalies, which point to a genetic proximity of one type region to the other.

Having further investigated primary magmas, we have noted that the three largest groups (ultrabasic, basic, and silicic) display no specific correlation with any particular type of deep-seated tectonics. An exception is silicic magma, which almost always is associated with geosynclines. But smaller groups of magmas (varieties) are associated with definite tectonic phenomena.

For extrageosynclinal regions and midocean ridges, basaltic magma is most widely distributed, both tholeiitic and alkalic-basaltic (olivine-basaltic). The latter is especially characteristic of continental areas. It appears most probable that basaltic magma melts out of mantle material in the region of the low-velocity channel. The occurrence of ultrabasic magmas in such regions requires supplementary heating, and this happens only rarely. A significant feature of basic magmas is the rapid rate at which they rise, so that it may be more accurate to speak of a stream of magma, and not about the slow rise of an asthenolith. As for these two types of region, so for geosynclines, there are no grounds for stating that very considerable masses of basic magmas get trapped at depth. For the most part, such magmas either pour out on the surface or are trapped at shallow depths in the crust.

Geosynclinal regions are characterized by the presence of a deep-seated zone of high-contrast movements and by increased heat flow. The melting of basic magmas in such regions

123

takes place through increased rise of energy from deep regions of the earth. This kind of deep-seated zone, a distinctive conductor of energy, determines the occurrence of a geosyncline above it and the special role of this geosyncline in the evolution of the earth's surface. Such deep-seated structure differs fundamentally from all other structures known to us, and we have called it a tectonofer.

The special position of silicic magmas is due to the fact that they form in the upper parts of the crust above the tectonofer, when this structure, in its growth from below, reaches shallow depths and heats the region above its upper part. At the same time, in causing increased mobility in this zone, it is responsible for folding, approximately simultaneous with the formation of silicic magmas, further rise of the magma, and other phenomena of the end of the geosynclinal process. We are convinced that the view that there is no connection between granitic magma and the development of geosynclines is unjustified, and that the fact that silicic magmas may rise in zones outside fold belts may be explained readily by noting the occurrence of fractures above the tectonofer, conductors of heat, passing to the surface somewhat to the side of the fold belt. This type of mechanism permits us also to explain the occurrence of volcanic-plutonic belts.

The uppermost part of the mantle (above the low-velocity channel) and the lower levels of the continental crust are sterile or almost sterile in regard to the generation of magmas.

In examining the role of oceanic subsidence in the formation of magmas, it was noted that the appearance of oceanic crust during extensive subsidence does not always attest to the formation of new oceanic areas. Such crust exists also beneath geosynclinal seas, such as the Sea of Japan, the seas of Indonesia, or some of the Mediterranean basins. We should distinguish ocean formation proper, not associated genetically with geosynclines. Examples of this are the Gulf of Guinea, the Arabian Sea, and some others. For such ocean formation proper, magmatic activity is hardly characteristic, and it is nowhere associated with fractures separating the continent from the ocean, as indicated by the rather shallow depth of such fractures.

The conditions of formation and the evolution of magmas have been examined in some detail. To a considerable degree we have based our views on the most thorough work on this subject [Green and Ringwood, 1966]. The principal mechanism leading to change in the composition of a magma during its passage from one condition of temperature and pressure to another is thought to be separation and settling out of crystals. It should be noted that, apart from the settling of crystals, a considerable role in the evolution of magmas may be played by the separation of liquids (liquation). However, the almost complete absence of experimental data compels us, pending their appearance, to reduce the process of magmatic evolution to separation of liquid and solid phases. In the future we may understand this process better. The conditions for melting out of **the earliest tholeiitic magmas and of the very basic magmas and** the alteration of these magmas, depending on rate of ascent and their detention at different levels, either to alkalic-basaltic liquid or to typical tholeiitic **series were investigated in some** detail (detention of the magma at some depth intermediate between the two indicated leads to the appearance of the high-alumina basalts of Kuno). The mechanism leading to the change of tholeiites to the residual alkalic-basalt series and further to nepheline basalts (as typical of the Hawaiian Islands) has also been described. And we have noted the conditions for direct melting of alkalic basalts and nephelinites from the mantle. Attention was called to the great significance of the amount of liquid melted out, i.e., the impossibility of its separation from the solid phase under ordinary conditions if it constitutes less than half the total volume. The presence of small masses of magma within the rocks (migmatitic segregations, veins of silicic rocks in dolerites, etc.) points to the fact that, apart from the ordinary conditions of separating liquid from crystals, special specific conditions appear in nature (filter pressing), when great temporary drops in pressure occur, permitting insignificant amounts of liquid to be squeezed out of the solid framework.

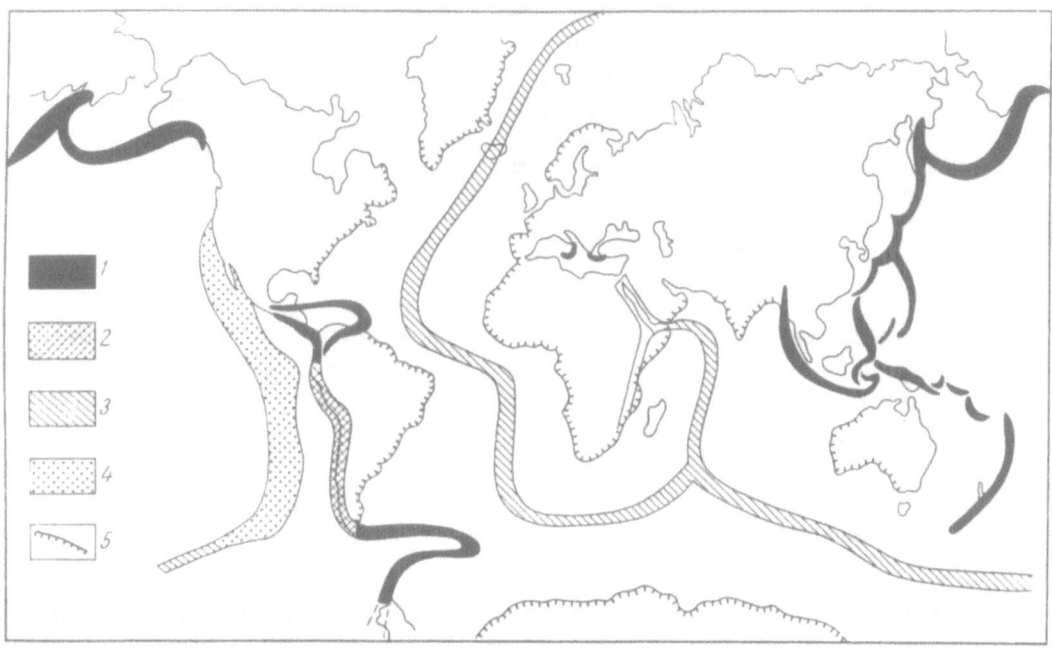

Fig. 47. Distribution of island arcs and midocean ridges. The "attraction" of the latter for the Atlantic type of coasts is clear, as is their absence where island arcs appear. 1) Island arc; 2) Andean region; 3) midocean ridge; 4) East Pacific rise; 5) coast of Atlantic type.

Having restricted ourselves to a very condensed repetition of what we have discussed in this book, let us pause further on some conclusions that seem to us to be important.

1. On the Nature of Midocean Ridges. It seems to us that there is great significance in the established fact of the shallow depth of high-contrast movements that cause the formation of waterways at the surface and large uplifts on continents, not associated with folding or with the formation of midocean ridges. In any case, earthquakes beneath such features do not extend to great depths. Strictly speaking, this indicates a limiting depth of sufficiently high-contrast movements (for the creation of shear displacements); it is not proof of immobility at great depths. More than this, the very presence of near-surface foci indicates some activity at depth. Therefore we cannot maintain that ocean formation and tectonic activation are purely near-surface in character. There is no doubt that slow and low-contrast movements take place here at great depths, which cannot now be measured. Earthquakes in the upper stage are merely reflections of these quiet displacements. Therefore, when we speak of the shallow occurrence of ocean-forming processes, activation, or the rise of midocean ridges, we have in mind the depth only of those movements that have sufficient contrast and strength for the production of earthquakes.

Not long ago [Sheinmann, 1965₁] our attention was turned to the peculiar "antipathy" between midocean ridges and zones of present-day geosynclines and the close connection between midocean ridges and coasts of Atlantic type (Fig. 47). The latter may serve as an indicator of the fundamental dependence of the one phenomenon on the other. This dependence can hardly be defined accurately at present, but we may attempt to approach it approximately, however crudely.

Magmas of the tholeiitic series dominate in midocean ridges. It is possible that they are the only type here (small residual masses of olivine basalt do not interest us here). We know

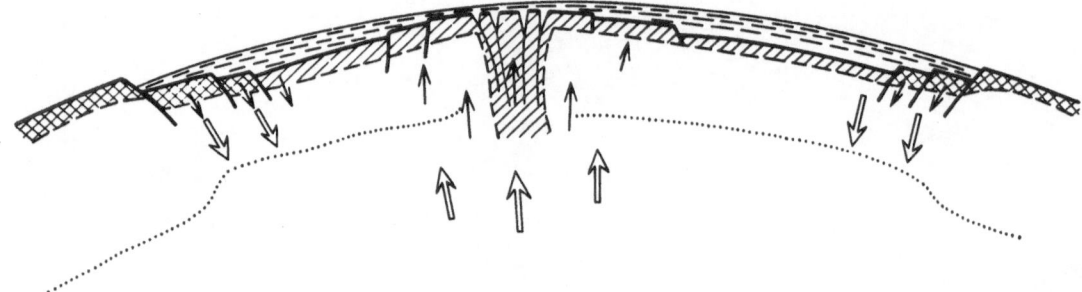

Fig. 48. Scheme of compensating uplift and subsidence, creating simultaneously a mid-ocean ridge and regions of thalassogenesis at the margin of the continent (open arrows indicate quiet movements, not causing earthquakes; solid arrows indicate high-contrast movements. The hachured zone is one of potential earthquakes).

that tholeiites may appear at a pressure of about 18 kbar and more, i.e., at depths no less than 60 km. Such depths correspond approximately to the upper boundary of the low-velocity channel of Gutenberg beneath the ocean.

We must recall, however, that this is a minimum value, and that, if we judge from the data of physical chemistry of basaltic melts [Green and Ringwood, 1966$_3$], this depth is the upper limit for the appearance of tholeiitic magma. Therefore, there are grounds for assuming that magmatic activity begins at somewhat greater depth. Since earthquake foci do not appear at such depths, we have no basis for seeking the formation of magma at considerably lower levels. We shall probably make no mistake if we set at 60-100 km the depth where the process that leads to formation of midocean ridges occurs. At the same time, this depth is that from which material is removed to the surface. It is probable that the great amount of crystallized basaltic melt at the uppermost levels of the mantle leads to the appearance beneath the ocean ridge of the "cushion" with low velocities (below 8.0 km/sec for transverse waves), about which so many people have written in recent years.

We may thus state that the removal of material to the near-surface zone in midocean ridges takes place most probably from a depth of 60-100 km, and no evidence is yet available to indicate the figure should be increased. Such rise of matter must inevitably lead to restoration of the loss by inflow from neighboring regions. This shift of mass is so slow, however, that it causes no seismic disturbance and we do not record it at the surface. This inflow (from the side) might lead to some compensatory settling of segments adjacent to the ridge, but this could hardly involve a very large area, most likely being restricted to the zones immediately adjacent to the ridge.

Volcanic phenomena are not the principal features in the growth of midocean ridges, and the indicated compensation of rising material by overflow from the side plays no leading role. Apparently the upward bulging of the ridge is caused primarily not by the outflow of lava but by general tectonic uplift of the central belt of the ocean. The moving force of this uplift must be sought, but it is not altogether clear where; the cause is about as uncertain as the cause of large uplifts on the continents. It is possible that the immediate cause of uplift of the midocean ridges is decompression of the matter below it, the rise of the less dense cushion of basaltic liquid perhaps not being the only, or even the chief, cause of the decompression.

It is thought that whereas lateral inflow to the place from which basalt has departed upward cannot be the cause of thalassogenesis at the edge of the ocean; compensatory subsidence, forming, as it were, a pair in conjunction with bulging of the ridge (but not with outpouring of basalt), may lead to an explanation of the expansion of the oceanic region with formation of

Atlantic-type shores. Perhaps it is just here that the explanation lies for the connection, which at first glance appears singular, between these two phenomena (Fig. 48). The tectonofer with a geosyncline above it is a phenomenon of a different order. Compensation of uplift and subsidence (since it takes place here) occurs over much smaller areas, and the entire process, with the accompanying upward movement of material, prevents expansion of the ocean through "chipping off" parts of the continent. The impossibility of this kind of thalassogenesis should prevent uplift of the midocean ridge.

The last view awaits confirmation, however, though one can hardly bring himself to expect it. With this approach to the problem it becomes incomprehensible why thalassogenesis is not accompanied by increased transfer of energy from the depths to the surface.

2. On Closed Convection Currents in the Mantle. Like the above discussion concerning the nature of midocean ridges, the linkage of these ridges with advance of the ocean (thalassogenesis) permits us once again to speak decisively against such views, common in our time, concerning circulation of matter in the mantle, the presence of normal and closed convection currents, ordinarily pictured in the upper branch from the center of the ocean (from the crest of the midocean ridge) to the shores, in particular to island arcs. There the picture changes to the branch descending to the depths, turning back toward the ocean, passing through the low point to the midocean ridge and then upward along the rising branch to emerge at the surface, closing the circle.

Apparently the idea that subcrustal currents played a leading role in the tectonic process first advanced by Ampferer [1906], has gradually changed in essence. In view of the absence of a continuous magmatic reservoir beneath the crust, it has become necessary to seek confirmation in the capacity of extremely viscous (as compared with magma) mantle material to create such rotation. It is true that the possibility of this has been demonstrated by no one, and confirmation now rests on faith, not on investigation. If such a current is possible, it is important to know its velocity, since movement of a few kilometers per 2-3 billion years can scarcely solve anything.

We can easily record uplift and subsidence. Horizontal displacements of mass, even within continents, where direct observation of structures can be made, are much more difficult to detect. The problem in the ocean is even worse. As a result, in this hypothetical convection ring, the length of the horizontal leg has been observed by no one. Even the upward leg, rising to the surface, cannot be seen; it is not necessary to say anything concerning the lower, deep leg.

The problem is the same with the presence of convection "cells." The necessity of the hypothesis itself, since the time it was proposed in 1906, has become very doubtful. At the beginning of the twentieth century it might have replaced other theories in explaining folding. Then the question was raised merely in relation to the folding itself, which was thought of as a crumpling of a layered plastic sequence through the action of two rigid masses between which the sequence lay. The approach of these masses toward each other was the cause of the folding and the accompanying orogeny. The hypothesis of subcrustal currents moving these masses one toward the other might have replaced the contraction theory, which was being dangerously fractured. The later works of Ampferer [1941, 1942] are an echo of this situation.

The position now is fundamentally different. It became clear that, in order to explain folding, it was necessary to prepare a model, reproducing the bending of the layers and their crumpling by means of some specific forces. The primitive concept of a geosyncline as a weakened zone, subsiding as a result of accumulating readily yielding sediments, gave way to the idea of a complex process, in which the formation of folds is by no means the main event. All the concepts, therefore, starting from Ampferer, have lost ground. That which we know

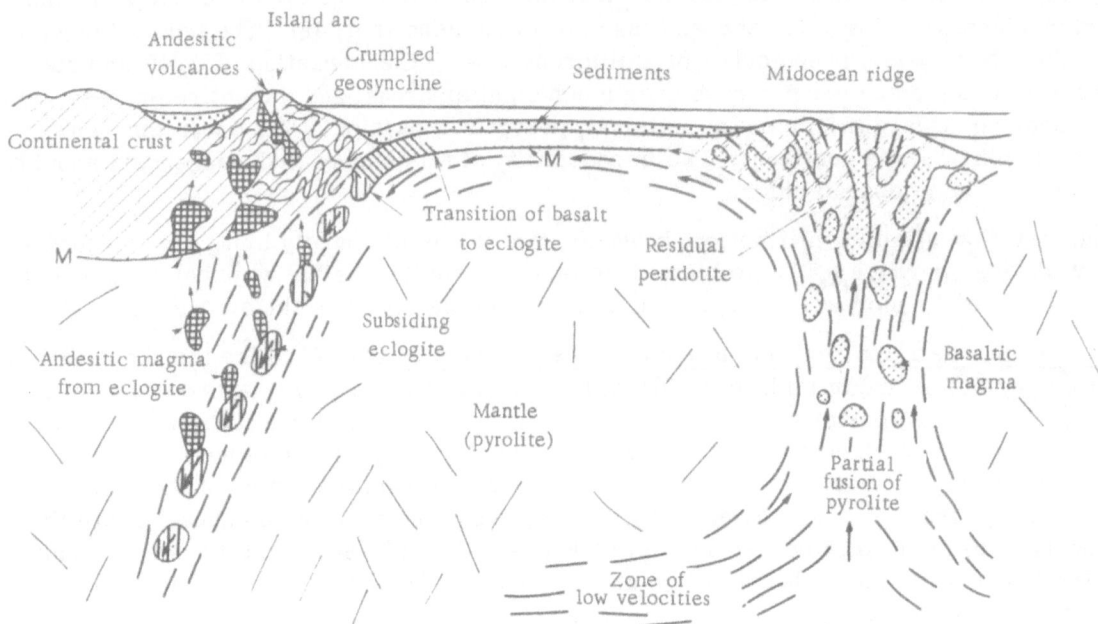

Fig. 49. Comparison of some peculiarities of magmas with convection currents (attention should be given to the combination in this scheme of the midocean ridge and the island arc into a single system and to the complex mechanism by which deep-seated andesites are generated) (after Ringwood and Green [1966]).

about magma, about the composition of the initial material, the conditions of temperature and pressure, about the environment in which it is converted to igneous rocks proves to be the most important part of the present model of the geosynclinal process. To this same model we must refer the ceaselessly increasing store of information concerning the physics and chemistry of the earth. Under these conditions, whatever needs justified the appearance of a hypothesis about subcrustal currents producing crustal deformation have disappeared.

Thus, the objective that the hypothesis should have met became superfluous, and the assumed currents in the hypothesis proved to be doubtful. The time obviously arrived for trying to approach the problem from other points of view, discarding, even if for a time, the view of convection as the first cause of the geosynclinal process and of tectonics as a whole. Only after physics has demonstrated the limits of possible currents in the mantle should we turn to the question of whether they play a role, and to what extent, in tectonic and magmatic processes.

Let us consider the last variant of the convection model proposed by Ringwood and Green [1966]. In this model, as in the many preceding models, the same defects are repeated. In this scheme a geosyncline is placed at the edge of the ocean, opposite the active zone of the midocean ridge. But, first, present-day midocean ridges and present-day geosynclines are antagonistic; second, we well know that geosynclines are by no means associated only with the oceanic margins (this view is a legacy from the old view of tectonic geologists in America); and, third, the descending branch strikes the geosynclinal zone, i.e., where material is observed to rise. It is true that in this scheme, as it is generally pictured, the region of ascending material is placed side by side with a zone where material is drawn into the interior of the earth (Fig. 49).

The distinctive feature of the scheme of Ringwood and Green is their treatment of the occurrence of andesites in the geosyncline (island arc). The necessity of treating andesites as products of very deep-seated magma apparently forced the authors to complicate their scheme extremely. In their opinion, masses of basaltic magma, separated in the mantle, float up in the midocean ridge. Entrapped by the circular current, they are crystallized near the surface as eclogite and are transported in this form to the edge of the ocean, where they follow the descending branch of the current and come to heated deep-seated zones, where the eclogite is partially melted, giving off andesitic liquid. The latter, floating upward toward the surface, is responsible for the andesitic vulcanism of the island arc. It is far from clear why such a complex mechanism for generating andesite is required when the presence of the andesite thermal trough at sufficiently high pressures guarantees its appearance by direct development of ascending deep-seated magma. Let us recall that the transition from basaltic to andesitic extrusions in the geosyncline are not associated with any rapid change, which would inevitably be the case in the model of Ringwood.

3. On a Single Magma Chamber Moving from Depth to the Surface. There are no grounds for assuming the existence of some single magma chamber, gradually rising from depth to the crust and, according to the distance it has ascended, generating different types of magma, beginning with ultrabasic, then basaltic, and, lastly, granitic. Dunite-harzburgite rocks can hardly be the products of some independently forming magma, the composition of which corresponds to the total composition of the rocks. For several years the author of the present work, like many others, believed that under special conditions the ultrabasic mass of the mantle could melt, resulting in the appearance of a special ultrabasic magma. Since the normal composition of the mantle might be that of dunite-harzburgite rocks (in which case the mantle could not yield basalt in any appreciable quantity), it was necessary to assume that ultrabasic magma must derive from the material of the mantle, previously deprived of its basaltic part. The author sought such material, the basalt already melted out, in the highest parts of the mantle, between the low-velocity channel of Gutenberg and the crust [Sheinmann, 1961]. The low-velocity channel itself, more properly its upper boundary, was interpreted as the boundary between the impoverished zone and the underlying still primary substance (with basalt) of the mantle (this idea was expressed somewhat later by Ringwood also [1962]).

At the present time it is necessary to reject this scheme. There are no apparent grounds for expecting the appearance, at such shallow depths, of temperatures sufficient for the fusion of ultrabasic material. More than that, it is rather doubtful that complete melting could take place even at greater depths. This problem is subject to verification. Present information on the conditions of melting compel us to believe that very basic magma, forming at depth, then ascending, might become trapped with its crystals of olivine (and pyroxenes ?) or, what is perhaps more likely, might adjust to the conditions of the higher levels and separate from these minerals. Whatever, even under the conditions that the trapped or segregated minerals after their separation from the liquid could be placed in an environment favorable for melting, the material of the ultrabasic rock is brought up together with the basalt and is separated from it later also at higher levels. We cannot therefore speak as if ultrabasic magma corresponds to the deeper half of the chamber and basalt appears after it is removed. Both are merely the result of the latest separation of magma or magmatic mush.

The one thing we must accept is the increase in silicic character of the magmatic material as it ascends in the mantle. We need also to realize that the top of the magma chamber rises, but the initial phase of the process is a silicic harzburgite. There is an upper limit of the chamber: in approaching the upper layers of the mantle it stops its advance. A magmatically dead zone appears here, within which the zone of heating rises, but there is no magma chamber.

Upward, in the upper half of the crust, when the zone of heating reaches this level, a new chamber appears, containing silicic magmas at this time.

Thus, the concept of a single chamber, three-staged, so to speak, and, in addition, rapidly rising [Izokh, 1965], does not appear to be justified.

4. The Tectonofer and the Geosyncline. One conclusion we have reached in the process of our discussion is that the geosyncline is the upper stage of a structural unit, rising from great depths, which we propose to call a tectonofer. This concept was formulated earlier [Sheinmann, 1964₂]. It emphasizes the special position of the geosyncline and of the geosynclinal process among other tectonic phenomena. This special position of the geosyncline was clear long ago to E. Haug, but only now, in connection with establishment of the fact that such great depths are involved in the process, has its significance become apparent. The geosyncline itself is something of a superstructure on the tectonofer. Whereas its frontal zone is directly related to emergence of the tectonofer at the surface (marginal depression, foredeep), the greater part is related to the tectonofer **indirectly by deep intrageosynclinal fractures.** These latter connect the surface with the hanging wall of the tectonofer, which here approaches rather close to the surface (from tens of kilometers to 200 km).

It is difficult to connect individual parts of the geosynclinal zone with the life of the tectonofer. It is possible to present different schemes, showing the transformation kinematically within the tectonofer into movements observable in the geosyncline. The relations there might be far from simple, and the author has set himself the task of investigating them. But the connection itself between the geosyncline and phenomena of the tectonofer appears to be convincingly demonstrated.

We note that, from the points of view we have analyzed, the geosynclinal process proves to be the near-surface expression of energy flowing from the depths of the mantle outward, especially powerful and concentrated in a narrow zone.

5. Evolution of the Tectonofer. The tectonofer is a distinctive vent, discharging upon the earth the excess energy from the interior. The energy of a tectonofer is both mechanical (tectonics) and thermal (heat flow, loss of heat by magma). We should not forget the energy of chemical reactions, however, which also favor outward loss of excess energy. The tectonofer is born at depth together with the initial ascent of energy. We do not know the mechanics of the appearance and upward growth of such a narrow and generally inclined zone. It is one of the many tasks for the new science of deep-seated tectonics. The rise of the zone of heating and the strong and high-contrast movements have been recorded nowhere, and observations of them are extremely difficult. Two types of complications occur: 1) difficulties in the observations themselves, when the phenomena are not perceptibly manifested near the surface, and 2) the exceptional quality of the conditions, since it is necessary, for the process to get a start, that the way be prepared for the appearance of a new geosyncline (a **geosynclinal cycle**). We have not learned to detect this process at the surface, to which we may add that no new geosynclines have been born anywhere in our time. Even the young post-Alpine residual geosynclines, which might be assumed to exist in the Mediterranean with adequate reasons, have already been formed. It appears most plausible that the preparatory stage takes place rather quickly, and the period when intense tectonic disturbances are confined to deep-seated regions, and the magma that has formed has not yet reached the surface, is relatively short, much shorter than the epoch of the active tectonofer.

Later, the store of excess energy, which may give rise to a tectonofer, declines, and finally becomes zero. Correspondingly, the aspect of the tectonofer alters. Its activity at depth declines, and then ceases altogether. Its depth gradually lessens. Apparently (if we judge from modern phenomena in tectonofers) the magmas that have melted out at appropriate depths cease first to form and, after this, considerable mobility (earthquakes) is maintained for an extended time. Finally, the tectonofer dies entirely; the melting of granitic magma ceases, and only very shallow earthquakes, common to the adjacent regions also, continue to occur.

6. On the Controversy between "Horizontalists" and "Vertical-
ists." All that we have said above forces us to believe that the controversy between so-
called "horizontalists" and "verticalists" cannot at this time play any significant role. More
than this, the argument is unnecessary. The controversy began many decades ago and was
useful at that time, when the most important task of investigators was to reduce folding to a
simple dynamic scheme. At present, this task has become but a part of the matter, and "folding" can no longer be reduced merely to the formation of a fold. It embraces other important
phenomena as well. In this work, we have therefore used the term "geosynclinal process." In
applying the ideal of mobilism to it, even if we do not stand on the very fashionable but, as it
seems to us, unpromising platform of this concept, the question of primarily vertical or horizontal forces proves to be secondary. Recently it has acquired a new freshness. The controversy has been revived, chiefly on the side of the "horizontalists." From all that we have said,
it is clear that movements within a geosyncline have a secondary character and are generated
by phenomena in a much deeper zone, in what we have called the tectonofer. To speak of horizontal or vertical directions in it seems senseless. There is no doubt that the principal direction here is from the depths toward the surface, since the entire process is associated with the
outward disposal of "excess" stores of intraterrestrial energy. Without doubt, this outward
discharge proceeds along the easiest path, usually the shortest. Consequently, "slippage"
along the tectonofer in a direction near the horizontal cannot be the principal movement in its
life.

However, this does not indicate that the paths are necessarily vertical. As we know, the
tectonofer is normally appreciably inclined. At times it is almost vertical, but its inclination
may be 25-30° (beneath the Andes). But the significance of the phenomena does not change because of this, and the direction of energy flux remains the same, from the depths outward, and
this is also the primary direction of tectonic structures at depth.

Thus, from the viewpoint of the present work, there are no grounds for stirring up a controversy over the primary character of vertical or horizontal forces. This question has been
essentially removed from the news of the day as a result of developments in science. There
are now many more burning and important questions, a few of which we have tried to touch on
in these pages. Whether we have been successful we leave to the judgment of the reader. Re-
birth of the controversy of vertical versus horizontal forces cannot only fail to aid in development of our science, it may retard it. It therefore seems proper to withdraw from the controversy.

7. Two Types of Magma Formation. The formation of magma, regardless of
the composition of the magmas that form, takes place in two fundamentally different ways. In
one, in extrageosynclinal regions and midocean ridges, conditions near those for the melting
of basalt exist in the deep zones of the earth. Furthermore, it is possible that basaltic liquid
in minor quantities is always present there. If this is so, then drops or thin intergranular films
of liquid exist in the material of the mantle. There is no doubt that the composition of the
liquid corresponds to the easily fusible components, under these conditions, or very near to
these. According to the latest data (Ringwood and Green) this material is andesite. However,
the amount of this liquid is small, and it cannot be separated from the solid matrix, even by
filter pressing. But conditions for this do not exist anyway, because a specific environment is
necessary, incompatible with tectonic calm, in order to create the required pressure drop
(even if very short-lived).

Under conditions of tectonic calm at depth, this type of phenomenon is either impossible
or extremely rare. As a result, if intergranular liquid is present in the low-velocity channel of
Gutenberg, as assumed by Belousov, it cannot be removed, and magma will not be formed.
There is no supplementary influx of energy sufficient to melt the supplementary masses (more

Fig. 50. Differences in conditions for melting and ascent of magma (outside the tectono-fer). I) Regions of low activity, shallow-focus earthquakes, relatively shallow fractures: a) beginning of asthenolith; b) breaking away of asthenolith, magmatic mush not in state for separation of magma; c) opening of asthenolith by a fracture, facilitating fusion and, as a consequence, separation of liquid and rapid rise of this liquid to the surface. This represents the probable conditions for small fields of chiefly olivine-basalt magma. Emergence of the asthenolith at the surface is relatively improbable. II) Regions of heightened activity and of the corresponding deep fractures: a) fracture opens up zone of potential melting, beginning of massive melting, separation of magma and its ascent; b) eruption of magma on the surface. Assumed conditions of midocean ridges and trap fields. 1) Zone of potential melting; 2) basalt.

difficult to melt than andesite). Thus, in the zone of the low-velocity channel, the liquid that appears proves to be a film, suppressed, as it were, by the great amount of solid phase. Massive fusion of material may begin there only if conditions favorable for it appear.

Thus, for magma to form in the first way, a change of conditions is essential, produced by high-contrast movements upward, which lead to the formation of fractures that extend to the deep regions of potential melting. There are no specific supplementary streams of energy from depth for this purpose. In any case, if some energy is supplied thus, the amount is small, and its role in forming magma is negligible at best (Fig. 50).

The second type of magma formation refers to the magma of the tectonofer. The depth at which the magma forms is at least two or three times the depth of magma generation in extrageosynclinal regions, and the chamber is independent of fractures that originate in the crust. The formation of magma here is associated with supplementary and, probably, strong heating, i.e., with a supply of additional deep-seated energy, which characterizes the tectonofer. The same picture holds also for high-level zones (silicic magmas). Melting here is directly associated with the influence of the tectonofer, with its energy.

Consequently, there are two types of relations between tectonics and the formation of primary magmas. In one case, the tectonic phenomena facilitate melting and create conditions for ready passage of the melt to the surface. In this process there is no significant increase in energy level in the zone of melting by transfer from deeper zones. This refers to extra-geosynclinal magmas. In the other case, the melting is due to the energy of a deep-seated tectonic process, and tectonics thus determines the melting to a great extent. It is, so to speak, the active factor, not some secondary phenomenon aiding the process. Subsequently, tectonics affects the possibility of magma ascending and its upward path to a lesser degree than the first type. In other words, there are magmas of relatively peaceful regions and magmas of active, especially deep-seated zones, charged with abundant energy. This difference appears to be the most important factor in the future development of the geology of deep-seated regions: abyssal geology.

CONCLUDING REMARKS FOR ENGLISH-LANGUAGE READERS

In the time between the writing of this book and its translation into English, much that is new has appeared, in both the world of facts and the world of ideas. Some of the supplementary material could be introduced into the text without hurting it. But some must be placed here in these last few lines. In them I must touch briefly on what has been published in the Russian language to supplement the book.

A significant addition for understanding processes at depth is a rough estimate of the energy required for the formation of magmas. This estimate shows not only the energy changes for the depth of formation (the shallower the depth, other things remaining the same, the greater the energy required for any particular magma) but also the changes during transition from one magma to another. In particular, the amount of energy declines from tholeiitic magma to alkalic and, farther, to alkalic-ultrabasic magma.

These data [Sheinmann, 1970] allow us to understand why tholeiitic magmas are restricted to tectonically more active zones, alkalic-basalt magmas to quieter regions. They also lead us to believe that in regions where alkalic basalts are developed (without tholeiites) there are better grounds for believing that alkalic-basaltic magma is primary, not a differentiate of tholeiitic magma.

The formation of andesitic magma in the crust proves to be relatively improbable, not only from the viewpoint of geology, but also from the viewpoint of the physics of the process: the melting of andesitic magma in the crust, even in the presence of abundant water, requires very much energy, much more than at greater depths.

An unusual solution was found for the question of why alkalic basalts and alkalic-ultra-basic magmas are absent or very rare in the oceans. Higher temperatures in the upper mantle beneath the ocean strongly assists the melting of magmas in comparison with the continents. Comparatively small amounts of additional heat are sufficient for not only alkalic but tholeiitic magmas also to begin melting under the oceans. This leads, in most cases, perhaps all, during formation of the magma chamber, to a change in the process after a short time, to the stage at which alkalic basalt is melted out, and the chamber then furnishes tholeiitic magma. This stage is by no means always reached beneath continents.

It seems to me that, of the questions discussed in the book, special attention today belongs to the problem of changes in conditions leading to the formation of magma. It is known that melting may begin as a result of one of the following phenomena: elevation of temperature, lowering of pressure, or enrichment in easily melted components, in particular the volatile substances. The last factor plays a very large role in melting crustal material, but in the general case it is of little interest for deep-seated magmas. Therefore, in discussing melting in the mantle, we must choose between elevation of temperature and decline of pressure. In the first case it is necessary to find a source of the heat, one that does not affect the entire earth but only some individual segments of it. Such a source might be local accumulation of radioactive elements. However, with any reasonable view concerning a primary inhomogeneity of the distribution of these elements we cannot obtain the necessary effect: in order to raise the temperature to the melting point it would be necessary to have implausibly large local concentrations of heat-forming substances. This has made it necessary to search for the cause of melting in decline of pressure. It has already been shown some time ago by Uffen [1959, 1963] and Gzovskii [1962] that this kind of process is possible, but 1) it is short-lived, and 2) it occurs only under special conditions. These have been the principal difficulties, but they are still smaller than that to be overcome in proposing a great inhomogeneity in the primary distribution of radioactive elements. The result of this kind of speculation was the view, defended in this book, that, outside of tectonofers, magma is generated as a result of local drop in pressure near a deep-seated fracture.

In the adopted hypothesis, two difficulties of a high order have been established. The first, particular, difficulty involves the fact that it was necessary to assume that matter in the asthenosphere is at the melting boundary, and that slight heating or a small drop in pressure is sufficient to cause melting to begin. It became necessary to close one's eyes to the difficulty involved in this assumption, since there was no other escape.

The second circumstance, of a more philosophical order, is found in the fact that magmatism, at least outside of tectonofers, reduces to accidental sporadic events. However, geology shows that this process is constant, of great power, and always operating in the earth (active since the Archean, up to the present). Its cause, therefore, should not be a random one, excesses that occur from time to time, but it should be a more important and invariable phenomenon, part of the life of the earth, determining this life.

We have seen that the choice of causes of deep-seated melting is limited, reducing merely to this: do we yield to a preference for local drop in pressure or for a local rise in temperature? It consequently becomes necessary to seek further. For tectonofers, the cause was accepted from the very first, and consisted of "tubes" along which the "excess" of energy from depth is expelled outward. In 1968 a new hypothesis was advanced [Artyushkov, 1968], and it is extremely important for understanding the overall mechanics of the earth as a whole. Basing this hypothesis on the fact that thermal convection in the mantle (on which most modern hypotheses have been constructed) proves to be impossible because of too rapid increase in density with depth, Artyushkov proposes a new mechanism of carrying energy to the surface. He suggests that substances of the earth's core continue to separate out in the depths of the mantle. These relatively easy-melting substances flow downward and in their place there remains a relatively infusible light differentiate. The disturbed equilibrium is restored when this differentiate breaks through to the upper mantle, adding to its mass. The process must continue until differentiation of the primary substance of the earth (lower mantle) ceases. All this is accompanied by the freeing and upward transfer of large amounts of energy, both kinetic and thermal.

The hypothesis of Artyushkov opens up great possibilities for geologists. We obtain a process for magmatism that secures local heating and removes the second, philosophical difficulty we just noted.

The rise of material from depth requires no thermal convection, and the difficulty in connection with this problem is removed. Ascending hot masses of light substance form "their customary paths," i.e., the ascent takes place most easily along the heated channel. Thus, tectonofers take their place in the general picture, the earth. The intermittent character of the process becomes understandable: accelerated ascent along the channel leads to depletion of the reserves of light differentiate. Renewal of the ascent of material is possible only after a new accumulation of reserves of this substance.

There is not room here to enumerate all the possible geological consequences of Artyushkov's hypothesis. We shall note only that it permits a satisfactory explanation both of the geosynclinal process and of the behavior of midocean ridges, as well as of movements on platforms.

In the framework of this book, the hypothesis must be of interest to us primarily because it gives a deeper understanding of the essence of tectonofers and it makes it possible for us to avoid the treacherous path of local drop in pressure to explain magmatism. Now we have every possibility of drawing on local rises in temperature, since these are inevitable in light of the idea that light and hot differentiate from depth rises to the upper part of the mantle.

We note that these new ideas do not alter our conclusion concerning the existence of two groups of magma: one connected directly with a tectonofer, and the other arising under conditions of many weaker deep-seated effects. The magmas of midocean ridges, if the almost universally accepted view of abundant magmatism on these ridges is justified, are probably nearer the first group.

The author would like to emphasize the fact that, although the conclusions discussed above represent the most probable solutions to him of the questions raised in these pages, he realizes that for most of his comrades in the work many of the conclusions will appear controversial. He understands that the vigorous development of our sciences concerning the behavior of substances at high temperatures and pressures may change many of the ideas on which he has based his conclusions. Therefore, however assured he may be now in what has been said, he knows that many things may change even within the coming year. It is hoped, however, that the thoughts expressed may aid in analyzing in their very basic features some of the problems of this new branch of science, being born before our eyes: the geology of deep-seated regions, abyssal geology.

# TABLES OF CHEMICAL ANALYSES

# TABLE 13. Ethiopia (Mohr [1960])*

## Trapp Series
## Eritrea and Tigre

| Analysis No. | 1 | | 2 | | 3 | | 4 | | 5 | | 6 | | 7 | |
|---|---|---|---|---|---|---|---|---|---|---|---|---|---|---|
| Component | % | M.Q. | % | M.Q. | % | M.Q. | % | M.Q. | % | M.Q. | % | M.Q. | % | M.Q. |
| $SiO_2$ | 46.96 | 782 | 49.03 | 817 | 55.38 | 922 | 57.81 | 968 | 61.28 | 1021 | 63.74 | 1061 | 67.03 | 1116 |
| $TiO_2$ | 1.45 | 18 | 2.48 | 31 | — | — | — | — | — | — | — | — | — | — |
| $Al_2O_3$ | 15.19 | 149 | 12.10 | 119 | 18.46 | 181 | 18.74 | 183 | 21.94 | 215 | 17.86 | 176 | 14.25 | 140 |
| $Fe_2O_3$ | 8.35 | 52 | 7.64 | 48 | 4.72 | 29 | 5.76 | 36 | 0.48 | 3 | 4.27 | 27 | 1.96 | 12 |
| $FeO$ | 7.30 | 102 | 3.87 | 54 | 6.28 | 88 | 0.42 | 6 | | | 0.30 | 4 | 1.70 | 24 |
| $MnO$ | 0.12 | 1 | — | — | 0.19 | 3 | — | — | — | — | 0.19 | 3 | — | — |
| $MgO$ | 6.63 | 165 | 6.56 | 163 | 1.19 | 30 | — | — | — | — | 0.10 | 3 | — | — |
| $CaO$ | 8.34 | 149 | 9.02 | 161 | 3.93 | 70 | 1.28 | 23 | 0.61 | 11 | 0.83 | 15 | 1.05 | 19 |
| $Na_2O$ | 4.53 | 73 | 4.21 | 68 | 7.55 | 122 | 9.35 | 151 | 3.37 | 54 | 7.23 | 117 | 3.85 | 62 |
| $K_2O$ | 0.59 | 6 | 1.67 | 18 | 2.94 | 31 | 4.52 | 48 | 2.58 | 28 | 5.19 | 55 | 3.90 | 41 |
| $P_2O_5$ | 0.17 | 11 | 0.22 | 1 | 0.42 | 3 | — | — | — | — | — | — | — | — |
| $qz$ | | −38 | | −30 | | −47 | | −65 | | +122 | | −13 | | +127 |

| Analysis No. | 8 | | 9 | | 10 | | 11 | | 12 | | 13 | | 14 | |
|---|---|---|---|---|---|---|---|---|---|---|---|---|---|---|
| Component | % | M.Q. | % | M.Q. | % | M.Q. | % | M.Q. | % | M.Q. | % | M.Q. | % | M.Q. |
| $SiO_2$ | 70.36 | 1171 | 71.98 | 1199 | 72.95 | 1214 | 73.44 | 1223 | 73.46 | 1223 | 74.72 | 1244 | 76.01 | 1256 |
| $TiO_2$ | — | — | — | — | 0.07 | 1 | — | — | — | — | — | — | — | — |
| $Al_2O_3$ | 16.90 | 166 | 14.48 | 142 | 13.85 | 136 | 14.35 | 140 | 12.47 | 123 | 13.38 | 131 | 11.96 | 118 |
| $Fe_2O_3$ | 0.91 | 6 | 0.88 | 6 | 2.50 | 9 | 0.50 | 3 | 3.64 | 23 | 1.29 | 8 | 2.06 | 13 |
| $FeO$ | | | 0.92 | 12 | 0.36 | 5 | | | | | 0.77 | 11 | | |
| $MnO$ | — | — | — | — | 0.08 | 1 | — | — | — | — | — | — | — | — |
| $MgO$ | — | — | 0.28 | 7 | 0.28 | 7 | 0.39 | 10 | — | — | 0.35 | 8 | — | — |
| $CaO$ | 0.48 | 9 | 0.76 | 14 | 0.20 | 4 | 0.53 | 10 | 0.32 | 5 | 0.18 | 4 | 0.26 | 5 |
| $Na_2O$ | 5.92 | 95 | 5.98 | 97 | 5.49 | 89 | 5.74 | 93 | 5.63 | 91 | 4.27 | 69 | 4.46 | 72 |
| $K_2O$ | 5.04 | 53 | 3.71 | 39 | 3.23 | 34 | 3.82 | 40 | 4.03 | 43 | 3.94 | 41 | 4.73 | 50 |
| $P_2O_5$ | — | — | — | — | 0.11 | 1 | — | — | — | — | — | — | — | — |
| $qz$ | | +72 | | +140 | | +148 | | +131 | | +122 | | +187 | | +187 |

Note: 1) olivine basalt (Marahano, Hamasen); 2) camptonite (Mai Enda Marugla, Komaile); 3) tinguaite (Azemo, Hasemo); 4) tinguaite (Elda, Georgis, Adua); 5) rhyolitic tuff (Mai Metere, Addi Ugri); 6) selvsbergite (Edda, Georgis, Adua); 7) obsidian (Amba Bera, Adula); 8) bostonite (Senafe); 9) paisanite (Amba Tokale); 10) rhyolite (Mehra Seitan, Demberguina); 11) quartz bostonite (Senafe); 12) grorudite (Amba Sebat, Adua); 13) quartz bostonite (Barakite); 14) paisanite (Amba Sheloda, Adua).

*In the following tables M.Q. stands for molecular quantity [equal to wt.% of oxide / molecular wt.% of oxide · 1000]—transl.

## TABLE 14. Central Ethiopia

| Analysis No. | 1 | | 2 | | 3 | | 4 | | 5 | | 6 | | 7 | |
|---|---|---|---|---|---|---|---|---|---|---|---|---|---|---|
| Component | % | M.Q. | % | M.Q. | % | M.Q. | % | M.Q. | % | M.Q. | % | M.Q. | % | M.Q. |
| $SiO_2$ | 43.60 | 726 | 45.14 | 752 | 45.59 | 759 | 46.66 | 777 | 46.69 | 778 | 61.78 | 1029 | 62.68 | 1044 |
| $TiO_2$ | 2.22 | 28 | 1.42 | 18 | 2.16 | 27 | 0.78 | 10 | 1.80 | 23 | 1.10 | 14 | 0.99 | 13 |
| $Al_2O_3$ | 10.65 | 104 | 18.64 | 182 | 14.23 | 139 | 21.86 | 215 | 14.88 | 146 | 13.23 | 129 | 13.44 | 131 |
| $Fe_2O_3$ | 3.53 | 22 | 9.26 | 58 | 3.33 | 21 | 0.39 | 3 | 0.09 | 1 | 7.98 | 50 | 5.25 | 33 |
| $FeO$ | 7.15 | 100 | 4.77 | 66 | 9.33 | 130 | 7.77 | 108 | 6.73 | 94 | — | — | 1.00 | 14 |
| $MnO$ | — | — | 14 | 2 | 0.20 | 3 | 0.16 | 2 | — | — | 0.12 | 1 | 0.15 | 2 |
| $MgO$ | 12.62 | 312 | 5.85 | 145 | 8.72 | 216 | 7.61 | 189 | 5.63 | 140 | 0.76 | 19 | 0.46 | 11 |
| $CaO$ | 17.42 | 310 | 10.54 | 188 | 10.96 | 195 | 9.60 | 171 | 8.97 | 160 | 3.18 | 57 | 2.72 | 48 |
| $Na_2O$ | 1.30 | 21 | 2.09 | 34 | 2.17 | 35 | 2.57 | 41 | 2.97 | 48 | 4.27 | 69 | 3.06 | 49 |
| $K_2O$ | 0.55 | 6 | 1.52 | 16 | 0.72 | 7 | 1.18 | 13 | 1.48 | 16 | 4.39 | 47 | 4.43 | 47 |
| $P_2O_5$ | — | | 0.40 | 3 | 0.42 | 3 | — | | — | | — | | — | |
| $qz$ | | —31 | | —26 | | —23 | | —23 | | —14 | | +39 | | +79 |

| Analysis No. | 8 | | 9 | | 10 | | 11 | |
|---|---|---|---|---|---|---|---|---|
| Component | % | M.Q. | % | M.Q. | % | M.Q. | % | M.Q. |
| $SiO_2$ | 63.13 | 1051 | 71.37 | 1188 | 71.78 | 1195 | 71.91 | 1197 |
| $TiO_2$ | 1.14 | 14 | 0.41 | 5 | 0.66 | 9 | 0.41 | 5 |
| $Al_2O_3$ | 15.72 | 154 | 10.87 | 107 | 10.29 | 101 | 11.44 | 112 |
| $Fl_2O_3$ | 2.50 | 16 | 3.87 | 24 | 4.43 | 28 | 2.67 | 17 |
| $FeO$ | 1.77 | 25 | 2.37 | 33 | 1.67 | 24 | 2.76 | 39 |
| $MnO$ | 0.21 | 3 | 0.09 | 1 | 0.21 | 3 | 0.14 | 2 |
| $MgO$ | 0.81 | 20 | 0.27 | 6 | — | — | 0.18 | 5 |
| $CaO$ | 1.72 | 30 | 0.38 | 7 | 1.02 | 18 | 0.21 | 4 |
| $Na_2O$ | 4.55 | 74 | 3.90 | 63 | 3.70 | 60 | 4.38 | 71 |
| $K_2O$ | 4.09 | 44 | 4.29 | 46 | 5.37 | 57 | 4.63 | 49 |
| $P_2O_5$ | 0.26 | 2 | 0.15 | 1 | — | — | 0.04 | 1 |
| $qz$ | | +52 | | +142 | | +128 | | +125 |

Note: 1) tokeite (Toke Mt., Shoa); 2) basalt (Mt. Meti); 3) olivine basalt (Addis Ababa); 4) basalt (Addis Ababa); 5) basalt (Barga Riv.); 6) anorthoclase trachyte (Addis Ababa); 7) anorthoclase trachyte (Addis Ababa); 8) trachyte (Addis Ababa); 9) pantellerite (Addis Ababa); 10) comendite (Addis Ababa); 11) pantellerite (Addis Ababa).

## TABLE 15. Southwestern Ethiopia

| Analysis No. | 1 | | 2 | | 3 | | 4 | | 5 | | 6 | | 7 | |
|---|---|---|---|---|---|---|---|---|---|---|---|---|---|---|
| Component | % | M.Q. | % | M.Q. | % | M.Q. | % | M.Q. | % | M.Q. | % | M.Q. | % | M.Q. |
| $SiO_2$ | 42.04 | 700 | 42.47 | 707 | 44.29 | 738 | 44.99 | 749 | 46.18 | 769 | 48.08 | 801 | 50.24 | 837 |
| $TiO_2$ | 4.18 | 53 | 3.25 | 40 | 2.51 | 31 | 2.64 | 33 | 3.07 | 39 | 2.86 | 36 | 2.39 | 30 |
| $Al_2O_3$ | 13.66 | 134 | 14.45 | 142 | 14.76 | 145 | 16.20 | 159 | 13.12 | 128 | 17.52 | 172 | 16.12 | 158 |
| $Fe_2O_3$ | 6.43 | 40 | 3.55 | 22 | 6.36 | 40 | 6.24 | 39 | 4.26 | 27 | 3.37 | 21 | 11.14 | 69 |
| FeO | 9.51 | 132 | 9.48 | 132 | 6.12 | 85 | 5.86 | 82 | 10.33 | 144 | 7.87 | 110 | 1.43 | 19 |
| Mn | — | — | 0.79 | 11 | 0.43 | 6 | 0.41 | 6 | 0.19 | 3 | 0.05 | 1 | 0.06 | 1 |
| Mg | 5.98 | 139 | 9.42 | 233 | 9.12 | 226 | 8.89 | 221 | 5.73 | 142 | 5.58 | 139 | 2.32 | 57 |
| Ca | 11.53 | 202 | 10.40 | 185 | 11.74 | 209 | 9.51 | 169 | 8.07 | 144 | 9.17 | 163 | 5.60 | 100 |
| $Na_2O$ | 2.53 | 41 | 2.65 | 43 | 2.07 | 34 | 1.82 | 29 | 3.36 | 54 | 3.19 | 52 | 2.26 | 36 |
| $K_2O$ | 2.46 | 27 | 2.00 | 21 | 1.16 | 2 | 1.69 | 7 | 2.10 | 22 | 1.46 | 16 | 5.50 | 58 |
| *qz* | | —91 | | —44 | | —24 | | —19 | | —34 | | —24 | | —18 |

| Analysis No. | 8 | | 9 | | 10 | | 11 | | 12 | | 13 | | 14 | |
|---|---|---|---|---|---|---|---|---|---|---|---|---|---|---|
| Component | % | M.Q. | % | M.Q. | % | M.Q. | % | M.Q. | % | M.Q. | % | M.Q. | % | M.Q. |
| $SiO_2$ | 50.74 | 845 | 53.31 | 888 | 53.77 | 895 | 54.22 | 902 | 55.62 | 926 | 56.19 | 936 | 58.22 | 969 |
| $TiO_2$ | 2.27 | 29 | 1.10 | 11 | 0.57 | 7 | 0.92 | 11 | 0.30 | 4 | 1.42 | 18 | 0.44 | 5 |
| $Al_2O_3$ | 15.31 | 150 | 16.98 | 167 | 21.47 | 211 | 18.90 | 185 | 17.81 | 175 | 17.15 | 168 | 17.08 | 168 |
| $Fe_2O_3$ | 9.88 | 62 | 5.50 | 34 | 3.54 | 22 | 2.89 | 18 | 1.46 | 9 | 1.72 | 11 | 1.84 | 11 |
| FeO | 5.40 | 75 | 3.88 | 54 | 1.80 | 25 | 5.02 | 70 | 5.49 | 77 | 6.48 | 90 | 6.09 | 85 |
| Mn | 0.03 | 0 | 0.87 | 4 | — | — | 0.22 | 3 | 1.42 | 20 | 0.85 | 12 | 0.07 | 1 |
| MgO | 2.66 | 66 | 2.37 | 58 | 0.78 | 20 | 1.80 | 45 | 0.55 | 14 | 0.98 | 25 | 0.22 | 5 |
| CaO | 6.04 | 108 | 3.90 | 70 | 2.15 | 38 | 4.80 | 86 | 2.32 | 41 | 3.70 | 66 | 3.28 | 59 |
| $Na_2O$ | 3.58 | 58 | 4.91 | 79 | 8.03 | 130 | 4.11 | 66 | 6.89 | 111 | 3.86 | 62 | 6.57 | 106 |
| $K_2O$ | 2.60 | 28 | 3.21 | 34 | 4.24 | 45 | 4.17 | 45 | 7.08 | 75 | 6.91 | 73 | 4.28 | 46 |
| *qz* | | —18 | | —19 | | —62 | | —14 | | —65 | | —24 | | —27 |

| Analysis No. | 15 | | 16 | | 17 | | 18 | | 19 | | 20 | |
|---|---|---|---|---|---|---|---|---|---|---|---|---|
| Component | % | M.Q. | % | M.Q. | % | M.Q. | % | M.Q. | % | M.Q. | % | M.Q. |
| $SiO_2$ | 59.70 | 994 | 59.96 | 998 | 60.08 | 1001 | 63.17 | 1052 | 63.50 | 1057 | 64.48 | 1074 |
| $TiO_2$ | 0.20 | 3 | 0.50 | 6 | 0.10 | 1 | 0.32 | 4 | 0.57 | 7 | 0.86 | 11 |
| $Al_2O_3$ | 17.14 | 168 | 16.55 | 162 | 16.47 | 162 | 18.01 | 177 | 14.51 | 142 | 10.28 | 101 |
| $Fe_2O_3$ | 5.87 | 37 | 3.43 | 21 | 5.05 | 32 | } 1.83 | 11 | 3.87 | 24 | 3.15 | 20 |
| FeO | 1.29 | 18 | 2.55 | 36 | 1.22 | 17 | | | 2.50 | 35 | 3.88 | 54 |
| MnO | 0.17 | 2 | 0.20 | 3 | 0.16 | 2 | 0.06 | 1 | 0.12 | 1 | 0.16 | 2 |
| MgO | 0.09 | 3 | 0.55 | 14 | 0.18 | 5 | 0.65 | 16 | 0.14 | 4 | 0.78 | 20 |
| CaO | 1.00 | 18 | 1.60 | 29 | 1.45 | 26 | 1.33 | 24 | 0.48 | 9 | 2.16 | 39 |
| $Na_2O$ | 7.36 | 117 | 6.19 | 100 | 5.90 | 95 | 7.69 | 124 | 6.36 | 103 | 3.13 | 51 |
| $K_2O$ | 5.82 | 62 | 6.69 | 71 | 6.96 | 74 | 7.17 | 76 | 5.41 | 57 | 5.24 | 55 |
| *qz* | | —40 | | —31 | | —27 | | —43 | | +5 | | +80 |

Note: 1) augitite (R. Laka Kalled, Jimma); 2) olivine basalt (Yubdo, Wallega); 3) basalt (R. Dabu, Wallega); 4) basalt (50 km northeast of Gore (Sodda, Wallega); 5) basalt (R. Karta, Wallega); 6) basalt (Achevo, Jimma); 7) porphyrite (basalt) (crest of Tulu Dauku, Wallega); 8) Gore rock (Gore, Illubabor); 9) trachyandesite (Katta Yorgo, Wallega); 10) kenyte (Rillo, Wallega); 11) trachyandesite (Katta Yorgo, Wallega); 12) phonolitic trachybasalt (Belmodo, Beni Shangul); 13) phonolite (Belmodo, Beni Shangul); 14) sodic trachyte (Gore Hills, Illubabor); 15) bostonite (crest of Tulu, Ergo, 50 km south of Ghimbi, Wallega); 16) alkalic trachyte (Bube, Wallega); 17) bostonite (crest of Tulu, Ergo, 50 km south of Ghimbi, Wallega); 18) anorthite trachyte (Bun, Illubabor); 19) comendite (Baro, Southwestern Gore, Illubabor); 20) rhyolite (Baliza, Wallega?).

## TABLE 16. Somali Plateau

| Analysis No. Component | 1 % | 1 M.Q. | 2 % | 2 M.Q. | 3 % | 3 M.Q. | 4 % | 4 M.Q. | 5 % | 5 M.Q. | 6 % | 6 M.Q. | 7 % | 7 M.Q. |
|---|---|---|---|---|---|---|---|---|---|---|---|---|---|---|
| $SiO_2$ | 45.50 | 758 | 46.64 | 777 | 46.99 | 783 | 47.42 | 789 | 43.43 | 790 | 47.86 | 797 | 50.41 | 839 |
| $TiO_2$ | 2.28 | 29 | 3.83 | 48 | 2.68 | 34 | 3.48 | 44 | 0.44 | 5 | 2.39 | 30 | 2.58 | 33 |
| $Al_2O_3$ | 14.42 | 102 | 13.23 | 129 | 18.07 | 178 | 13.40 | 131 | 16.83 | 165 | 14.24 | 139 | 15.94 | 156 |
| $Fe_2O_3$ | 4.82 | 30 | 4.80 | 30 | 3.21 | 20 | 4.95 | 30 | 5.40 | 34 | 4.24 | 26 | 1.26 | 8 |
| $FeO$ | 4.03 | 56 | 10.51 | 146 | 9.32 | 129 | 9.77 | 136 | 4.31 | 60 | 8.81 | 122 | 11.53 | 160 |
| $MnO$ | 0.14 | 2 | 0.19 | 3 | 0.21 | 3 | 0.18 | 3 | 0.25 | 4 | 0.49 | 7 | 0.22 | 3 |
| $MgO$ | 5.65 | 140 | 5.96 | 149 | 5.13 | 128 | 5.97 | 148 | 1.76 | 44 | 6.56 | 163 | 5.02 | 124 |
| $CaO$ | 12.24 | 218 | 10.17 | 181 | 8.50 | 152 | 9.84 | 176 | 6.31 | 112 | 10.86 | 194 | 9.28 | 166 |
| $Na_2O$ | 3.44 | 55 | 2.58 | 42 | 3.91 | 63 | 2.79 | 45 | 7.11 | 115 | 3.37 | 54 | 2.44 | 39 |
| $K_2O$ | 2.81 | 30 | 0.59 | 6 | 1.04 | 11 | 0.40 | 4 | 3.00 | 32 | 91 | 10 | 1.12 | 12 |
| $P_2O_5$ | 0.62 | 4 | 0.47 | 4 | 53 | 4 | 0.47 | 4 | 0.66 | 5 | 0.22 | 1 | — | — |
| qz | | −37 | | | | −31 | | −16 | | −66 | | −27 | | −6 |

| Analysis No. Component | 8 % | 8 M.Q. | 9 % | 9 M.Q. | 10 % | 10 M.Q. | 11 % | 11 M.Q. | 12 % | 12 M.Q. | 13 % | 13 M.Q. | 14 % | 14 M.Q. | 15 % | 15 M.Q. |
|---|---|---|---|---|---|---|---|---|---|---|---|---|---|---|---|---|
| $SiO_2$ | 50.69 | 844 | 50.92 | 848 | 51.02 | 849 | 54.33 | 905 | 64.42 | 1072 | 65.14 | 1085 | 67.08 | 1117 | 71.98 | 1199 |
| $TiO_2$ | 0.46 | 6 | 3.07 | 39 | 0.98 | 13 | — | — | 0.18 | 3 | — | — | — | — | 0.36 | 5 |
| $Al_2O_3$ | 20.18 | 198 | 21.15 | 208 | 15.40 | 151 | 20.24 | 198 | 14.54 | 143 | 15.80 | 155 | 16.18 | 159 | 10.15 | 100 |
| $Fe_2O_3$ | 2.49 | 16 | 4.33 | 27 | 2.72 | 17 | 3.40 | 21 | 5.34 | 33 | 3.16 | 20 | 1.23 | 7 | 5.15 | 32 |
| $FeO$ | 2.16 | 30 | 8.44 | 117 | 10.21 | 142 | 2.59 | 36 | 1.53 | 21 | 1.92 | 26 | 2.90 | 40 | 1.78 | 25 |
| $MnO$ | 0.15 | 2 | 0.22 | 3 | 0.19 | 3 | 0.23 | 3 | — | — | 0.26 | 4 | 0.16 | 2 | 0.14 | 2 |
| $MgO$ | 0.73 | 18 | 5.66 | 140 | 5.04 | 125 | 0.87 | 21 | 0.34 | 8 | — | — | — | — | 0.16 | 4 |
| $CaO$ | 2.23 | 40 | 10.78 | 193 | 8.94 | 160 | 1.65 | 30 | 0.52 | 9 | 0.60 | 11 | 0.46 | 8 | 0.22 | 4 |
| $Na_2O$ | 9.34 | 151 | 2.63 | 43 | 2.73 | 44 | 8.10 | 131 | 8.03 | 130 | 8.18 | 132 | 6.83 | 110 | 4.37 | 70 |
| $K_2O$ | 4.77 | 55 | 0.77 | 8 | 1.32 | 14 | 5.19 | 55 | 4.32 | 46 | 4.54 | 48 | 4.58 | 49 | 5.07 | 54 |
| $P_2O_5$ | 0.20 | 1 | 0.43 | 3 | 0.76 | 6 | — | — | — | — | — | — | — | — | 0.14 | 1 |
| qz | | −96 | | −16 | | −8 | | −64 | | −13 | | −6 | | +26 | | +119 |

Note: 1) trachybasalt (Dodola, Arusi); 2) augite-olivine basalt (Sheik Gure, Bulo Burti, Somalia); 3) basalt (Mt. Kululu, Arusi); 4) olivine basalt (Sheik Gure, Somalia); 5) nepheline tinguaite (Dodola, Arusi); 6) augite-olivine basalt (Kuretka, southeast of Lukh Ferandi, Somalia); 7) basalt (Farso, Harar); 8) nepheline tinguaite (Mt. Ladjo, Arusi); 9) basalt (Gasara, Arusi); 10) basalt (Garamulata, Harar); 11) tinguaite (Alengo, Bardara, Somalia); 12) selvsbergite (Karsa, Harar); 13) selvsbergite (Karsa, Harar); 14) selvsbergite (Karsa, Harar); 15) pantellerite (Mt. Kakka, Arusi).

## TABLE 17. Quaternary Lavas of Ethiopia

### The Volcanic Aden Series of the Ethiopian Rift

| Analysis No. | 1 | | 2 | | 3 | | 4 | | 5 | | 6 | | 7 | |
|---|---|---|---|---|---|---|---|---|---|---|---|---|---|---|
| Component | % | M.Q. | % | M.Q. | % | M.Q. | % | M.Q. | % | M.Q. | % | M.Q. | % | M.Q. |
| $SiO_2$ | 44.42 | 739 | 45.57 | 759 | 46.30 | 771 | 46.67 | 777 | 47.08 | 801 | 48.50 | 808 | 48.84 | 813 |
| $TiO_2$ | 1.70 | 21 | — | — | — | — | — | — | 1.59 | 20 | 2.23 | 28 | 3.00 | 38 |
| $Al_2O_3$ | 14.05 | 138 | 13.07 | 128 | 13.44 | 131 | 12.64 | 124 | 17.59 | 173 | 12.32 | 121 | 11.00 | 108 |
| $Fe_2O_3$ | 6.54 | 41 | 6.72 | 42 | 4.11 | 26 | 6.13 | 38 | 10.49 | 66 | 10.87 | 68 | 13.12 | 82 |
| $FeO$ | 5.69 | 78 | 12.43 | 173 | 12.61 | 175 | 10.07 | 141 | 1.45 | 20 | 5.33 | 74 | 4.06 | 57 |
| $MnO$ | 0.17 | 2 | 0.28 | 4 | 0.22 | 3 | 0.19 | 3 | 0.27 | 4 | — | | 0.20 | 3 |
| $MgO$ | 8.44 | 209 | 2.80 | 69 | 4.42 | 109 | 5.64 | 140 | 4.11 | 102 | 5.36 | 133 | 6.01 | 149 |
| $CaO$ | 13.37 | 238 | 6.79 | 121 | 11.88 | 212 | 11.48 | 205 | 9.85 | 176 | 10.42 | 185 | 10.75 | 192 |
| $Na_2O$ | 2.32 | 37 | 2.04 | 33 | 2.13 | 34 | 1.64 | 26 | 2.72 | 44 | 3.45 | 56 | 1.56 | 25 |
| $K_2O$ | 0.80 | 9 | 3.36 | 36 | 1.94 | 20 | 2.31 | 24 | 1.83 | 19 | 1.74 | 18 | 1.64 | 16 |
| $P_2O_5$ | — | | 0.52 | 4 | 0.59 | 4 | 0.74 | 5 | 0.23 | 1 | 0.24 | 1 | — | |
| *qz* | | −26 | | −25 | | −24 | | −22 | | −33 | | −29 | | −9 |

| Analysis No. | 8 | | 9 | | 10 | | 11 | | 12 | | 13 | | 14 | |
|---|---|---|---|---|---|---|---|---|---|---|---|---|---|---|
| Component | % | M.Q. | % | M.Q. | % | M.Q. | % | M.Q. | % | M.Q. | % | M.Q. | % | M.Q. |
| $SiO_2$ | 49.05 | 817 | 49.28 | 821 | 49.28 | 821 | 49.95 | 832 | 50.19 | 836 | 51.95 | 848 | 56 55 | 942 |
| $TiO_2$ | 1.65 | 20 | 0.33 | 4 | 1.83 | 23 | 1.35 | 17 | 2.24 | 28 | 52.12 | 26 | 0.67 | 9 |
| $Al_2O_3$ | 8.50 | 83 | 8.10 | 79 | 14.36 | 141 | 12.95 | 126 | 15.34 | 150 | 14.12 | 138 | 17.74 | 174 |
| $Fe_2O_3$ | 10.80 | 68 | 8.08 | 51 | 6.38 | 40 | 7.53 | 47 | 7.43 | 46 | 5.52 | 34 | 3.76 | 24 |
| $FeO$ | 5.45 | 76 | 5.61 | 78 | 8.16 | 114 | 7.41 | 103 | 4.65 | 64 | 10.23 | 142 | 2.82 | 39 |
| $MnO$ | 0.18 | 3 | 0.32 | 4 | 0.57 | 8 | 0.30 | 4 | Traces | — | 0.35 | 5 | 0.19 | 3 |
| $MgO$ | 5.18 | 128 | 6.88 | 171 | 5.46 | 135 | 5.54 | 137 | 6.69 | 166 | 4.58 | 121 | 0.98 | 25 |
| $CaO$ | 16.70 | 298 | 17.55 | 313 | 10.56 | 188 | 11.12 | 193 | 11.32 | 201 | 9.22 | 164 | 4.39 | 79 |
| $Na_2O$ | 1.70 | 27 | 1.80 | 29 | 1.67 | 27 | 2.06 | 33 | 1.82 | 29 | 1.47 | 24 | 7.55 | 122 |
| $K_2O$ | 0.78 | 9 | 1.03 | 11 | 0.58 | 6 | 1 15 | 12 | 0.63 | 6 | 0.49 | 5 | 1.74 | 18 |
| $P_2O_5$ | — | | — | | — | | — | | 0.14 | 1 | | | 0.81 | 6 |
| *qz* | | −11 | | −16 | | −1 | | −7 | | −2 | | +10 | | −25 |

| Analysis No. | 15 | | 16 | | 17 | | 18 | | 19 | | 20 | | 21 | |
|---|---|---|---|---|---|---|---|---|---|---|---|---|---|---|
| Component | % | M.Q. | % | M.Q. | % | M.Q. | % | M.Q. | % | M.Q. | % | M.Q. | % | M.Q. |
| $SiO_2$ | 57.74 | 961 | 57.94 | 965 | 59.60 | 992 | 62.70 | 1044 | 65.68 | 1094 | 67.30 | 1121 | 68.69 | 1144 |
| $TiO_2$ | 1.06 | 14 | 0.58 | 9 | 1.19 | 15 | 0.82 | 10 | 0.55 | 7 | 0.79 | 10 | 0.40 | 5 |
| $Al_2O_3$ | 14.40 | 141 | 18.22 | 178 | 15.82 | 155 | 13.42 | 131 | 13.81 | 135 | 10.30 | 101 | 9.98 | 98 |
| $Fe_2O_3$ | 3.99 | 25 | 3.56 | 23 | 4.93 | 30 | 3.36 | 21 | 5.51 | 34 | 3.09 | 19 | 8.52 | 53 |
| $FeO$ | 6.82 | 95 | 2.46 | 34 | 3.05 | 42 | 6.37 | 89 | 0.38 | 6 | 5.99 | 83 | 0.68 | 10 |
| $MnO$ | 0.13 | 2 | 0.19 | 3 | 1.02 | 14 | 0.40 | 6 | 0.06 | 1 | 0.32 | 4 | 0.14 | 2 |
| $MgO$ | 2.09 | 52 | 0.75 | 18 | 1.08 | 27 | 0.11 | 3 | 0.14 | 4 | 0.36 | 9 | 0.03 | 1 |
| $CaO$ | 4.41 | 79 | 3.59 | 64 | 5.05 | 90 | 3.06 | 55 | 0.47 | 8 | 1.92 | 34 | 0.76 | 14 |
| $Na_2O$ | 4.49 | 73 | 8.53 | 138 | 3.36 | 54 | 5.79 | 94 | 6.26 | 101 | 5.94 | 96 | 5 84 | 94 |
| $K_2O$ | 4.38 | 47 | 2.08 | 22 | 4.41 | 47 | 3.28 | 35 | 4.72 | 50 | 3.59 | 38 | 4.23 | 45 |
| $P_2O_5$ | 0.32 | 2 | 0.25 | 2 | 0.29 | 2 | 0.10 | 1 | 0.06 | 1 | 0.12 | 1 | 0.03 | 1 |
| *qz* | | −11 | | −36 | | +20 | | +15 | | +44 | | +65 | | +92 |

| Analysis No. | 22 | | 23 | | 24 | | 25 | | 26 | | 27 | | 28 | |
|---|---|---|---|---|---|---|---|---|---|---|---|---|---|---|
| Component | % | M.Q. | % | M.Q. | % | M.Q. | % | M.Q. | % | M.Q. | % | M.Q. | % | M.Q. |
| $SiO_2$ | 69.06 | 1150 | 69.41 | 1155 | 69.86 | 1163 | 70.52 | 1174 | 70.76 | 1178 | 70.99 | 1182 | 71.30 | 1187 |
| $TiO_2$ | 0.61 | 8 | 0.20 | 3 | 0.52 | 6 | 0.60 | 8 | 0.59 | 8 | 0.38 | 5 | 0.61 | 8 |
| $Al_2O_3$ | 6.30 | 62 | 13.31 | 130 | 11.41 | 112 | 9.30 | 91 | 8.31 | 81 | 9.18 | 90 | 7.42 | 73 |
| $Fe_2O_3$ | 4.98 | 31 | 3.85 | 24 | 3.48 | 22 | 1.38 | 9 | 3.96 | 25 | 3.15 | 20 | 1.96 | 12 |
| $FeO$ | 6.26 | 87 | 0.92 | 12 | 3.35 | 46 | 5.72 | 79 | 4.61 | 64 | 4.67 | 65 | 6.79 | 93 |
| $MnO$ | 0.17 | 2 | 0.09 | 1 | 0.39 | 6 | 0.24 | 3 | 0.22 | 3 | 0.25 | 4 | 0.36 | 5 |
| $MgO$ | 0.27 | 6 | 0.17 | 4 | — | | — | | — | | 0.19 | 5 | — | |
| $CaO$ | 0.94 | 17 | 0.74 | 13 | 0.92 | 16 | 1.92 | 34 | 1.58 | 29 | 1.71 | 13 | 0.60 | 11 |
| $Na_2O$ | 7.37 | 119 | 6.13 | 99 | 5.72 | 92 | 5.98 | 97 | 5.64 | 91 | 6.07 | 98 | 6.63 | 107 |
| $K_2O$ | 3.39 | 36 | 4.53 | 48 | 4.22 | 45 | 4.01 | 42 | 4.06 | 43 | 4.24 | 45 | 4.26 | 46 |
| $P_2O_5$ | 0.07 | 1 | — | | — | | 0.06 | 1 | — | | 0.28 | 2 | — | |
| *qz* | | +107 | | +70 | | +91 | | +111 | | +122 | | +114 | | +127 |

## TABLE 17. (continued)

| Analysis No. / Component | 29 % | 29 M.Q. | 30 % | 30 M.Q. | 31 % | 31 M.Q. | 32 % | 32 M.Q. | 33 % | 33 M.Q. | 34 % | 34 M.Q. | 35 % | 35 M.Q. | 36 % | 36 M.Q. |
|---|---|---|---|---|---|---|---|---|---|---|---|---|---|---|---|---|
| $SiO_2$ | 71.63 | 1193 | 72.02 | 1199 | 72.46 | 1206 | 72.88 | 1214 | 73.28 | 1220 | 75.86 | 1263 | 78.56 | 1292 | 78.72 | 1311 |
| $TiO_2$ | 0.36 | 5 | — | — | — | — | — | — | 0.47 | 6 | 0.39 | 5 | 0.20 | 3 | 0.31 | 4 |
| $Al_2O_3$ | 11.31 | 111 | 14.44 | 141 | 13.96 | 137 | 13.86 | 136 | 8.33 | 81 | 9.75 | 96 | 11.47 | 113 | 7 78 | 76 |
| $Fe_2O_3$ | 2.41 | 15 | 1.32 | 8 | 1.65 | 10 | 1.52 | 9 | 5.72 | 36 | 2.0 | 12 | 0.64 | 4 | 4.39 | 28 |
| $FeO$ | 3.56 | 50 | 1.64 | 23 | 89 | 12 | 0.73 | 10 | 1.19 | 17 | 1.98 | 28 | 0.04 | 0 | 1.31 | 18 |
| $MnO$ | 0.20 | 3 | — | — | — | — | — | — | — | — | 0.08 | 1 | 0.03 | 0 | — | — |
| $MgO$ | 0.02 | — | 0.41 | 10 | — | — | 0.23 | 6 | 0.35 | 8 | — | — | 0.18 | 5 | — | — |
| $CaO$ | 0.41 | 7 | 0.98 | 18 | 0.81 | 14 | 1.82 | 32 | 0.72 | 13 | 0.58 | 11 | 0.19 | 4 | — | — |
| $Na_2O$ | 5.61 | 90 | 6.10 | 98 | 5.69 | 92 | 3.98 | 65 | 4.34 | 70 | 3.32 | 53 | 3.62 | 58 | 4.24 | 68 |
| $K_2O$ | 4.56 | 49 | 4.26 | 46 | 5.16 | 55 | 4.40 | 47 | 4.03 | 43 | 4.72 | 50 | 4.58 | 49 | 3.0 | 32 |
| $P_2O_5$ | 0.07 | 1 | — | — | — | — | — | — | — | — | — | — | — | — | — | — |
| qz | | +112 | | +180 | | +97 | | +146 | | +192 | | +234 | | +260 | | +293 |

Note: 1) olivine basalt (Kandala, northern Somalia); 2-4) basalt (Assab, Eritrea); 5) basalt (Bahar Assol Gulf, Eritrea); 6) porphyritic olivine basalt (Alid Volcano, Eritrea); 7-9) basalt (Beilul, Maraho, Eritrea); 10) basalt (Assab, Eritrea); 11) basalt (Bahar Assol Gulf, Eritrea); 12) porphyritic basalt (Maraho Volcano, Eritrea); 13) basalt (Geleli Volcano, Eritrea); 14) sodic trachyandesite (Asakoma, Eritrea); 15) alkalic trachyte (Adama, Shoa); 16) dankalite (Asakoma, Eritrea); 17) trachyte (Bahar Assol Gulf, Eritrea); 18) sodic olivine trachyte (Fontale Volcano, Shoa); 19) sodic trachyte (Gahare, Beiluk, Eritrea); 20) granophyre (Fontale Volcano, Shoa); 21) pantellerite (Metahara, Shoa); 22) obsidian (Fontale Volcano, Shoa); 23) pantellerite (Bahar Assol Gulf, Eritrea); 24) granophyre (Fontale Volcano, Shoa); 25) porphyritic obsidian (Fontale Volcano, Shoa); 26) aegerine pantellerite (Fontale Volcano, Shoa); 27) pantellerite (Fontale Volcano, Shoa); 28) obsidian (Fontale Volcano, Shoa); 29) pantelleritic obsidian (Moyo, Shoa); 30) amphibole felsodacite (Alid Volcano, Eritrea); 31) violet obsidian (Alid Volcano, Eritrea); 32) porphyritic rhyolite with tridymite (Badda Samoti, Eritrea); 33) pantellerite (Dabita, French Somaliland); 34) comendite (Mabla Mt., French Somaliland); 35) comendite (Bahar Assol Gulf, Eritrea); 36) pantellerite (Hol-Hol, French Seomaliland).

## TABLE 18. Volcanic Field South of Lake Rudolf

| Analysis No. / Component | 1 % | 1 M.Q. | 2 % | 2 M.Q. | 3 % | 3 M.Q. | 4 % | 4 M.Q. | 5 % | 5 M.Q. |
|---|---|---|---|---|---|---|---|---|---|---|
| $SiO_2$ | 45.27 | 754 | 45.41 | 756 | 46.25 | 770 | 50.27 | 837 | 50.58 | 843 |
| $TiO_2$ | 2.40 | 30 | 2.72 | 34 | 1.98 | 25 | 2.88 | 36 | 2.73 | 34 |
| $Al_2O_3$ | 13.91 | 136 | 13.06 | 128 | 13.28 | 130 | 18.09 | 178 | 17.20 | 169 |
| $Fe_2O_3$ | 3.41 | 21 | 2.94 | 18 | 2.55 | 16 | 1.65 | 10 | 1.22 | 7 |
| $FeO$ | 7.17 | 100 | 7.64 | 106 | 7.65 | 106 | 8.64 | 120 | 8.49 | 118 |
| $MnO$ | 0.15 | 2 | 0.28 | 4 | 0.12 | 1 | 0.14 | 2 | 0.27 | 4 |
| $MgO$ | 10.76 | 267 | 12.05 | 299 | 12.36 | 306 | 2.18 | 55 | 4.06 | 101 |
| $CaO$ | 12.68 | 227 | 12.01 | 214 | 12.62 | 225 | 8.74 | 156 | 8.05 | 144 |
| $Na_2O$ | 2.14 | 34 | 2.26 | 37 | 2.00 | 32 | 4.32 | 69 | 4.57 | 74 |
| $K_2O$ | 0.97 | 11 | 0.91 | 10 | 0.53 | 5 | 2.04 | 21 | 1.88 | 20 |
| $P_2O_5$ | 0.26 | 2 | 0.36 | 3 | 0.14 | 1 | 0.95 | 6 | 0.79 | 6 |
| qz | | -30 | | -32 | | -26 | | -23 | | -26 |

Note: 1) olivine basalt with nepheline in groundmass (Neangoil Riv. Likayo); 2) glassy basalt (lapilli) (northwest of the Likayo cone); 3) olivine basalt with rare nepheline in the groundmass (late flow at Likayo); 4) sodic trachybasalt (older flow, Teleki Volcano); 5) glassy sodic trachybasalt (late flow on north side of Teleki Volcano). This group of recent flows (basanites and olivine basalts) rests on a volcanic shield of phonolite-trachytes, which covers the basement complex. Elsewhere (west of Teleki) young lavas rest on older olivine basalts, olivine nephelinites, and phonolites. This region (south of Turkany) resembles Kenya, Tanganyika, and Uganda.

## TABLE 19.  Lavas of Lake Tana

### (described by Komuchi 1950. Probable age:
### T represents the Trapp Series, A the Aden Series)

| Analysis No. / Component | 1A % | 1A M.Q. | 2A % | 2A M.Q. | 3T % | 3T M.Q. | 4T % | 4T M.Q. | 5T % | 5T M.Q. | 6A % | 6A M.Q. | 7T % | 7T M.Q. |
|---|---|---|---|---|---|---|---|---|---|---|---|---|---|---|
| $SiO_2$ | 42.32 | 704 | 43.42 | 723 | 43.66 | 727 | 44.43 | 740 | 45.00 | 749 | 45.17 | 752 | 45.41 | 756 |
| $TiO_2$ | 2.53 | 32 | 2.11 | 26 | 2.57 | 32 | 1.16 | 15 | 3.70 | 46 | 1.05 | 14 | 1.41 | 18 |
| $Al_2O_3$ | 16.13 | 158 | 13.01 | 127 | 14.30 | 140 | 14.27 | 140 | 8.80 | 86 | 14.86 | 146 | 13.98 | 137 |
| $Fe_2O_3$ | 14.13 | 88 | 10.74 | 67 | 4.46 | 28 | 8.28 | 52 | 5.87 | 37 | 3.27 | 21 | 5.36 | 34 |
| $FeO$ | 0.21 | 3 | 3.49 | 49 | 7.41 | 103 | 3.49 | 49 | 10.11 | 141 | 6.82 | 95 | 6.62 | 92 |
| $MnO$ | 0.17 | 2 | 0.13 | 2 | 0.14 | 2 | 0.15 | 2 | 0.16 | 2 | 0.12 | 1 | 0.14 | 2 |
| $MgO$ | 6.99 | 174 | 9.81 | 243 | 8.89 | 221 | 8.81 | 218 | 10.89 | 270 | 12.53 | 311 | 8.93 | 222 |
| $CaO$ | 8.90 | 159 | 10.21 | 182 | 9.80 | 175 | 9.60 | 171 | 12.05 | 215 | 11.06 | 197 | 9.38 | 168 |
| $Na_2O$ | 1.40 | 23 | 4.31 | 69 | 3.02 | 48 | 1.51 | 24 | 1.87 | 30 | 2.38 | 39 | 1.29 | 37 |
| $K_2O$ | 1.25 | 14 | 1.29 | 14 | 3 07 | 33 | 0.31 | 3 | 0.73 | 8 | 2 05 | 22 | 0.96 | 11 |
| $qz$ |  | −21 |  | −52 |  | −48 |  | −11 |  | −25 |  | −41 |  | −23 |

| Analysis No. / Component | 8T % | 8T M.Q. | 9T % | 9T M.Q. | 10A % | 10A M.Q. | 11A % | 11A M.Q. | 12A % | 12A M.Q. | 13T % | 13T M.Q. |
|---|---|---|---|---|---|---|---|---|---|---|---|---|
| $SiO_2$ | 45.57 | 759 | 45.64 | 760 | 45.77 | 762 | 45.78 | 763 | 46.23 | 770 | 46.29 | 771 |
| $TiO_2$ | 1.64 | 20 | 3.07 | 39 | 1.58 | 20 | 1.13 | 14 | 1.63 | 20 | 1.61 | 20 |
| $Al_2O_3$ | 16.25 | 160 | 16.02 | 157 | 16.05 | 158 | 11.16 | 110 | 14.34 | 140 | 13.93 | 136 |
| $Fe_2O_3$ | 3.21 | 20 | 6.65 | 42 | 3.63 | 23 | 4.95 | 30 | 4.95 | 30 | 6.87 | 43 |
| $FeO$ | 6.91 | 96 | 7.54 | 105 | 7.12 | 99 | 7.47 | 104 | 8.47 | 118 | 6.20 | 86 |
| $MnO$ | 0.16 | 2 | 0.18 | 3 | 0.19 | 3 | 0.08 | 1 | 0.18 | 3 | 0.18 | 3 |
| $MgO$ | 9.19 | 228 | 4.76 | 118 | 8.34 | 207 | 16.24 | 403 | 9.20 | 228 | 6.19 | 154 |
| $CaO$ | 10.76 | 192 | 7.39 | 132 | 10.29 | 184 | 8.43 | 151 | 9.76 | 174 | 9.72 | 173 |
| $Na_2O$ | 2.24 | 36 | 4.24 | 68 | 3.65 | 57 | 1.01 | 16 | 2.33 | 38 | 2.97 | 48 |
| $K_2O$ | 1.17 | 13 | 2.38 | 25 | 2.21 | 23 | 1.23 | 13 | 0.73 | 8 | 2.60 | 28 |
| $qz$ |  | −27 |  | −44 |  | −43 |  | −25 |  | −24 |  | −35 |

| Analysis No. / Component | 14A % | 14A M.Q. | 15T % | 15T M.Q. | 16A % | 16A M.Q. | 17A % | 17A M.Q. | 18 % | 18 M.Q. | 19 % | 19 M.Q. |
|---|---|---|---|---|---|---|---|---|---|---|---|---|
| $SiO_2$ | 46.40 | 773 | 46.47 | 774 | 47.10 | 784 | 48.33 | 805 | 48.43 | 807 | 48.54 | 808 |
| $TiO_2$ | 2.06 | 26 | 1.87 | 24 | 0.86 | 11 | 0.98 | 13 | 2.22 | 28 | 1.14 | 14 |
| $Al_2O_3$ | 16.89 | 162 | 15.17 | 149 | 14.42 | 141 | 17.58 | 153 | 18.62 | 182 | 16.83 | 165 |
| $Fe_2O_3$ | 6.00 | 38 | 4.67 | 29 | 4.21 | 26 | 2.38 | 15 | 6.63 | 41 | 3.53 | 22 |
| $FeO$ | 5.48 | 77 | 8.40 | 117 | 7.90 | 110 | 7.90 | 110 | 3.98 | 56 | 7.83 | 109 |
| $MnO$ | 0.14 | 2 | 0.16 | 2 | 0.17 | 2 | 0.11 | 1 | 0.25 | 4 | 0.13 | 2 |
| $MgO$ | 5.76 | 143 | 7.50 | 186 | 10.46 | 259 | 7.75 | 192 | 3.35 | 83 | 7.05 | 175 |
| $CaO$ | 9.48 | 169 | 9.45 | 168 | 9.12 | 162 | 11.49 | 187 | 6.90 | 123 | 10.74 | 192 |
| $Na_2O$ | 4.12 | 66 | 3.46 | 56 | 2.95 | 48 | 2.90 | 47 | 5.36 | 86 | 2.08 | 34 |
| $K_2O$ | 1.54 | 16 | 1.15 | 12 | 1.19 | 13 | 0,13 | 1 | 2.12 | 22 | 1.09 | 12 |
| $qz$ |  | −40 |  | −33 |  | −31 |  | −15 |  | −37 |  | −15 |

## TABLE 19. (continued)

| Analysis No. / Component | 20T % | 20T M.Q. | 21 % | 21 M.Q. | 22 % | 22 M.Q. | 23 % | 23 M.Q. | 24 % | 24 M.Q. | 25 % | 25 M.Q. | 26 % | 26 M.Q. |
|---|---|---|---|---|---|---|---|---|---|---|---|---|---|---|
| $SiO_2$ | 49.06 | 817 | 53.93 | 898 | 55.04 | 916 | 56.56 | 942 | 60.57 | 1008 | 61.85 | 1030 | 62.27 | 1037 |
| $TiO_2$ | 2.57 | 32 | 1.24 | 15 | 0.50 | 6 | 0.67 | 9 | 0.51 | 6 | 0.11 | 1 | 1.08 | 13 |
| $Al_2O_3$ | 15.40 | 151 | 17.82 | 175 | 19.57 | 192 | 20.80 | 204 | 17.45 | 172 | 17.86 | 176 | 13.02 | 127 |
| $Fe_2O_3$ | 4.34 | 27 | 5.08 | 32 | 3.63 | 23 | 2.63 | 16 | 4.61 | 29 | 4.61 | 29 | 2.55 | 16 |
| $FeO$ | 6.34 | 88 | 3.70 | 51 | 0.93 | 13 | 1.07 | 15 | 0.50 | 7 | 0.49 | 7 | 6.75 | 94 |
| $MnO$ | 0.15 | 2 | 10 | 1 | 0.22 | 3 | 0.25 | 4 | 0.06 | 1 | 0.11 | 1 | 0.16 | 2 |
| $MgO$ | 5.81 | 144 | 2.43 | 60 | 0.44 | 11 | — | — | 0.84 | 21 | 0.61 | 15 | 0.72 | 17 |
| $CaO$ | 8.84 | 157 | 4.87 | 87 | 1.78 | 34 | 2.61 | 36 | 1.88 | 34 | 0.54 | 10 | 4.11 | 73 |
| $Na_2O$ | 3.62 | 58 | 4.66 | 75 | 8.86 | 143 | 9.33 | 151 | 5.29 | 85 | 5.66 | 91 | 4.02 | 65 |
| $K_2O$ | 1.83 | 19 | 4.20 | 45 | 7.62 | 81 | 5.05 | 54 | 5.94 | 63 | 5.42 | 57 | 3.49 | 37 |
| $qz$ | | −24 | | −25 | | −78 | | −74 | | −6 | | −6 | | +40 |

| Analysis No. / Component | 27T % | 27T M.Q. | 28? % | 28? M.Q. | 29T % | 29T M.Q. | 30? % | 30? M.Q. | 31a % | 31a M.Q. | 32T % | 32T M.Q. | 33T % | 33T M.Q. |
|---|---|---|---|---|---|---|---|---|---|---|---|---|---|---|
| $SiO_2$ | 66.19 | 1102 | 68.55 | 1141 | 70.63 | 1176 | 72.07 | 1200 | 73.87 | 1230 | 75.25 | 1251 | 75.50 | 1262 |
| $TiO_2$ | 0.91 | 11 | 0.67 | 9 | 0.24 | 3 | 0.36 | 5 | 26 | 4 | 0.43 | 5 | 0.39 | 5 |
| $Al_2O_3$ | 12.95 | 126 | 13.82 | 135 | 11.95 | 118 | 13.97 | 127 | 12.56 | 124 | 10.58 | 104 | 9.56 | 94 |
| $Fe_2O_3$ | 5.10 | 32 | 5.73 | 36 | 4.91 | 30 | 1.02 | 6 | 3.28 | 21 | 2.60 | 16 | 6.18 | 39 |
| $FeO$ | 0.99 | 14 | 0.30 | 4 | 0.30 | 4 | 1.00 | 14 | 0.36 | 5 | 1.64 | 23 | 0.20 | 3 |
| $MnO$ | 0.07 | 1 | 0.03 | 0 | — | — | 0.08 | 1 | — | — | 0.04 | — | — | — |
| $MgO$ | 0.55 | 14 | 0.41 | 10 | 0.10 | 3 | 0.38 | 10 | 0.79 | 20 | — | — | 0.15 | 4 |
| $CaO$ | 3.01 | 54 | 0.68 | 13 | 0.25 | 4 | 1.25 | 22 | 1.59 | 29 | 0.43 | 8 | 0.36 | 6 |
| $Na_2O$ | 3.81 | 61 | 3.58 | 58 | 4.78 | 77 | 3.80 | 61 | 2.45 | 40 | 3.76 | 61 | 3.20 | 52 |
| $K_2O$ | 4.24 | 45 | 3.20 | 34 | 5.31 | 56 | 3.70 | 39 | 4.60 | 49 | 5.06 | 54 | 3.08 | 33 |
| $qz$ | | +78 | | +136 | | +115 | | +177 | | +182 | | +193 | | +241 |

Note: 1) glassy basalt (Devani, Bahar Dar); 2) glassy basalt (Daga Is., Lake Tana); 3) porphyritic basalt (Goanch); 4) porphyritic basalt (Gondar); 5) porphyritic olivine basalt (Araii Libo ?); 6) porphyritic basalt (Zuma, Dangila); 7) doleritic basalt (Kertadi Michael?); 8) porphyritic basalt (Mani); 9) nonporphyritic basalt (Gidor); 10) glassy basalt (Enyabara); 11) altered hornblendite (Selki Mt., Simen); 12) porphyritic glassy basalt (Moskha ?); 13) porphyritic basalt (Libo Georgis, Ifag); 14) porphyritic basalt (Ismala Georgis); 15) porphyritic basalt (Umbera Ezus ?); 16) picritic basalt (Bahar Dar); 17) doleritic basalt (Kobetoa, Dangila); 18) nonporphyritic basalt (Ezus Tabor ?); 19) doleritic basalt (Kidane Meret); 20) porphyritic basalt (Konsela); 21) trachydolerite (Enyabara); 22) nepheline phonolite (Enyabara); 23) nepheline phonolite (Enyabara); 24) nepheline trachyte (Amba Libo, Ifag); 25) nepheline trachyte (Amba Libo, Ifag); 26) sodic trachyte (Vashit ?, Gondar); 27) trachydolerite (Amba, Georgis); 28) rhyolite (Machuda ?); 29) quartz porphyry (southwest of Ifag); 30) rhyolite (Kicha ?); 31) rhyolite (Anguasek); 32) rhyolite (Mai Shaha, Simen); 33) rhyolite (Keddis Arit, Simen).

## TABLE 20. Southern Lake Kivu Field (Bowen [1938]; Holmes [1940])

| Analysis No. | 1 | | 2 | | 3 | | 4 | | 5 | |
|---|---|---|---|---|---|---|---|---|---|---|
| Component | % | M.Q. | % | M.Q. | % | M.Q. | % | M.Q. | % | M.Q. |
| $SiO_2$ | 49.84 | 830 | 50.19 | 836 | 43.56 | 725 | 57.45 | 957 | 60.25 | 1003 |
| $TiO_2$ | 0.71 | 9 | 0.96 | 12 | 2.82 | 35 | 1.26 | 16 | 0.84 | 10 |
| $Al_2O_3$ | 14.13 | 138 | 14.56 | 143 | 15.04 | 147 | 19.46 | 191 | 19.55 | 192 |
| $Fe_2O_3$ | 4.19 | 26 | 3.19 | 20 | 2.74 | 17 | 4.65 | 29 | 3.60 | 23 |
| FeO | 9.63 | 134 | 8.05 | 112 | 8.84 | 123 | 0.91 | 12 | 0.60 | 8 |
| MnO | — | — | 0.13 | 1 | 0.23 | 2 | 0.12 | 1 | 0.16 | 2 |
| MgO | 7.78 | 194 | 7.22 | 179 | 8.01 | 198 | 1.18 | 30 | 0.50 | 12 |
| CaO | 10.07 | 179 | 9.88 | 177 | 10.95 | 195 | 2.99 | 54 | 0.61 | 11 |
| $Na_2O$ | 1.76 | 28 | 2.51 | 40 | 3.28 | 53 | 5.47 | 88 | 5.68 | 92 |
| $K_2O$ | 0.61 | 6 | 0.54 | 5 | 0.92 | 10 | 4.11 | 44 | 5.33 | 56 |
| *qz* | | —5 | | —6 | | —38 | | —10 | | —2 |

Note: 1) basalt, 15 km northwest of Mwenga; 2) olivine basalt, Mukaba R. south of Kahusi; 3) essexitic basalt, at Kahusi; 4) trachyte, 5 km south of Costersmanville; 5) trachyte, south end of Lake Kivu.

## TABLE 21. Isle of Mull (Northwestern Scotland) (Bailey, Thomas, and others [1924]). Plateau Type Magma (Stage I)

| Analysis No. | 1 | | 2 | | 3 | | 4 | | 5 | | 6 | | 7 | | 8 | |
|---|---|---|---|---|---|---|---|---|---|---|---|---|---|---|---|---|
| Component | % | M.Q. | % | M.Q. | % | M.Q. | % | M.Q. | % | M.Q. | % | M.Q. | % | M.Q. | % | M.Q. |
| $SiO_2$ | 43.94 | 732 | 45.24 | 753 | 45.37 | 755 | 45.48 | 758 | 45.52 | 758 | 46.46 | 774 | 46.61 | 776 | 47.64 | 793 |
| $TiO_2$ | 2.45 | 26 | 2.26 | 24 | 2.87 | 31 | 3.48 | 37 | 2.85 | 31 | 2.07 | 22 | 1.81 | 19 | 1.27 | 14 |
| $Al_2O_3$ | 14.03 | 137 | 15.63 | 157 | 15.16 | 149 | 15.66 | 154 | 14.30 | 137 | 15.48 | 152 | 15.32 | 150 | 14.15 | 139 |
| $Fe_2O_3$ | 1.95 | 12 | 5.56 | 35 | 3.38 | 21 | 3.64 | 23 | 3.43 | 21 | 3.63 | 23 | 3.49 | 48 | 5.15 | 33 |
| FeO | 11.65 | 162 | 7.19 | 100 | 11.58 | 161 | 10.56 | 147 | 9.00 | 125 | 10.23 | 142 | 7.71 | 107 | 7.96 | 111 |
| MnO | 0.32 | 4 | 0.23 | 3 | 0.31 | 4 | 0.20 | 3 | 0.19 | 3 | 0.48 | 7 | 0.13 | 1 | 7.38 | 183 |
| MgO | 10.46 | 259 | 7.82 | 194 | 6.72 | 166 | 6.99 | 174 | 10.65 | 264 | 6.80 | 169 | 8.66 | 215 | 11.71 | 209 |
| CaO | 8.99 | 161 | 9.38 | 168 | 8.11 | 144 | 8.24 | 147 | 9.54 | 170 | 9.05 | 162 | 10.08 | 180 | 2.38 | 39 |
| $Na_2O$ | 2.68 | 44 | 2.01 | 32 | 2.90 | 47 | 2.68 | 44 | 2.21 | 35 | 3.01 | 48 | 2.43 | 39 | 0.71 | 7 |
| $K_2O$ | 0.33 | 3 | 0.72 | 7 | 0.44 | 4 | 0.49 | 5 | 0.42 | 4 | 0.68 | 7 | 0.67 | 7 | | |
| *qz* | | —63 | | —18 | | —23 | | —22 | | —23 | | —24 | | —26 | | —19 |

### Nonporphyritic Type of Central Intrusion. Olivine Poor (Stage II)

| Analysis No. | 9 | | 10 | | 11 | | 12 | | 13 | | 14 | | 15 | | 16 | | 17 | | 18 | |
|---|---|---|---|---|---|---|---|---|---|---|---|---|---|---|---|---|---|---|---|---|
| Component | % | M.Q. | % | M.Q. | % | M.Q. | % | M.Q. | % | M.Q. | % | M.Q. | % | M.Q. | % | M.Q. | % | M.Q. | % | M.Q. |
| $SiO_2$ | 47.35 | 785 | 47.80 | 796 | 49.76 | 829 | 52.13 | 868 | 50.54 | 842 | 53.78 | 896 | 51.53 | 858 | 51.63 | 860 | 52.16 | 869 | 53.97 | 898 |
| $TiO_2$ | 1.75 | 19 | — | — | 0.94 | 10 | — | — | 2.80 | 30 | 2.28 | 24 | 1.57 | 17 | 2.00 | 21 | 3.25 | 35 | 1.24 | 13 |
| $Al_2O_3$ | 13.90 | 136 | 14.80 | 145 | 14.42 | 141 | 14.87 | 146 | 12.86 | 126 | 12.69 | 125 | 11.05 | 109 | 11.77 | 116 | 11.95 | 118 | 14.65 | 14 |
| $Fe_2O_3$ | 5.87 | 37 | — | — | 3.95 | 25 | — | — | 4.13 | 26 | 3.44 | 21 | 2.73 | 17 | 3.23 | 20 | 4.86 | 30 | 3.62 | 24 |
| FeO | 8.96 | 125 | 13.08 | 182 | 7.77 | 108 | 11.40 | 159 | 8.75 | 122 | 8.94 | 124 | 10.98 | 153 | 10.47 | 146 | 9.92 | 138 | 6.32 | 83 |
| MnO | 0.23 | 3 | 0.09 | 1 | 0.20 | 3 | 0.32 | 4 | 0.32 | 4 | 0.53 | 7 | 0.45 | 7 | 0.35 | 6 | 0.18 | 3 | 0.30 | 8 |
| MgO | 5.97 | 148 | 6.84 | 170 | 5.30 | 132 | 6.46 | 160 | 4.63 | 115 | 2.58 | 64 | 5.21 | 129 | 5.02 | 124 | 3.77 | 93 | 4.49 | 114 |
| CaO | 10.65 | 190 | 12.89 | 230 | 10.22 | 182 | 10.56 | 188 | 8.74 | 156 | 6.36 | 113 | 9.68 | 170 | 9.34 | 167 | 7.14 | 127 | 7.98 | 142 |
| $Na_2O$ | 2.73 | 44 | 2.48 | 40 | 2.49 | 40 | 2.60 | 42 | 2.89 | 47 | 2.74 | 44 | 3.48 | 56 | 2.90 | 47 | 2.36 | 38 | 2.54 | 43 |
| $K_2O$ | 0.54 | 5 | 0.86 | 10 | 1.83 | 19 | 0.69 | 7 | 1.43 | 15 | 2.27 | 24 | 0.86 | 10 | 0.91 | 10 | 1.74 | 18 | 1.52 | 11 |
| *qz* | | —19 | | —23 | | —13 | | —5 | | —7 | | +8 | | —11 | | —4 | | +8 | | +16 |

## TABLE 21. (continued)
### Stage III

| Component | 19 % | 19 M.Q. | 20 % | 20 M.Q. | 21 % | 21 M.Q. | 22 % | 22 M.Q. | 23 % | 23 M.Q. | 24 % | 24 M.Q. |
|---|---|---|---|---|---|---|---|---|---|---|---|---|
| $SiO_2$ | 55.82 | 929 | 59.21 | 986 | 61.69 | 1027 | 62.37 | 1038 | 64.13 | 1068 | 66.27 | 1104 |
| $TiO_2$ | 1.62 | 17 | 1.06 | 12 | 1.00 | 11 | 1.06 | 12 | 1.19 | 13 | 0.87 | 10 |
| $Al_2O_3$ | 11.47 | 113 | 14.06 | 138 | 14.43 | 141 | 12.04 | 118 | 13.15 | 129 | 11.92 | 117 |
| $Fe_2O_3$ | 3.68 | 23 | 2.66 | 17 | 1.23 | 7 | 1.87 | 12 | 1.08 | 7 | 3.09 | 19 |
| $FeO$ | 7.66 | 107 | 4.87 | 68 | 5.86 | 82 | 5.81 | 81 | 6.31 | 88 | 3.18 | 44 |
| $MnO$ | 0.40 | 6 | 0.24 | 3 | 0.30 | 4 | 0.24 | 3 | 0.27 | 4 | 0.31 | 4 |
| $MgO$ | 4.08 | 101 | 3.71 | 92 | 2.81 | 69 | 0.97 | 24 | 1.08 | 27 | 1.44 | 36 |
| $CaO$ | 7.88 | 141 | 5.95 | 106 | 4.91 | 88 | 3.51 | 62 | 3.62 | 64 | 3.30 | 59 |
| $Na_2O$ | 2.58 | 42 | 2.06 | 33 | 3.20 | 52 | 3.47 | 56 | 3.64 | 59 | 2.89 | 47 |
| $K_2O$ | 2.00 | 21 | 2.83 | 30 | 1.72 | 18 | 2.34 | 24 | 2.32 | 24 | 4.03 | 43 |
| $qz$ | | +17 | | +46 | | +59 | | +83 | | +80 | | +92 |

### Stage IV Ring Structures
### (Silicic magma)

| Component | 25 % | 25 M.Q. | 26 % | 26 M.Q. | 27 % | 27 M.Q. | 28 % | 28 M.Q. | 29 % | 29 M.Q. |
|---|---|---|---|---|---|---|---|---|---|---|
| $SiO_2$ | 70.70 | 1177 | 71.30 | 1187 | 72.66 | 1210 | 73.12 | 1217 | 73.32 | 1220 |
| $TiO_2$ | 1.27 | 3 | 0.58 | 6 | 0.34 | 3 | 0.39 | 4 | 0.51 | 5 |
| $Al_2O_3$ | 11.78 | 116 | 11.24 | 118 | 12.00 | 118 | 12.44 | 122 | 12.25 | 121 |
| $Fe_2O_3$ | 1.32 | 8 | 1.80 | 11 | 2.03 | 12 | 2.09 | 13 | 2.77 | 17 |
| $FeO$ | 3.45 | 48 | 2.84 | 39 | 2.04 | 28 | 1.65 | 23 | 2.20 | 81 |
| $MnO$ | 0.07 | 1 | 0.31 | 4 | 0.18 | 3 | 0.17 | 2 | 0.12 | 1 |
| $MgO$ | 0.53 | 13 | 0.61 | 15 | 0.07 | 2 | 0.14 | 4 | 0.11 | 3 |
| $CaO$ | 1.30 | 23 | 1.56 | 28 | 1.25 | 22 | 0.88 | 16 | 1.65 | 30 |
| $Na_2O$ | 2.48 | 40 | 3.44 | 55 | 3.26 | 53 | 3.90 | 63 | 3.92 | 63 |
| $K_2O$ | 4.71 | 50 | 4.66 | 50 | 5.26 | 56 | 4.67 | 50 | 2.34 | 24 |
| $qz$ | | +166 | | +138 | | +153 | | +148 | | +184 |

### Field of Plateau and Dike Magmas

| Component | Allivalite 30 % | 30 M.Q. | Eucrite 31 % | 31 M.Q. | Dolerite 32 % | 32 M.Q. | Dolerite 33 % | 33 M.Q. | Olivine gabbro 34 % | 34 M.Q. | Gabbro 35 % | 35 M.Q. | Gabbro 36 % | 36 M.Q. |
|---|---|---|---|---|---|---|---|---|---|---|---|---|---|---|
| $SiO_2$ | 42.20 | 703 | 46.66 | 777 | 48.05 | 800 | 45.54 | 758 | 46.39 | 773 | 47.28 | 788 | 48.34 | 805 |
| $TiO_2$ | 0.09 | 1 | 0.47 | 6 | 0.49 | 6 | 1.06 | 12 | 0.26 | 3 | 0.28 | 3 | 0.95 | 11 |
| $Al_2O_3$ | 17.56 | 173 | 16.71 | 164 | 15.35 | 151 | 23.39 | 229 | 26.34 | 458 | 21.11 | 207 | 20.10 | 197 |
| $Fe_2O_3$ | 1.20 | 7 | 2.69 | 17 | 1.86 | 12 | 1.98 | 12 | 2.02 | 12 | 3.52 | 22 | 1.97 | 12 |
| $FeO$ | 6.33 | 88 | 5.87 | 82 | 7.53 | 105 | 6.98 | 97 | 3.15 | 44 | 3.91 | 54 | 6.62 | 92 |
| $MnO$ | 0.18 | 3 | 0.12 | 1 | 0.28 | 4 | 0.27 | 4 | 0.14 | 2 | 0.15 | 2 | 0.32 | 5 |
| $MgO$ | 20.38 | 507 | 12.36 | 306 | 12.53 | 311 | 4.60 | 114 | 4.82 | 119 | 8.06 | 200 | 5.49 | 136 |
| $CaO$ | 9.61 | 171 | 12.57 | 224 | 11.02 | 196 | 11.82 | 210 | 15.29 | 273 | 13.42 | 239 | 13.16 | 235 |
| $Na_2O$ | 1.11 | 18 | 1.16 | 19 | 1.26 | 20 | 2.50 | 40 | 1.63 | 26 | 1.52 | 24 | 1.66 | 27 |
| $K_2O$ | 0.11 | 1 | 0.27 | 3 | 0.19 | 2 | 0.44 | 4 | 0.20 | 2 | 0.29 | 3 | 0.98 | 11 |
| $qz$ | | −36 | | −16 | | −12 | | −19 | | −11 | | −12 | | −10 |

| Component | Basalt 37 % | 37 M.Q. | Basalt 38 % | 38 M.Q. | Basalt 39 % | 39 M.Q. | Mugearite 40 % | 40 M.Q. | Mugearite 41 % | 41 M.Q. | Mugearite 42 % | 42 M.Q. | Mugearite 43 % | 43 M.Q. | Syenite 44 % | 44 M.Q. | Trachyte 45 % | 45 M.Q. | Trachyte 46 % | 46 M.Q. |
|---|---|---|---|---|---|---|---|---|---|---|---|---|---|---|---|---|---|---|---|---|
| $SiO_2$ | 47.24 | 787 | 47.49 | 791 | 48.51 | 808 | 49.24 | 820 | 49.92 | 831 | 50.70 | 844 | 55.76 | 928 | 58.81 | 979 | 60.13 | 1101 | 63.12 | 1051 |
| $TiO_2$ | 1.46 | 16 | 0.93 | 10 | 1.46 | 16 | 1.84 | 19 | 2.04 | 21 | 1.89 | 20 | 1.78 | 19 | 0.76 | 8 | 0.73 | 7 | 0.51 | 5 |
| $Al_2O_3$ | 18.55 | 182 | 21.46 | 211 | 19.44 | 190 | 15.84 | 155 | 12.83 | 125 | 14.60 | 143 | 16.55 | 163 | 14.81 | 145 | 16.53 | 162 | 15.44 | 151 |
| $Fe_2O_3$ | 6.02 | 38 | 1.72 | 11 | 5.66 | 36 | 6.09 | 38 | 6.96 | 44 | 5.23 | 33 | 3.10 | 19 | 4.58 | 29 | 2.86 | 18 | 1.73 | 11 |
| $FeO$ | 4.06 | 57 | 4.80 | 67 | 4.00 | 56 | 7.18 | 100 | 6.21 | 86 | 7.68 | 107 | 6.02 | 83 | 4.21 | 58 | 2.55 | 36 | 3.53 | 49 |
| $MnO$ | 0.31 | 5 | 0.15 | 2 | 0.23 | 4 | 0.29 | 4 | 0.52 | 7 | 0.42 | 6 | 0.22 | 3 | 0.27 | 4 | 0.48 | 7 | 0.27 | 4 |
| $MgO$ | 5.24 | 130 | 4.59 | 114 | 5.12 | 127 | 3.02 | 74 | 3.78 | 94 | 4.15 | 103 | 1.08 | 27 | 0.80 | 20 | 1.20 | 30 | 0.62 | 5 |
| $CaO$ | 11.72 | 209 | 13.24 | 236 | 12.03 | 214 | 5.28 | 95 | 7.25 | 129 | 7.20 | 128 | 3.23 | 58 | 2.33 | 42 | 1.61 | 29 | 1.31 | 23 |
| $Na_2O$ | 2.42 | 39 | 2.17 | 35 | 2.53 | 41 | 5.21 | 84 | 3.72 | 60 | 3.71 | 60 | 6.28 | 102 | 5.60 | 90 | 8.06 | 130 | 5.81 | 94 |
| $K_2O$ | 0.15 | 2 | 0.42 | 4 | 0.25 | 3 | 2.10 | 22 | 1.73 | 19 | 1.33 | 14 | 3.87 | 41 | 4.96 | 53 | 3.99 | 42 | 5.36 | 57 |
| $qz$ | | −11 | | −8 | | −16 | | −35 | | −15 | | −13 | | −31 | | −16 | | −27 | | +8 |

## TABLE 21. (continued)

| Component | At Cruach Choireadail | | | | | | | | At Coire an t' Salen | | | |
|---|---|---|---|---|---|---|---|---|---|---|---|---|
| | 47 | | 48 | | 49 | | 50 | | 51 | | 52 | |
| | % | M.Q. | % | M.Q. | % | M.Q. | % | M.Q. | % | M.Q. | % | M.Q. |
| $SiO_2$ | 49.90 | 831 | 51.32 | 854 | 56.22 | 936 | 68.12 | 1134 | 50.04 | 833 | 57.18 | 952 |
| $TiO_2$ | 2.56 | 28 | 0.98 | 11 | 2.74 | 29 | 1.26 | 14 | 2.56 | 28 | 3.25 | 35 |
| $Al_2O_3$ | 12.70 | 125 | 13.96 | 137 | 12.45 | 123 | 13.08 | 128 | 13.32 | 130 | 10.75 | 106 |
| $Fe_2O_3$ | 4.20 | 26 | 2.48 | 16 | 3.09 | 19 | 1.02 | 6 | 4.71 | 29 | 4.96 | 31 |
| FeO | 7.88 | 110 | 7.10 | 99 | 7.58 | 106 | 3.26 | 46 | 8.07 | 112 | 6.24 | 87 |
| MnO | 0.36 | 5 | 0.34 | 5 | 0.43 | 6 | 0.39 | 6 | 0.33 | 4 | 0.32 | 4 |
| MgO | 5.58 | 146 | 5.78 | 144 | 2.78 | 69 | 0.71 | 17 | 5.01 | 124 | 2.15 | 53 |
| CaO | 10.39 | 185 | 11.51 | 205 | 5.93 | 106 | 1.81 | 32 | 10.02 | 178 | 5.73 | 102 |
| $Na_2O$ | 2.86 | 46 | 3.50 | 50 | 3.82 | 61 | 4.15 | 67 | 3.28 | 53 | 4.62 | 74 |
| $K_2O$ | 0.95 | 11 | 1.16 | 13 | 2.67 | 29 | 4.47 | 48 | 1.08 | 12 | 2.67 | 29 |
| qz | | −11 | | −9 | | +7 | | +88 | | −16 | | +4 |

Note: 1) crinanite (dike, Sloc na Sgarth, Jura); 2) dolerite (sill, Ben Lee, Skye); 3) basalt (lava, Pennycross House, Mull); 4) basalt (lava, Lochaline Pier, Morven); 5) basalt (lava, Lochaline, Morven); 6) basalt (lava, Orval, Rum); 7) basalt (lava, Drynoch, Skye); 8) dolerite (dike, Cuillins, Skye); 9) tholeiite (dike, Kintallen, Mull); 10) basalt (lava, Staffa); 11) basalt (lava, Rudha na h' Uamha, Mull); 12) basalt (lava, Giant's Causeway, Ireland); 13) basalt (lava, Monadh Bheag, Mull); 14) basalt (lava, Loch Ba, Mull); 15) basalt (dike, Arla, Mull); 16) basalt (dike, Kintallen, Mull); 17) quartz dolerite (Cruach an Dearg, Mull); 18) basalt (margin of sill, Rudh' a Chromain, Mull); 19) craignurite (Dubh Choir, Mull); 20) dellenite (sill, Leuldach, Glach an t' Smehde, Mull); 21) glassy leidleite (sill, Leuldach, Glach an t' Smehda, Mull); 22-23) inninmorite ("layer," Tom a Choilich, Mull); 24) craignurite (Allt an Dubh Goir, Mull); 25) felsite (sill, Coire Buidhe, Mull); 26) granophyre (Craignure Bay, Mull); 27) rhyolite (ring dike, Beinn a' Ghraig, Mull); 28) granophyre (dike, Loch Ba, Mull); 29) granophyre (ring dike, Knock, Mull); 30) allivalite (intrusion, Allival, Inverness-shire); 31) eucrite (intrusion, Ben Buie, Mull); 32) eucrite (intrusion, Allt Mor na h' Uamha); 33) dolerite (sill, Coire Buidhe, Mull); 34-35) olivine gabbro (intrusion, Skye); 36) olivine gabbro (intrusion, Beinn na Duatharach, Mull); 37) basalt (lava, Derrynaculen, Mull); 38) basalt (pillow lava, Cruach Choireadail, Mull); 39) basalt (lava, Cruach Doin, Mull); 40) mugearite (sill, Druim na Criche, Skye); 41) mugearite (sill, Eiden a Bheird, Canna); 42) mugearite (sill, Fionchra, Rum); 43) mugearite (lava, Kein Loch, Mull); 44) syenite (intrusion, Gamnach Mor, Mull); 45) trachytic dome (Salen-Tobermory, Mull); 46) trachytic dome (Salen, Mull); 47) quartz gabbro (Mull); 48) quartz gabbro (Mull); 49) rock near craignurite (Mull); 50) rock near silicic craignurite (Mull); 51) quartz gabbro (Mull); 52) rock near craignurite (Mull).

## TABLE 22. Antrim Plateau (Northern Ireland)
### (Patterson and Swaine [1955]; Patterson [1952]) Middle Series

| Analysis No. Component | 1 | | 2 | | 3 | | 4 | | 5 | | 6 | | 7 | | 8 | | 9 | |
|---|---|---|---|---|---|---|---|---|---|---|---|---|---|---|---|---|---|---|
| | % | M.Q. | % | M.Q. | % | M.Q. | % | M.Q. | % | M.Q. | % | M.Q. | % | M.Q. | % | M.Q. | % | M.Q. |
| $SiO_2$ | 51.6 | 859 | 52.0 | 866 | 51.9 | 848 | 51.6 | 859 | 51.6 | 859 | 52.6 | 876 | 54.9 | 914 | 54.8 | 912 | 48.0 | 799 |
| $TiO_2$ | 1.1 | 14 | 1.0 | 13 | 1.0 | 13 | 0.9 | 11 | 0.9 | 11 | 1.1 | 14 | 1.1 | 14 | 1.2 | 15 | 0.7 | 9 |
| $Al_2O_3$ | 14.8 | 145 | 15.0 | 147 | 15.4 | 151 | 14.0 | 137 | 14.5 | 142 | 14.3 | 140 | 14.8 | 145 | 14.3 | 140 | 14.9 | 146 |
| $Fe_2O_3$ | 2.7 | 17 | 2.5 | 16 | 2.8 | 17 | 2.2 | 14 | 3.0 | 19 | 3.9 | 24 | 5.2 | 33 | 2.5 | 16 | 3.4 | 21 |
| FeO | 8.3 | 115 | 8.4 | 117 | 7.2 | 99 | 8.1 | 113 | 6.8 | 95 | 7.4 | 103 | 5.8 | 81 | 8.6 | 120 | 9.0 | 125 |
| MnO | 0.1 | 1 | 0.2 | 2 | 0.2 | 2 | 0.1 | 1 | 0.2 | 2 | 0.2 | 2 | 0.2 | 2 | 0.2 | 2 | 0.2 | 2 |
| MgO | 6.4 | 159 | 6.0 | 149 | 6.8 | 169 | 7.6 | 189 | 7.4 | 183 | 5.8 | 144 | 4.6 | 114 | 5.4 | 134 | 8.7 | 216 |
| CaO | 11.0 | 196 | 10.7 | 191 | 11.3 | 201 | 12.0 | 214 | 11.7 | 209 | 10.1 | 180 | 8.6 | 153 | 9.2 | 164 | 12.2 | 218 |
| $Na_2O$ | 2.5 | 40 | 2.7 | 44 | 3.0 | 48 | 2.7 | 44 | 3.1 | 50 | 3.0 | 48 | 2.4 | 39 | 2.6 | 42 | 2.4 | 39 |
| $K_2O$ | 1.0 | 11 | 0.9 | 10 | 0.4 | 4 | 0.5 | 5 | 0.4 | 4 | 0.8 | 9 | 1.5 | 16 | 1.1 | 12 | 0.1 | 1 |
| qz | | −7 | | −6 | | −10 | | −9 | | −12 | | −4 | | +12 | | +8 | | −19 |

### Lower Series

| Component | Olivine basalt | | | | | |
|---|---|---|---|---|---|---|
| | 10 | | 11 | | 12 | |
| | % | M.Q. | % | M.Q. | % | M.Q. |
| $SiO_2$ | 44.88 | 748 | 44.87 | 747 | 46.65 | 777 |
| $TiO_2$ | 1.26 | 16 | 1.39 | 18 | 1.03 | 13 |
| $Al_2O_3$ | 13.79 | 135 | 14.88 | 146 | 15.27 | 150 |
| $Fe_2O_3$ | 2.52 | 16 | 2.23 | 14 | 2.67 | 17 |
| FeO | 7.78 | 109 | 10.57 | 147 | 7.39 | 103 |
| MnO | 0.15 | 2 | 0.20 | 2 | 0.16 | 2 |
| MgO | 11.88 | 295 | 11.51 | 285 | 9.69 | 241 |
| CaO | 12.54 | 224 | 9.79 | 175 | 10.37 | 185 |
| $Na_2O$ | 1.74 | 28 | 2.20 | 35 | 1.92 | 31 |
| $K_2O$ | 0.20 | 2 | 0.12 | 1 | 0.36 | 4 |
| $P_2O_5$ | 0.09 | | 0.23 | | 0.14 | |
| qz | | −24 | | −26 | | −15 |

TABLE 22. (continued)
## Upper Series

| Component | Olivine basalt 13 % | M.Q. | Olivine basalt 14 % | M.Q. | Olivine basalt 15 % | M.Q. | Tholeiitic basalt 16 % | M.Q. | Tholeiitic basalt 17 % | M.Q. | Quartz trachyte 18 % | M.Q. | Quartz trachyte 19 % | M.Q. | Rhyolite 20 % | M.Q. | Obsidian 21 % | M.Q. | Rhyolite 22 % | M.Q. | T. Magee 23 % |
|---|---|---|---|---|---|---|---|---|---|---|---|---|---|---|---|---|---|---|---|---|---|
| $SiO_2$ | 45.93 | 765 | 45.34 | 753 | 45.09 | 751 | 50.36 | 839 | 53.76 | 879 | 63.91 | 1064 | 65.07 | 1083 | 75.80 | 1262 | 74.55 | 1241 | 76.03 | 1266 | 44.60 |
| $TiO_2$ | 1.54 | 19 | 1.13 | 14 | 0.63 | 8 | 1.06 | 14 | 1.20 | 15 | 0.59 | 8 | 0.47 | 6 | — | — | 0.12 | 1 | 0.10 | 1 | 1.25 |
| $Al_2O_3$ | 15.22 | 149 | 14.67 | 144 | 14.21 | 139 | 14.51 | 142 | 14.00 | 137 | 14.27 | 140 | 14.34 | 140 | 12.45 | 123 | 11.56 | 114 | 13.01 | 127 | 12.15 |
| $Fe_2O_3$ | 3.06 | 19 | 2.40 | 15 | 3.42 | 21 | 2.61 | 16 | 2.46 | 16 | 4.95 | 31 | 4.10 | 26 | 1.47 | 9 | 0.92 | 6 | 0.06 | 1 | 3.41 |
| $FeO$ | 9.26 | 129 | 9.15 | 128 | 7.37 | 103 | 8.09 | 113 | 8.55 | 119 | 2.97 | 41 | 2.64 | 37 | 0.44 | 5 | 0.80 | 11 | 0.09 | 12 | 6.57 |
| $MnO$ | 0.18 | 2 | 0.22 | 2 | 0.19 | 2 | 0.12 | 1 | 0.23 | 2 | 0.16 | 2 | 0.15 | 2 | 0.02 | 0 | 0.03 | 0 | 0.01 | 0 | 0.15 |
| $MgO$ | 10.30 | 256 | 13.32 | 330 | 12.24 | 304 | 6.26 | 155 | 5.37 | 133 | 0.30 | 7 | 1.30 | 32 | 0.08 | 2 | 0.17 | 4 | — | — | 16.84 |
| $CaO$ | 8.55 | 153 | 9.12 | 162 | 9.61 | 171 | 10.77 | 192 | 9.14 | 163 | 2.34 | 42 | 1.79 | 32 | 1.00 | 18 | 1.06 | 19 | 0.36 | 6 | 7.84 |
| $Na_2O$ | 2.63 | 43 | 1.86 | 30 | 1.99 | 32 | 2.48 | 40 | 2.63 | 43 | 4.27 | 69 | 4.15 | 67 | 2.30 | 37 | 3.49 | 56 | 2.25 | 36 | 0.66 |
| $K_2O$ | 0.19 | 2 | 0.24 | 2 | 0.21 | 2 | 0.99 | 11 | 1.13 | 12 | 3.66 | 39 | 4.73 | 50 | 4.17 | 45 | 4.46 | 48 | 5.02 | 53 | 0.62 |
| $P_2O_5$ | 0.28 | | 0.09 | | 0.52 | | 0.45 | | 0.14 | | 0.17 | | 0.08 | | | | | | 0.02 | | |
| $qz$ | | −24 | | −24 | | −23 | | −7 | | +3 | | +57 | | +50 | | +270 | | +213 | | +286 | |

Note: 1-7) flows 1 to 7 (uppermost) in section; 8) from the second flow 10-15 km from 1-7 (it is the second flow, but also the top flow, so that its place in the section is not clear); 9) from the sixth flow (also from another district and not correlated with 1-7).

TABLE 23. Faeroe Islands

### (Noe-Nygaard and Rasmussen [1957])

| Analysis No. / Component | 1 % | M.Q. | 2 % | M.Q. | 3 % | M.Q. | 4 % | M.Q. | 5 % | M.Q. | 6 % | M.Q. | 7 % | M.Q. |
|---|---|---|---|---|---|---|---|---|---|---|---|---|---|---|
| $SiO_2$ | 4.97 | 828 | 48.02 | 799 | 47.43 | 790 | 47.34 | 788 | 45.40 | 756 | 42.62 | 709 | 44.40 | 739 |
| $TiO_2$ | 4.83 | 60 | 2.20 | 28 | 1.36 | 17 | 1.97 | 25 | 3.20 | 40 | 1.98 | 25 | 0.75 | 10 |
| $Al_2O_3$ | 10.98 | 108 | 14.17 | 139 | 15.81 | 155 | 13.22 | 129 | 14.54 | 142 | 12.11 | 119 | 12.25 | 121 |
| $Fe_2O_3$ | 3.02 | 19 | 3.62 | 23 | 5.35 | 34 | 6.27 | 39 | 1.96 | 12 | 6.56 | 41 | 1.50 | 9 |
| $FeO$ | 11.19 | 156 | 8.37 | 116 | 7.07 | 98 | 8.00 | 111 | 10.45 | 146 | 3.40 | 47 | 8.92 | 122 |
| $MnO$ | 0.22 | 3 | 0.16 | 2 | 0.06 | 2 | 0.18 | 3 | 0.25 | 4 | 0.24 | 3 | 0.18 | 3 |
| $MgO$ | 5.34 | 133 | 7.08 | 176 | 5.92 | 147 | 5.98 | 149 | 6.25 | 155 | 17.54 | 435 | 17.66 | 438 |
| $CaO$ | 9.71 | 173 | 11.02 | 196 | 11.27 | 201 | 11.25 | 201 | 11.72 | 209 | 9.75 | 174 | 10.00 | 178 |
| $Na_2O$ | 2.22 | 35 | 1.73 | 28 | 1.99 | 16 | 2.41 | 39 | 2.08 | 34 | 1.12 | 18 | 0.95 | 16 |
| $K_2O$ | 0.63 | 6 | 0.27 | 3 | 0.16 | 2 | 0.16 | 2 | 0.49 | 5 | 0.34 | 3 | 0.25 | 3 |
| $qz$ | | +2 | | −4 | | +4 | | −12 | | −17 | | −29 | | −28 |

| Analysis No. / Component | 8 % | M.Q. | 9 % | M.Q. | 10 % | M.Q. | 11 % | M.Q. | 12 % | M.Q. | 13 % | M.Q. | 14 % | M.Q. |
|---|---|---|---|---|---|---|---|---|---|---|---|---|---|---|
| $SiO_2$ | 46.13 | 768 | 46.40 | 773 | 47.58 | 793 | 48.06 | 800 | 48.09 | 801 | 48.10 | 810 | 53.30 | 888 |
| $TiO_2$ | 1.05 | 14 | 3.05 | 39 | 2.72 | 34 | 4.42 | 55 | 3.58 | 45 | 2.50 | 31 | 1.70 | 21 |
| $Al_2O_3$ | 13.73 | 134 | 16.30 | 160 | 17.12 | 168 | 13.64 | 133 | 13.23 | 129 | 16.01 | 157 | 9.37 | 92 |
| $Fe_2O_3$ | 3.03 | 19 | 3.60 | 23 | 3.03 | 19 | 8.43 | 53 | 10.77 | 68 | 3.47 | 22 | 5.09 | 32 |
| $FeO$ | 9.65 | 135 | 7.17 | 100 | 8.73 | 121 | 3.70 | 51 | 2.90 | 40 | 7.16 | 100 | 8.11 | 113 |
| $MnO$ | 0.63 | 9 | 0.23 | 3 | 0.14 | 2 | 0.24 | 3 | 0.26 | 4 | 0.15 | 2 | 0.20 | 3 |
| $MgO$ | 10.32 | 256 | 6.00 | 149 | 5.18 | 129 | 6.45 | 160 | 6.49 | 161 | 6.00 | 149 | 8.60 | 213 |
| $CaO$ | 12.49 | 223 | 11.04 | 197 | 10.12 | 180 | 10.91 | 194 | 10.32 | 184 | 10.55 | 188 | 8.99 | 161 |
| $Na_2O$ | 2.40 | 39 | 2.14 | 34 | 3.89 | 63 | 2.21 | 35 | 2.27 | 36 | 2.50 | 40 | 4.12 | 66 |
| $K_2O$ | 0.05 | 1 | 0.29 | 3 | 0.86 | 10 | 0.61 | 6 | 0.83 | 9 | 0.30 | 3 | 0.37 | 4 |
| $qz$ | | −27 | | −10 | | −29 | | −7 | | −11 | | −7 | | −15 |

Note: 1) nonporphyritic basalt; 2) olivine basalt; 3) olivine basalt; 4) olivine basalt; 5) olivine basalt; 6) oceanite; 7) olivine basalt; 8) olivine basalt; 9) basalt with feldspar phenocrysts; 10) basalt with feldspar phenocrysts; 11) porphyritic (plagioclase) basalt; 12) porphyritic (feldspar) basalt; 13) porphyritic (feldspar) basalt; 14) light gray olivine basalt.

## TABLE 24. Iceland

(Tyrrell and Peacock [1926, 1927]; Tyrrell [1949]; Hoppe [1938]; Barth [1950]; Tryggvason [1943, 1955]; Washington [1917, 1922]; Noe-Nygaard [1940]; Wolff [1931]; Cargill, Hawkes, and others [1928]; Thorarinsson [1950]; Sigvaldson [1958])

| Analysis No. | 1 | | 2 | | 3 | | 4 | | 5 | | 6 | | 7 | |
|---|---|---|---|---|---|---|---|---|---|---|---|---|---|---|
| Component | % | M.Q. | % | M.Q. | % | M.Q. | % | M.Q. | % | M.Q. | % | M.Q. | % | M.Q. |
| $SiO_2$ | 35.34 | 588 | 39.35 | 655 | 43.83 | 730 | 45.10 | 751 | 45.22 | 753 | 45.37 | 755 | 45.82 | 763 |
| $TiO_2$ | 2.10 | 26 | 3.10 | 39 | 0.46 | 6 | 2.17 | 27 | 3.56 | 45 | 1.85 | 24 | 2.61 | 33 |
| $Al_2O_3$ | 11.15 | 110 | 8.74 | 85 | 14.70 | 144 | 9.58 | 94 | 11.86 | 117 | 14.51 | 142 | 15.30 | 150 |
| $Fe_2O_3$ | 10.28 | 64 | 3.63 | 23 | 1.47 | 9 | 5.37 | 34 | 6.43 | 40 | 5.63 | 35 | 0.89 | 6 |
| $FeO$ | 2.19 | 31 | 14.85 | 207 | 6.81 | 95 | 10.59 | 148 | 8.81 | 122 | 6.09 | 85 | 11.97 | 167 |
| $MnO$ | 0.22 | 3 | 0.28 | 4 | 0.12 | 1 | 0.16 | 2 | 0.19 | 3 | 0.21 | 3 | 0.38 | 6 |
| $MgO$ | 6.52 | 162 | 19.57 | 485 | 19.08 | 473 | 14.24 | 353 | 5.77 | 145 | 7.44 | 184 | 7.86 | 195 |
| $CaO$ | 7.01 | 125 | 8.94 | 160 | 11.85 | 211 | 10.18 | 182 | 10.83 | 193 | 12.46 | 222 | 10.81 | 193 |
| $Na_2O$ | 0.16 | 2 | 0.76 | 12 | 1.18 | 19 | 2.31 | 37 | 2.23 | 36 | 1.89 | 31 | 1.96 | 32 |
| $K_2O$ | 0.19 | 2 | 0.08 | 1 | 0.14 | 1 | 0.19 | 2 | Traces | — | 0.09 | 1 | 1.46 | 16 |
| $qz$ | | — | | −40 | | −33 | | −33 | | −12 | | −15 | | −26 |

| Analysis No. | 8 | | 9 | | 10 | | 11 | | 12 | | 13 | | 14 | |
|---|---|---|---|---|---|---|---|---|---|---|---|---|---|---|
| Component | % | M.Q. | % | M.Q. | % | M.Q. | % | M.Q. | % | M.Q. | % | M.Q. | % | M.Q. |
| $SiO_2$ | 45.94 | 765 | 46.07 | 767 | 46.08 | 768 | 46.26 | 770 | 46.28 | 771 | 46.38 | 773 | 46.50 | 774 |
| $TiO_2$ | 0.82 | 10 | 3.83 | 48 | 1.55 | 20 | 2.97 | 37 | 2.00 | 25 | 3.68 | 46 | 2.25 | 28 |
| $Al_2O_3$ | 17.63 | 173 | 13.70 | 134 | 12.14 | 119 | 14.04 | 138 | 14.64 | 144 | 15.06 | 148 | 14.00 | 137 |
| $Fe_2O_3$ | 1.67 | 11 | 7.53 | 47 | 4.89 | 30 | 5.20 | 33 | 2.39 | 15 | 2.76 | 17 | 4.03 | 25 |
| $FeO$ | 5.24 | 73 | 7.09 | 99 | 9.04 | 126 | 11.30 | 157 | 10.79 | 150 | 9.73 | 135 | 9.38 | 131 |
| $MnO$ | 0.37 | 5 | 0.32 | 4 | 0.17 | 2 | — | — | 0.33 | 5 | 0.42 | 6 | 0.10 | 1 |
| $MgO$ | 8.05 | 200 | 5.09 | 127 | 9.80 | 243 | 5.64 | 140 | 9.46 | 235 | 6.75 | 167 | 8.00 | 198 |
| $CaO$ | 16.26 | 290 | 11.79 | 210 | 12.23 | 218 | 10.11 | 180 | 11.89 | 212 | 9.74 | 174 | 11.83 | 211 |
| $Na_2O$ | 1.67 | 27 | 1.64 | 26 | 2.07 | 33 | 2.81 | 45 | 1.19 | 19 | 3.07 | 49 | 2.09 | 34 |
| $K_2O$ | 0.24 | 2 | 0.98 | 11 | 0.31 | 3 | 0.68 | 7 | 0.37 | 4 | 1.02 | 11 | 0.09 | 1 |
| $qz$ | | −18 | | −12 | | −23 | | −7 | | −15 | | −27 | | −17 |

| Analysis No. | 15 | | 16 | | 17 | | 18 | | 19 | | 20 | | 21 | |
|---|---|---|---|---|---|---|---|---|---|---|---|---|---|---|
| Component | % | M.Q. | % | M.Q. | % | M.Q. | % | M.Q. | % | M.Q. | % | M.Q. | % | M.Q. |
| $SiO_2$ | 46.84 | 780 | 46.95 | 782 | 46.97 | 783 | 46.97 | 782 | 47.27 | 787 | 47.27 | 787 | 47.27 | 787 |
| $TiO_2$ | 2.01 | 25 | 1.32 | 16 | 1.99 | 25 | 1.99 | 25 | 3.37 | 42 | 2.00 | 25 | 2.36 | 30 |
| $Al_2O_3$ | 13.65 | 134 | 14.56 | 143 | 13.87 | 136 | 13.87 | 136 | 13.37 | 131 | 15.23 | 149 | 14.44 | 141 |
| $Fe_2O_3$ | 3.68 | 23 | 1.41 | 9 | 8.27 | 52 | 8.27 | 52 | 2.35 | 14 | 3.21 | 20 | 0.97 | 6 |
| $FeO$ | 9.39 | 131 | 10.57 | 147 | 5.76 | 80 | 5.76 | 80 | 13.53 | 188 | 8.77 | 122 | 10.15 | 142 |
| $MnO$ | 0.20 | 3 | 0.25 | 4 | 1.20 | 17 | 0.20 | 17 | 0.18 | 3 | 0.07 | 1 | 0.19 | 3 |
| $MgO$ | 8.73 | 216 | 10.47 | 259 | 4.76 | 119 | 4.76 | 118 | 5.25 | 130 | 9.30 | 231 | 10.30 | 256 |
| $CaO$ | 12.46 | 222 | 12.19 | 217 | 14.46 | 258 | 14.46 | 258 | 10.00 | 178 | 11.17 | 199 | 11.73 | 209 |
| $Na_2O$ | 1.60 | 26 | 1.75 | 28 | 1.75 | 28 | 1.75 | 28 | 2.37 | 38 | 2.12 | 34 | 1.86 | 30 |
| $K_2O$ | 0.35 | 4 | 0.25 | 3 | 0.45 | 5 | 0.45 | 4 | 0.46 | 5 | 0.42 | 4 | 0.18 | 2 |
| $qz$ | | −16 | | −20 | | −13 | | −12 | | −12 | | −18 | | −17 |

| Analysis No. | 22 | | 23 | | 24 | | 25 | | 26 | | 27 | | 28 | |
|---|---|---|---|---|---|---|---|---|---|---|---|---|---|---|
| Component | % | M.Q. | % | M.Q. | % | M.Q. | % | M.Q. | % | M.Q. | % | M.Q. | % | M.Q. |
| $SiO_2$ | 47.30 | 788 | 47.35 | 788 | 47.67 | 794 | 47.68 | 796 | 47.96 | 799 | 48.04 | 800 | 48.06 | 800 |
| $TiO_2$ | 0.58 | 8 | 2.86 | 36 | 2.17 | 27 | 5.01 | 63 | 1.40 | 18 | 3.00 | 38 | 1.55 | 20 |
| $Al_2O_3$ | 25.04 | 245 | 13.65 | 134 | 15.81 | 155 | 12.54 | 123 | 14.05 | 138 | 12.26 | 120 | 12.14 | 119 |
| $Fe_2O_3$ | 0.93 | 6 | 2.27 | 14 | 1.70 | 11 | 3.44 | 21 | 3.02 | 19 | 2.07 | 13 | 4.89 | 30 |
| $FeO$ | 3.40 | 47 | 13.76 | 192 | 10.41 | 145 | 12.34 | 172 | 8.52 | 118 | 10.78 | 150 | 9.04 | 126 |
| $MnO$ | Traces | — | — | — | 0.16 | 2 | — | — | 0.20 | 3 | 0.31 | 4 | 0.17 | 2 |
| $MgO$ | 4.00 | 99 | 5.03 | 125 | 8.08 | 200 | 5.25 | 131 | 7.75 | 192 | 9.18 | 228 | 9.80 | 243 |
| $CaO$ | 15.92 | 284 | 9.63 | 172 | 12.23 | 218 | 9.58 | 171 | 13.38 | 139 | 12.04 | 215 | 12.23 | 218 |
| $Na_2O$ | 1.56 | 25 | 3.00 | 48 | 1.98 | 32 | 2.43 | 39 | 1.90 | 31 | 1.82 | 30 | 2.07 | 34 |
| $K_2O$ | 0.42 | 4 | 0.91 | 10 | 1.27 | 3 | 0.88 | 10 | 0.29 | 3 | 0.38 | 4 | 0.31 | 3 |
| $qz$ | | −6 | | −21 | | −17 | | −13 | | −13 | | −14 | | −18 |

## TABLE 24. (continued)

| Analysis No. Component | 29 % | 29 M.Q. | 30 % | 30 M.Q. | 31 % | 31 M.Q. | 32 % | 32 M.Q. | 33 % | 33 M.Q. | 34 % | 34 M.Q. | 35 % | 35 M.Q. |
|---|---|---|---|---|---|---|---|---|---|---|---|---|---|---|
| SiO$_2$ | 48.15 | 802 | 48.38 | 806 | 48.52 | 808 | 48.56 | 809 | 48.60 | 810 | 48.79 | 813 | 48.80 | 813 |
| TiO$_2$ | 2.00 | 25 | 1.26 | 16 | 2.68 | 34 | 1.25 | 16 | 2.20 | 28 | 4.17 | 52 | 2.08 | 26 |
| Al$_2$O$_3$ | 16.24 | 159 | 16.88 | 166 | 13.33 | 130 | 16.06 | 158 | 12.62 | 124 | 11.96 | 118 | 13.89 | 136 |
| Fe$_2$O$_3$ | 2.12 | 13 | 2.31 | 14 | 0.51 | 3 | 2.58 | 16 | 4.34 | 27 | 2.51 | 16 | 4.11 | 26 |
| FeO | 8.74 | 121 | 6.10 | 85 | 11.92 | 166 | 8.69 | 121 | 7.34 | 102 | 12.10 | 168 | 11.88 | 166 |
| MnO | 0.20 | 3 | 0.04 | — | 0.28 | 4 | 0.11 | 1 | 0.12 | 1 | 0.21 | 3 | 0.32 | 4 |
| MgO | 5.84 | 145 | 7.28 | 180 | 7.06 | 175 | 7.36 | 182 | 7.33 | 182 | 5.60 | 139 | 5.28 | 131 |
| CaO | 13.66 | 244 | 13.64 | 243 | 12.24 | 218 | 12.99 | 232 | 11.85 | 211 | 10.15 | 181 | 10.27 | 183 |
| Na$_2$O | 2.20 | 35 | 2.36 | 38 | 1.62 | 26 | 1.91 | 31 | 1.85 | 30 | 2.40 | 39 | 2.18 | 35 |
| K$_2$O | 0.33 | 3 | 0.19 | 2 | 0.16 | 2 | 0.09 | 1 | 0.31 | 3 | 0.70 | 7 | 0.33 | 3 |
| qz | | —12 | | —12 | | —14 | | —10 | | —4 | | —8 | | —7 |

| Analysis No. Component | 36 % | 36 M.Q. | 37 % | 37 M.Q. | 38 % | 38 M.Q. | 39 % | 39 M.Q. | 40 % | 40 M.Q. | 41 % | 41 M.Q. | 42 % | 42 M.Q. |
|---|---|---|---|---|---|---|---|---|---|---|---|---|---|---|
| SiO$_2$ | 48.98 | 816 | 49.24 | 820 | 49.56 | 825 | 49.60 | 826 | 49.67 | 827 | 50.46 | 840 | 50.52 | 841 |
| TiO$_2$ | 1.39 | 18 | 2.46 | 31 | 4.16 | 52 | 2.46 | 31 | 1.50 | 19 | 2.68 | 34 | 0.43 | 5 |
| Al$_2$O$_3$ | 15.32 | 150 | 14.12 | 138 | 10.36 | 102 | 13.49 | 132 | 13.57 | 133 | 13.86 | 136 | 16.31 | 160 |
| Fe$_2$O$_3$ | 0.63 | 4 | 1.51 | 9 | 7.04 | 44 | 3.43 | 21 | 7.79 | 49 | 1.56 | 10 | 8.76 | 55 |
| FeO | 11.33 | 158 | 12.68 | 177 | 8.19 | 114 | 11.38 | 159 | 7.21 | 100 | 12.28 | 171 | 3.72 | 51 |
| MnO | 0.22 | 3 | 0.08 | 1 | — | — | 0.17 | 2 | — | — | 0.23 | 3 | — | — |
| MgO | 7.31 | 181 | 5.79 | 144 | 6.04 | 150 | 5.07 | 126 | 5.53 | 137 | 5.14 | 128 | 7.04 | 175 |
| CaO | 12.56 | 224 | 10.62 | 189 | 11.66 | 208 | 9.90 | 177 | 12.37 | 220 | 9.96 | 178 | 13.26 | 236 |
| Na$_2$O | 1.52 | 24 | 2.17 | 35 | 1.86 | 30 | 3.01 | 48 | 1.57 | 25 | 2.86 | 46 | | |
| K$_2$O | 0.14 | 1 | 0.37 | 4 | 0.63 | 6 | 0.52 | 5 | 1.20 | 13 | 0.50 | 5 | | |
| qz | | —4 | | —6 | | —2 | | —11 | | —7 | | —7 | | (+15?) |

| Analysis No. Component | 43 % | 43 M.Q. | 44 % | 44 M.Q. | 45 % | 45 M.Q. | 46 % | 46 M.Q. | 47 % | 47 M.Q. | 48 % | 48 M.Q. | 49 % | 49 M.Q. |
|---|---|---|---|---|---|---|---|---|---|---|---|---|---|---|
| SiO$_2$ | 53.39 | 889 | 53.55 | 892 | 54.25 | 903 | 55.41 | 922 | 60.60 | 1.009 | 64.74 | 1.078 | 66.66 | 1.110 |
| TiO$_2$ | 2.28 | 29 | 1.48 | 19 | 1.54 | 19 | 1.60 | 20 | 0.92 | 11 | 0.35 | 4 | 0.25 | 4 |
| Al$_2$O$_3$ | 13.37 | 131 | 17.45 | 172 | 16.34 | 160 | 15.84 | 155 | 17.55 | 172 | 17.04 | 167 | 14.89 | 146 |
| Fe$_2$O$_3$ | 2.07 | 13 | 3.60 | 23 | 2.24 | 14 | 2.43 | 15 | 1.17 | 7 | 1.49 | 9 | 2.74 | 17 |
| FeO | 10.62 | 148 | 6.98 | 97 | 10.05 | 140 | 9.10 | 127 | 6.08 | 85 | 3.87 | 54 | 3.22 | 44 |
| MnO | 0.20 | 3 | 0.25 | 4 | 0.26 | 4 | 0.23 | 3 | 0.35 | 5 | 0.06 | 1 | — | — |
| MgO | 3.30 | 82 | 3.47 | 86 | 3.39 | 84 | 2.82 | 70 | 2.04 | 51 | 1.33 | 33 | 1.04 | 26 |
| CaO | 6.90 | 123 | 6.53 | 117 | 7.09 | 127 | 6.93 | 124 | 4.02 | 71 | 3.45 | 62 | 3.20 | 57 |
| Na$_2$O | 3.84 | 62 | 3.23 | 52 | 3.41 | 55 | 3.84 | 62 | 3.97 | 64 | 4.17 | 67 | 4.44 | 72 |
| K$_2$O | 1.19 | 13 | 1.72 | 18 | 0.95 | 10 | 0.80 | 9 | 2.73 | 29 | 2.77 | 30 | 2.72 | 29 |
| qz | | 0 | | +3 | | +6 | | +10 | | +29 | | +60 | | +72 |

| Analysis No. Component | 50 % | 50 M.Q. | 51 % | 51 M.Q. | 52 % | 52 M.Q. | 53 % | 53 M.Q. | 54 % | 54 M.Q. | 55 % | 55 M.Q. | 56 % | 56 M.Q. |
|---|---|---|---|---|---|---|---|---|---|---|---|---|---|---|
| SiO$_2$ | 67.17 | 1.118 | 70.36 | 1.171 | 70.49 | 1.174 | 70.80 | 1.179 | 70.83 | 1.179 | 71.25 | 1.186 | 71.47 | 1.190 |
| TiO$_2$ | 0.29 | 4 | — | — | 0.31 | 4 | Traces | — | 0.18 | 3 | 0.27 | 4 | — | — |
| Al$_2$O$_3$ | 15.27 | 150 | 14.85 | 146 | 13.32 | 130 | 14.79 | 145 | 13.08 | 128 | 14.81 | 145 | 12.69 | 125 |
| Fe$_2$O$_3$ | 1.58 | 10 | 2.34 | 14 | 1.76 | 11 | 2.82 | 17 | 2.91 | 18 | 0.84 | 5 | 0.37 | 3 |
| FeO | 2.41 | 33 | 2.36 | 33 | 1.24 | 17 | 1.39 | 19 | 1.58 | 22 | 1.52 | 21 | 2.56 | 36 |
| MnO | 0.08 | 1 | 0.18 | 3 | 0.36 | 5 | Traces | — | 0.04 | — | Traces | — | — | — |
| MgO | 0.85 | 21 | 0.44 | 11 | 0.41 | 10 | 0.37 | 9 | 0.19 | 5 | 0.92 | 23 | 0.41 | 10 |
| CaO | 2.33 | 42 | 1.82 | 32 | 0.87 | 15 | 0.70 | 13 | 1.09 | 20 | 1.87 | 33 | 1.85 | 33 |
| Na$_2$O | 6.20 | 100 | 4.25 | 68 | 5.09 | 82 | 4.19 | 68 | 4.86 | 78 | 3.32 | 53 | 4.27 | 69 |
| K$_2$O | 3.09 | 33 | 2.91 | 31 | 4.53 | 48 | 4.17 | 45 | 3.34 | 24 | 3.63 | 38 | 1.59 | 17 |
| qz | | +47 | | +121 | | +100 | | +120 | | +128 | | +140 | | +183 |

## TABLE 24. (continued)

| Analysis No. | 57 % | 57 M.Q. | 58 % | 58 M.Q. | 59 % | 59 M.Q. | 60 % | 60 M.Q. | 61 % | 61 M.Q. |
|---|---|---|---|---|---|---|---|---|---|---|
| $SiO_2$ | 72.41 | 1,205 | 74.50 | 1,240 | 74.80 | 1,245 | 75.10 | 1,250 | 77.21 | 1,285 |
| $TiO_2$ | 0.37 | 5 | 0.34 | 3 | 0.21 | 3 | 0.33 | 4 | 0.26 | 4 |
| $Al_2O_3$ | 12.35 | 122 | 12.77 | 125 | 12.97 | 127 | 12.27 | 120 | 12.53 | 123 |
| $Fe_2O_3$ | 1.48 | 9 | 0.92 | 6 | 0.48 | 3 | 0.80 | 5 | 0.24 | 1 |
| FeO | 1.38 | 19 | 1.45 | 20 | 0.24 | 3 | 2.78 | 39 | 0.12 | 1 |
| MnO | 0.08 | 1 | — | — | 0 | 0 | 0.06 | 1 | 0 | 0 |
| MgO | 0.11 | 3 | 0.79 | 20 | 0.11 | 3 | 0.08 | 2 | 0.04 | 1 |
| CaO | 1.18 | 21 | 1.10 | 20 | 1.94 | 35 | 1.87 | 33 | 0.55 | 10 |
| $Na_2O$ | 4.45 | 72 | 3.21 | 52 | 4.17 | 67 | 3.36 | 54 | 4.43 | 72 |
| $K_2O$ | 2.78 | 30 | 3.75 | 40 | 2.70 | 29 | 2.80 | 30 | 1.83 | 19 |
| qz | | +178 | | +201 | | +219 | | +215 | | +301 |

Note: 1) palagonite (Hvalfjördr, Iceland); 2) gabbro-peridotite (Southeastern Iceland); 3) recent basalt (Grindavik, Gulbringu, Sisla); 4) doleritic basalt (Hafnafjördr); 5) basalt (Vatnajökull); 6) altered rock from borehole (Sudur Reykir); 7) olivine basalt, deluvial (Keflavik, Sisla); 8) bytownite gabbro (eucrite) (Snaefellsnes, Iceland); 9) basalt, Tertiary (Arkafjöll Hvalfjördr); 10) basalt, deluvial (Reykjavik); 11) basalt.(Sog, southwestern Iceland); 12) basalt (south of Reydarbarum); 13) basalt, recent (Raudimel, Snaefellsnes); 14) basalt (Ellidaarvag, Reykjavik); 15) lava (Sekjöllobreid, west-southwest of Ljaitra); 16) Ljodberg (at Almanadjo); 17) basalt (Raudnalar Volcano, Reykjanes Island); 18) basalt, recent (Raudnalar Volcano); 19) diabasic dike (Reykir); 20) diabasic dike (Hellisheidt); 21) olivine basalt, recent (Grundarfjörd); 22) anorthositic gabbro (southeast Iceland); 23) basalt (valley of Sog Riv., southwest Iceland); 24) basalt, recent (near Thingva); 25) basalt (Mirdalsjökull); 26) lava (Ikjöllbreid); 27) basalt (east of Hrafnabjorg); 28) pre-glacial basalt; 29) Ikjöllbreid block; 30) olivine gabbro (southeastern Iceland); 31) basaltic glass from breccia (Kirkjubjavklaustur); 32) basalt (from eastern part of Hengill); 33) basalt (Jackson Island, Greenland); 34) basalt (Eskifjördr); 35) basalt (Skifolt, Arnesisla); 36) basalt (Stjörnflos); 37) lava (from Laksarnes); 38) lava (Ikafraredhraun); 39) basalt (Vatnajökull); 40) post-glacial lava (Thjorsa); 41) basalt (Vatnajökull); 42) basalt; 43) basalt (Hekla); 44) hedenbergite diorite (Vestur Horn); 45) lava flow (Hekla crater, April 1948); 46) lava flow (Hekla crater, beginning of June 1947); 47) gabbro-andesite (Snaefelljökull); 48) diorite with granulitic texture (intrusion into gabbro at Vestur Horn); 49) graphic hedenbergitic granodiorite (Oster Horn); 50) recent lava (Torfajökull); 51) granophyre (Vestur Horn); 52) liparite (Hrafutiknuskar); 53) magnetite granite (Snaefellsnes, Iceland); 54) albite granophyre (Hornafjördr, Iceland); 55) granophyre (Lon Gulf, Iceland); 56) black dacitic obsidian (Hamardfjördr); 57) liparite (Moskordsnjörkar); 58) leucogranite (Slaifrudal); 59) liparite (Moskordsnjörkar); 60) obsidian (Krafla); 61) liparite (Moskordsnjörkar).

## TABLE 25. Jan Mayen

### (Tyrrell [1926]; Holmes [1918]; Carstens [1961])

| Analysis No. | 1 % | 1 M.Q. | 2 % | 2 M.Q. | 3 % | 3 M.Q. | 4 % | 4 M.Q. | 5 % | 5 M.Q. | 6 % | 6 M.Q. |
|---|---|---|---|---|---|---|---|---|---|---|---|---|
| $SiO_2$ | 41.89 | 698 | 45.08 | 751 | 46.04 | 767 | 46.15 | 768 | 46.25 | 770 | 46.30 | 771 |
| $TiO_2$ | 3.49 | 44 | 2.93 | 37 | 3.96 | 50 | 3.05 | 38 | 3.75 | 47 | 3.85 | 48 |
| $Al_2O_3$ | 16.89 | 166 | 14.15 | 138 | 16.39 | 161 | 17.44 | 171 | 16.07 | 158 | 16.25 | 160 |
| $Fe_2O_3$ | 4.95 | 30 | 2.28 | 14 | 4.01 | 25 | 4.94 | 30 | 1.88 | 12 | 1.94 | 12 |
| FeO | 6.54 | 91 | 8.89 | 124 | 7.89 | 110 | 6.19 | 86 | 10.08 | 141 | 8.54 | 119 |
| MnO | 0.24 | 3 | 0.22 | 3 | 0.30 | 4 | 0.33 | 5 | 0.19 | 3 | 0.43 | 6 |
| MgO | 4.32 | 107 | 7.01 | 174 | 5.29 | 132 | 5.40 | 134 | 5.37 | 133 | 5.67 | 140 |
| CaO | 10.55 | 188 | 10.20 | 182 | 9.75 | 174 | 11.05 | 197 | 9.33 | 167 | 10.45 | 186 |
| $Na_2O$ | 1.51 | 24 | 3.99 | 65 | 3.18 | 52 | 2.37 | 38 | 3.45 | 55 | 3.18 | 52 |
| $K_2O$ | 1.65 | 18 | 1.79 | 19 | 1.97 | 21 | 2.45 | 26 | 1.97 | 21 | 2.40 | 25 |
| qz | | —5 | | —44 | | —33 | | —29 | | —34 | | —35 |

## TABLE 25. (continued)

| Analysis No. Component | 7 % | 7 M.Q. | 8 % | 8 M.Q. | 9 % | 9 M.Q. | 10 % | 10 M.Q. | 11 % | 11 M.Q. | 12 % | 12 M.Q. |
|---|---|---|---|---|---|---|---|---|---|---|---|---|
| $SiO_2$ | 46.73 | 773 | 49.03 | 817 | 54.45 | 906 | 56.71 | 944 | 64.88 | 1080 | 65.85 | 1096 |
| $TiO_2$ | 1.90 | 24 | 2.79 | 35 | 1.78 | 23 | 1.80 | 23 | 0.60 | 8 | 0.25 | 3 |
| $Al_2O_3$ | 9.30 | 91 | 16.63 | 163 | 17.77 | 175 | 17.06 | 168 | 16.10 | 158 | 16.10 | 158 |
| $Fe_2O_3$ | 3.25 | 20 | 10.37 | 65 | 2.05 | 12 | 3.55 | 22 | 3.35 | 21 | 3.17 | 20 |
| $FeO$ | 5.37 | 75 | 0.53 | 7 | 5.46 | 76 | 3.68 | 51 | 0.70 | 10 | 0.42 | 6 |
| $MnO$ | 0.29 | 4 | 0.21 | 3 | 0.44 | 6 | 0.17 | 2 | 0.18 | 3 | 0.20 | 3 |
| $MgO$ | 14.93 | 370 | 4.20 | 104 | 2.57 | 64 | 2.57 | 64 | 0.32 | 8 | 0.98 | 24 |
| $CaO$ | 14.50 | 250 | 7.94 | 142 | 5.35 | 96 | 4.67 | 83 | 1.23 | 22 | 2.05 | 37 |
| $Na_2O$ | 1.32 | 21 | 4.09 | 66 | 4.65 | 75 | 4.94 | 80 | 5.61 | 90 | 5.58 | 90 |
| $K_2O$ | 1.14 | 12 | 2.11 | 22 | 3.25 | 34 | 3.43 | 36 | 5.33 | 56 | 5.20 | 55 |
| qz | | −26 | | −27 | | −15 | | −9 | | +27 | | +25 |

Note: 1) doleritic basalt with potassic feldspar; 2) olivine trachyandesite; 3) trachybasalt (Seilen); 4) trachytic basalt with olivine (Nunatak, Beerenberg); 5) trachybasalt (Sjörbukta); 6) olivine trachytic basanite (Turin Bey); 7) ankaramite (Cape Hope); 8) inclusion of hornblende basalt in trachyte (Skrukkefjöll); 9) trachyandesite (Eggjoija); 10) trachyandesite (Dollar Mt.); 11) trachyte (Jundalsveja); 12) quartz trachyte (Bombell Mt.).

## TABLE 26. Hawaiian Islands

### Honolulu Series
### (Winchell [1947]; Macdonald [1949]; Macdonald and Katsura [1964])

| Analysis No. Component | 1 % | 1 M.Q. | 2 % | 2 M.Q. | 3 % | 3 M.Q. | 4 % | 4 M.Q. | 5 % | 5 M.Q. | 6 % | 6 M.Q. | 7 % | 7 M.Q. |
|---|---|---|---|---|---|---|---|---|---|---|---|---|---|---|
| $SiO_2$ | 36.64 | 605 | 36.72 | 611 | 36.73 | 611 | 36.75 | 612 | 37.10 | 618 | 37.12 | 618 | 37.22 | 619 |
| $TiO_2$ | 2.87 | 36 | 2.82 | 35 | 2.84 | 35 | 2.41 | 30 | 2.90 | 36 | 2.64 | 33 | 2.02 | 25 |
| $Al_2O_3$ | 10.14 | 99 | 11.56 | 114 | 10.78 | 106 | 11.98 | 118 | 11.12 | 109 | 11.43 | 112 | 12.08 | 119 |
| $Fe_2O_3$ | 6.53 | 41 | 4.94 | 30 | 5.57 | 35 | 6.05 | 38 | 6.53 | 41 | 5.74 | 36 | 5.18 | 33 |
| $FeO$ | 10.66 | 149 | 8.17 | 114 | 8.76 | 122 | 7.45 | 104 | 7.31 | 102 | 8.19 | 114 | 7.88 | 110 |
| $MnO$ | 0.20 | 3 | 0.13 | 1 | 0.12 | 1 | 0.08 | 1 | 0.09 | 1 | 0.12 | 1 | 0.11 | 1 |
| $MgO$ | 10.68 | 264 | 13.27 | 329 | 12.74 | 315 | 12.08 | 300 | 12.81 | 317 | 12.44 | 308 | 12.71 | 315 |
| $CaO$ | 13.10 | 234 | 14.34 | 256 | 13.70 | 224 | 13.81 | 246 | 13.56 | 242 | 13.50 | 241 | 13.34 | 238 |
| $Na_2O$ | 4.54 | 73 | 3.93 | 64 | 3.88 | 63 | 4.75 | 77 | 4.56 | 74 | 4.52 | 73 | 5.12 | 82 |
| $K_2O$ | 1.78 | 19 | 0.62 | 6 | 0.91 | 10 | 0.91 | 10 | 1.20 | 13 | 1.07 | 12 | 0.71 | 7 |
| qz | | −76 | | −65 | | −66 | | −72 | | −71 | | −70 | | −71 |

| Analysis No. Component | 8 % | 8 M.Q. | 9 % | 9 M.Q. | 10 % | 10 M.Q. | 11 % | 11 M.Q. | 12 % | 12 M.Q. | 13 % | 13 M.Q. | 14 % | 14 M.Q. | 15 % | 15 M.Q. | 16 % | 16 M.Q. |
|---|---|---|---|---|---|---|---|---|---|---|---|---|---|---|---|---|---|---|
| $SiO_2$ | 38.57 | 642 | 38.66 | 644 | 39.44 | 656 | 42.86 | 714 | 43.98 | 733 | 43.94 | 733 | 45.00 | 749 | 45.13 | 752 | 45.30 | 754 |
| $TiO_2$ | 2.79 | 35 | 2.71 | 34 | 3.30 | 41 | 2.94 | 37 | 2.73 | 34 | 2.32 | 28 | 3.16 | 40 | 2.94 | 37 | 2.30 | 29 |
| $Al_2O_3$ | 11.71 | 115 | 10.85 | 106 | 10.24 | 100 | 11.46 | 113 | 13.49 | 132 | 12.60 | 124 | 13.00 | 127 | 16.40 | 161 | 11.76 | 116 |
| $Fe_2O_3$ | 5.21 | 33 | 5.82 | 36 | 6.52 | 41 | 3.34 | 21 | 3.53 | 22 | 3.84 | 24 | 4.04 | 25 | 3.42 | 21 | 3.98 | 25 |
| $FeO$ | 7.78 | 108 | 7.75 | 108 | 7.02 | 97 | 9.03 | 126 | 8.79 | 122 | 9.18 | 128 | 8.32 | 115 | 8.17 | 114 | 8.80 | 122 |
| $MnO$ | 0.11 | 1 | 0.10 | 1 | 0.09 | 1 | 0.13 | 1 | 0.10 | 1 | 0.09 | 1 | 0.11 | 1 | 0.07 | 1 | 0.11 | 1 |
| $MgO$ | 13.08 | 324 | 13.56 | 336 | 14.14 | 350 | 13.61 | 337 | 10.19 | 253 | 11.43 | 284 | 9.03 | 223 | 5.52 | 137 | 12.88 | 320 |
| $CaO$ | 12.84 | 229 | 13.04 | 232 | 12.32 | 219 | 11.24 | 200 | 11.11 | 198 | 10.78 | 192 | 10.16 | 181 | 11.30 | 201 | 10.72 | 191 |
| $Na_2O$ | 4.22 | 68 | 4.00 | 65 | 2.67 | 43 | 3.02 | 48 | 3.49 | 56 | 3.84 | 62 | 4.75 | 76 | 3.62 | 58 | 2.36 | 38 |
| $K_2O$ | 1.20 | 13 | 1.12 | 12 | 1.21 | 13 | 0.93 | 10 | 0.99 | 11 | 1.02 | 11 | 1.18 | 13 | 1.02 | 11 | 0.72 | 7 |
| qz | | −66 | | −64 | | −52 | | −40 | | −44 | | −48 | | −49 | | −35 | | −32 |

TABLE 26. (continued)

| Analysis No. Component | 17 | | 18 | | 19 | | 20 | | 21 | |
|---|---|---|---|---|---|---|---|---|---|---|
| | % | M.Q. | % | M.Q. | % | M.Q. | % | M.Q. | % | M.Q. |
| $SiO_2$ | 37.50 | 624 | 44.07 | 734 | 44.02 | 733 | 40.66 | 677 | 46.39 | 773 |
| $TiO_2$ | 3.21 | 40 | 2.16 | 17 | 2.05 | 26 | 2.28 | 29 | 2.13 | 27 |
| $Al_2O_3$ | 9.12 | 89 | 13.68 | 134 | 12.41 | 122 | 10.57 | 104 | 12.94 | 126 |
| $Fe_2O_3$ | 5.59 | 35 | 4.01 | 25 | 2.14 | 13 | 2.95 | 18 | 2.05 | 12 |
| FeO | 8.81 | 122 | 8.35 | 116 | 10.99 | 153 | 9.74 | 135 | 10.16 | 142 |
| MnO | 0.15 | 2 | 0.20 | 3 | 0.20 | 3 | 0.19 | 3 | 0.17 | 2 |
| MgO | 13.72 | 340 | 11.74 | 290 | 13.40 | 332 | 15.75 | 390 | 11.38 | 283 |
| CaO | 13.85 | 247 | 10.22 | 182 | 10.95 | 195 | 11.37 | 202 | 10.81 | 193 |
| $Na_2O$ | 2.69 | 44 | 3.45 | 56 | 2.39 | 39 | 1.48 | 24 | 2.43 | 39 |
| $K_2O$ | 0.63 | 6 | 1.35 | 14 | 0.62 | 6 | 0.49 | 5 | 0.93 | 10 |
| *qz* | | −53 | | − 46 | | −37 | | −39 | | −30 |

Note: Honolulu Series, Island of Oahu: 1) nepheline melilite basalt; 2) nepheline melilite basalt; 3) melilite nepheline basalt (average of six analyses); 4) melilite-nepheline basalt; 5) nepheline basalt; 6) nepheline basalt (average of six analyses); 7) nepheline-melilite basalt; 8) nepheline basalt; 9) nepheline basalt (average of three analyses); 10) nepheline ankaratritic basalt; 11) nepheline basanite; 12) nepheline basanite (average of three analyses); 13) nepheline basanite; 14) nepheline basanite (average of five analyses); 15) nosite (alkalic basalt); 16) analcime basanite. Lahaina Series, Island of Kauai (preceded by a long period of erosion); 17) nepheline-melilite basalt (Kilauea Gulf); 18-19 basanites (crater on slope). Koloa Series, Island of Maui: 20) basanite; 21) alkalic olivine basalt.

TABLE 27. East Caroline Islands

(Tsuboi [1932]; Hobbs [1953]; Stark and Hay [1963]; Yagi [1960])

| Analysis No. Component | 1 | | 2 | | 3 | | 4 | | 5 | |
|---|---|---|---|---|---|---|---|---|---|---|
| | % | M.Q. | % | M.Q. | % | M.Q. | % | M.Q. | % | M.Q. |
| $SiO_2$ | 36.79 | 613 | 38.99 | 649 | 42.74 | 712 | 47.62 | 793 | 47.92 | 798 |
| $TiO_2$ | 4.93 | 62 | 2.99 | 38 | — | — | — | — | — | — |
| $Al_2O_3$ | 13.84 | 135 | 11.80 | 116 | 11.13 | 109 | 15.20 | 149 | 13.59 | 133 |
| $Fe_2O_3$ | 10.94 | 68 | 8.96 | 56 | 9.66 | 61 | 4.50 | 55 | 9.67 | 61 |
| FeO | 3.37 | 47 | 9.48 | 132 | 3.96 | 55 | 8.85 | 63 | 2.80 | 39 |
| MnO | 0.51 | 7 | 0 | 0 | — | — | — | — | — | — |
| MgO | 8.79 | 218 | 7.42 | 183 | 11.63 | 289 | 4.90 | 121 | 6.04 | 150 |
| CaO | 12.74 | 227 | 11.34 | 202 | 13.04 | 233 | 8.42 | 150 | 9.20 | 164 |
| $Na_2O$ | 3.95 | 64 | 3.92 | 63 | 2.64 | 43 | 3.20 | 52 | 2.98 | 48 |
| $K_2O$ | 1.30 | 14 | 1.62 | 17 | 1.20 | 13 | 1.25 | 14 | 1.67 | 18 |
| *qz* | | −64 | | −60 | | −44 | | −18 | | −18 |

Note: 1) nepheline basalt (Truk); 2) nepheline basalt (Ponape); 3) nepheline basalt (Kusaie); 4) nepheline basalt (MOEK — Truk group); 5) basalt (Ponape).

## TABLE 28. Cameroons, Adamawa Plateau

### Wolff [1931]

| Analysis No. Component | 1 % | 1 M.Q. | 2 % | 2 M.Q. | 3 % | 3 M.Q. | 4 % | 4 M.Q. | 5 % | 5 M.Q. | 6 % | 6 M.Q. | 7 % | 7 M.Q. |
|---|---|---|---|---|---|---|---|---|---|---|---|---|---|---|
| $SiO_2$ | 48.12 | 801 | 46.48 | 774 | 39.36 | 654 | 40.10 | 668 | 39.97 | 665 | 40.15 | 668 | 38.39 | 639 |
| $TiO_2$ | 1.58 | 20 | 1.22 | 15 | 3.62 | 45 | 3.64 | 45 | 3.34 | 42 | 3.21 | 40 | 4.44 | 55 |
| $Al_2O_3$ | 16.21 | 159 | 19.00 | 186 | 13.66 | 134 | 15.27 | 150 | 17.30 | 170 | 17.32 | 170 | 12.64 | 124 |
| $Fe_2O_3$ | 2.94 | 18 | 4.74 | 29 | 7.42 | 46 | 10.13 | 63 | 7.41 | 46 | 7.25 | 46 | 7.40 | 46 |
| FeO | 7.56 | 105 | 2.30 | 32 | 4.45 | 62 | 1.85 | 26 | 3.05 | 42 | 4.00 | 56 | 6.15 | 86 |
| MnO | — | — | — | — | 0.08 | 1 | 0.08 | 1 | 0.08 | 1 | 0.09 | 1 | — | — |
| MgO | 7.24 | 180 | 2.49 | 62 | 4.46 | 111 | 4.59 | 114 | 3.82 | 95 | 4.43 | 110 | 6.46 | 160 |
| CaO | 8.71 | 155 | 4.35 | 78 | 11.37 | 202 | 12.08 | 216 | 10.53 | 216 | 11.78 | 210 | 14.17 | 252 |
| $Na_2O$ | 2.91 | 47 | 8.46 | 136 | 5.78 | 94 | 4.78 | 77 | 5.14 | 83 | 5.99 | 97 | 4.35 | 70 |
| $K_2O$ | 1.44 | 15 | 6.78 | 72 | 1.44 | 15 | 3.34 | 35 | 3.56 | 38 | 3.78 | 40 | 2.44 | 26 |
| $qz$ | | −21 | | −105 | | −72 | | −70 | | −74 | | −84 | | −68 |

| Analysis No. Component | 8 % | 8 M.Q. | 9 % | 9 M.Q. | 10 % | 10 M.Q. | 11 % | 11 M.Q. | 12 % | 12 M.Q. | 13 % | 13 M.Q. | 14 % | 14 M.Q. |
|---|---|---|---|---|---|---|---|---|---|---|---|---|---|---|
| $SiO_2$ | 45.99 | 766 | 44.74 | 745 | 46.16 | 769 | 46.90 | 781 | 45.77 | 762 | 45.89 | 764 | 38.30 | 804 |
| $TiO_2$ | 3.42 | 43 | 3.39 | 43 | 3.87 | 49 | 2.90 | 36 | 2.99 | 38 | 3.36 | 42 | 2.85 | 36 |
| $Al_2O_3$ | 16.45 | 162 | 6.85 | 68 | 16.62 | 159 | 19.60 | 192 | 13.77 | 135 | 15.32 | 150 | 16.90 | 166 |
| $Fe_2O_3$ | 1.40 | 9 | 14.74 | 92 | 2.51 | 16 | 4.60 | 29 | 4.82 | 30 | 4.52 | 28 | 5.95 | 38 |
| FeO | 10.28 | 143 | 1.08 | 15 | 8.52 | 118 | 8.90 | 124 | 8.60 | 120 | 7.06 | 98 | 8.35 | 116 |
| MnO | 0.15 | 2 | 0.39 | 6 | 0.15 | 2 | — | — | 0.16 | 2 | 0.18 | 3 | — | — |
| MgO | 6.03 | 150 | 13.31 | 330 | 5.11 | 127 | 4.25 | 106 | 6.85 | 170 | 7.99 | 198 | 9.20 | 228 |
| CaO | 9.83 | 176 | 12.56 | 224 | 9.90 | 177 | 6.95 | 124 | 11.14 | 199 | 10.32 | 184 | 10.40 | 185 |
| $Na_2O$ | 3.49 | 56 | 1.92 | 31 | 4.20 | 68 | 5.00 | 81 | 3.26 | 53 | 3.10 | 50 | 3.05 | 49 |
| $K_2O$ | 1.57 | 17 | 0.32 | 3 | 1.94 | 20 | 1.30 | 14 | 1.48 | 16 | 1.97 | 21 | 1.75 | 18 |
| $qz$ | | −35 | | −30 | | −41 | | −43 | | −33 | | −37 | | −36 |

Note: 1) dolerite (Mao Manduku, Adamawa); 2) leucitite (Etinde, Cameroons); 3) haüynophyre (Etinde, Cameroons); 4) leucite-nephelinite (Etinde, Cameroons); 5) leucite-nephelinite (Etinde, Cameroons); 6) nephelinite (Etinde, Cameroons); 7) nephelinite (Etinde, Cameroons); 8) porphyritic (feldspar) basalt (Cameroons); 9) porphyritic (feldspar) basalt (Cameroons); 10) porphyritic (feldspar) basalt (Buea Waterfall); 11) porphyritic (feldspar) basalt (Bonking); 12) porphyritic (feldspar) basalt (Mannus); 13) porphyritic (feldspar) basalt (Bibundi); 14) limburgite (Elephant Lake).

TABLE 29. São Tomé Island

(W. Boese [1912])

| Compo- nents | Analysis No. | | | | | |
|---|---|---|---|---|---|---|
| | 1 | | 2 | | 3 | |
| | % | M.Q. | % | M.Q. | % | M.Q. |
| $SiO_2$ | 44.96 | 749 | 46.76 | 779 | 60.34 | 1006 |
| $TiO_2$ | 1.94 | 24 | 2.05 | 26 | 0.62 | 8 |
| $Al_2O_3$ | 18.06 | 177 | 14.80 | 145 | 20.69 | 203 |
| $Fe_2O_3$ | 4.04 | 25 | 8.71 | 54 | 1.53 | 10 |
| FeO | 7.82 | 108 | 8.32 | 115 | 1.15 | 16 |
| MnO | 0.26 | 4 | 0.34 | 5 | | |
| MgO | 5.58 | 140 | 1.70 | 43 | 0.38 | 10 |
| CaO | 8.99 | 161 | 9.46 | 169 | 1.97 | 36 |
| $Na_2O$ | 4.56 | 74 | 2.75 | 45 | 3.94 | 64 |
| $K_2O$ | 2.54 | 27 | 2.48 | 27 | 8.15 | 87 |

Note: 1) trachydolerite (Mt. Coffee); 2) basalt (Mt. Coffee);
3) aegirine trachyte (Mt. Coffee).

TABLE 30. Fernando Poo

(W. Boese [1912])

| Compo- nents | Analysis No. | | | |
|---|---|---|---|---|
| | 1 | | 2 | |
| | % | M.Q. | % | M.Q. |
| $SiO_2$ | 45.73 | 762 | 47.81 | 787 |
| $TiO_2$ | 3.23 | 40 | 2.87 | 36 |
| $Al_2O_3$ | 11.20 | 110 | 13.71 | 134 |
| $Fe_2O_3$ | 6.46 | 41 | 6.31 | 39 |
| FeO | 5.53 | 76 | 5.83 | 81 |
| MnO | 0.54 | 8 | 0.11 | 2 |
| MgO | 11.36 | 284 | 7.47 | 187 |
| CaO | 10.45 | 187 | 9.28 | 166 |
| $Na_2O$ | 2.19 | 35 | 2.27 | 37 |
| $K_2O$ | 1.36 | 15 | 1.62 | 17 |

Note: 1) basalt (Santa Isabel); 2) basalt.

TABLE 31. The Canary Islands

Fuerteventura
(Bourcart and Jérémine [1938]; Hausen [1959])

| Analysis No. / Component | 1 | | 2 | | 3 | | 4 | | 5 | | 6 | | 7 | |
|---|---|---|---|---|---|---|---|---|---|---|---|---|---|---|
| | % | M.Q. | % | M.Q. | % | M.Q. | % | M.Q. | % | M.Q. | % | M.Q. | % | M.Q. |
| $SiO_2$ | 35.66 | 594 | 39.53 | 658 | 40.01 | 666 | 40.52 | 674 | 41.10 | 684 | 41.98 | 699 | 42.78 | 713 |
| $TiO_2$ | 6.40 | 80 | 2.68 | 34 | 2.18 | 28 | 4.65 | 58 | 3.28 | 41 | 3.74 | 47 | 3.62 | 45 |
| $Al_2O_3$ | 6.64 | 65 | 9.75 | 95 | 12.86 | 126 | 12.51 | 58 | 12.64 | 124 | 14.94 | 146 | 13.40 | 131 |
| $Fe_2O_3$ | 13.75 | 86 | 5.37 | 34 | 6.65 | 42 | 4.34 | 27 | 4.05 | 25 | 4.20 | 26 | 11.43 | 71 |
| FeO | 11.31 | 157 | 7.17 | 100 | 4.88 | 68 | 9.45 | 131 | 8.42 | 117 | 9.12 | 127 | 1.71 | 24 |
| MnO | 0.12 | 1 | 0.14 | 2 | 0.11 | 1 | 0.21 | 3 | 0.22 | 3 | 0.18 | 3 | 0.12 | 2 |
| MgO | 10.10 | 250 | 14.53 | 361 | 11.48 | 284 | 8.62 | 214 | 12.37 | 306 | 7.13 | 177 | 11.66 | 274 |
| CaO | 15.06 | 269 | 9.20 | 164 | 11.89 | 212 | 12.65 | 226 | 11.36 | 202 | 8.24 | 154 | 10.84 | 193 |
| $Na_2O$ | 0.49 | 8 | 1.84 | 30 | 3.09 | 50 | 3.67 | 59 | 3.14 | 51 | 2.40 | 39 | 2.39 | 39 |
| $K_2O$ | 0.20 | 2 | 0.78 | 9 | 0.69 | 7 | 1.24 | 13 | 1.58 | 17 | 1.42 | 15 | 1.21 | 19 |
| $qz$ | | −41 | | −39 | | −47 | | −49 | | −53 | | −34 | | −38 |

## TABLE 31. (continued)

| Analysis No. | 8 | | 9 | | 10 | | 11 | | 12 | | 13 | | 14 | |
|---|---|---|---|---|---|---|---|---|---|---|---|---|---|---|
| Component | % | M.Q. | % | M.Q. | % | M.Q. | % | M.Q. | % | M.Q. | % | M.Q. | % | M.Q. |
| $SiO_2$ | 42.85 | 713 | 42.85 | 714 | 43.21 | 719 | 44.10 | 734 | 44.62 | 743 | 44.06 | 734 | 44.62 | 743 |
| $TiO_2$ | 2.15 | 27 | 1.85 | 24 | 2.55 | 32 | 3.96 | 49 | 2.92 | 36 | 4.95 | 61 | 2.08 | 35 |
| $Al_2O_3$ | 14.21 | 139 | 7.52 | 74 | 8.00 | 78 | 11.65 | 114 | 14.13 | 138 | 12.66 | 125 | 22.12 | 217 |
| $Fe_2O_3$ | 5.35 | 34 | 5.68 | 36 | 3.99 | 25 | 2.19 | 19 | 1.58 | 10 | 7.22 | 45 | 2.63 | 16 |
| FeO | 7.52 | 104 | 10.40 | 145 | 9.75 | 136 | 9.71 | 135 | 8.67 | 121 | 5.44 | 76 | 5.62 | 78 |
| MnO | 0.21 | 3 | 0.27 | 4 | 0.20 | 3 | 0.15 | 2 | 0.20 | 3 | 0.14 | 2 | 0.08 | 1 |
| MgO | 8.94 | 222 | 16.47 | 408 | 16.12 | 400 | 11.37 | 282 | 12.50 | 310 | 6.49 | 161 | 4.80 | 119 |
| CaO | 9.59 | 171 | 12.28 | 219 | 12.55 | 224 | 11.42 | 203 | 10.65 | 190 | 12.66 | 219 | 15.56 | 277 |
| $Na_2O$ | 3.44 | 55 | 1.05 | 17 | 1.68 | 27 | 2.61 | 42 | 2.71 | 44 | 1.71 | 27 | 1.44 | 23 |
| $K_2O$ | 1.05 | 11 | 0.29 | 3 | 0.72 | 8 | 1.03 | 11 | 1.14 | 12 | 1.57 | 17 | 0.42 | 4 |
| qz | | −42 | | −33 | | −37 | | −37 | | −37 | | −22 | | −15 |

| Analysis No. | 15 | | 16 | | 17 | | 18 | | 19 | | 20 | |
|---|---|---|---|---|---|---|---|---|---|---|---|---|
| Component | % | M.Q. | % | M.Q. | % | M.Q. | % | M.Q. | % | M.Q. | % | M.Q. |
| $SiO_2$ | 44.78 | 746 | 46.60 | 776 | 46.76 | 779 | 49.30 | 821 | 50.94 | 848 | 51.50 | 858 |
| $TiO_2$ | 2.81 | 35 | 2.44 | 30 | 1.60 | 20 | 2.90 | 36 | 2.55 | 32 | 3.24 | 40 |
| $Al_2O_3$ | 12.87 | 126 | 6.18 | 61 | 17.33 | 170 | 14.96 | 147 | 17.88 | 176 | 17.43 | 171 |
| $Fe_2O_3$ | 2.69 | 17 | 4.64 | 29 | 1.89 | 12 | 5.77 | 36 | 2.52 | 16 | 1.44 | 9 |
| FeO | 9.38 | 131 | 5.80 | 81 | 5.98 | 83 | 4.56 | 64 | 5.74 | 80 | 6.70 | 93 |
| MnO | 0.18 | 3 | 0.18 | 3 | 0.10 | 1 | 0.18 | 3 | 0.23 | 3 | 0.12 | 1 |
| MgO | 12.02 | 298 | 12.80 | 317 | 8.12 | 201 | 3.15 | 78 | 3.55 | 88 | 3.19 | 79 |
| CaO | 9.49 | 169 | 19.48 | 348 | 16.34 | 291 | 8.03 | 143 | 7.38 | 132 | 8.12 | 144 |
| $Na_2O$ | 2.42 | 39 | 1.10 | 18 | 1.27 | 20 | 4.23 | 68 | 5.15 | 83 | 4.71 | 76 |
| $K_2O$ | 1.08 | 12 | 0.21 | 2 | 0.38 | 4 | 2.08 | 22 | 2.40 | 25 | 2.13 | 22 |
| qz | | −34 | | −22 | | −14 | | −23 | | −33 | | −23 |

| Analysis No. | 21 | | 22 | | 23 | | 24 | | 25 | | 26 | |
|---|---|---|---|---|---|---|---|---|---|---|---|---|
| Component | % | M.Q. | % | M.Q. | % | M.Q. | % | M.Q. | % | M.Q. | % | M.Q. |
| $SiO_2$ | 59.45 | 990 | 65.51 | 1091 | 65.84 | 1096 | 66.05 | 1100 | 66.28 | 1103 | 68.34 | 1138 |
| $TiO_2$ | 1.10 | 14 | 0.36 | 5 | 0.52 | 7 | 0.23 | 3 | 1.27 | 15 | | |
| $Al_2O_3$ | 18.53 | 181 | 17.50 | 172 | 16.07 | 158 | 18.06 | 178 | 13.27 | 130 | 13.68 | 134 |
| $Fe_2O_3$ | 2.46 | 16 | 2.64 | 16 | 2.58 | 16 | 2.07 | 13 | 5.60 | 35 | 4.78 | 30 |
| FeO | 1.50 | 21 | 0.27 | 4 | 1.08 | 25 | 0.18 | 3 | 0.88 | 12 | 0.15 | 2 |
| MnO | 0.13 | 2 | 0.12 | 1 | 0.17 | 2 | 0.02 | 0 | 0.13 | 2 | — | — |
| MgO | 0.51 | 12 | 0.06 | 1 | 0.22 | 5 | 0.01 | 0 | 0.04 | 1 | 0.15 | 4 |
| CaO | 2.75 | 49 | 0.06 | 1 | 0.92 | 16 | 1.32 | 23 | 0.70 | 13 | 0.93 | 17 |
| $Na_2O$ | 7.95 | 128 | 7.54 | 122 | 6.91 | 111 | 4.82 | 77 | 5.61 | 90 | 5.39 | 87 |
| $K_2O$ | 4.30 | 46 | 5.30 | 56 | 4.86 | 52 | 4.95 | 52 | 4.23 | 45 | 4.81 | 51 |
| qz | −37 | | | +2 | | +15 | | +63 | | +60 | | +68 |

Note: 1) pyroxenite; 2) picrite (Toto Woods); 3) spilite from trap formation (eastern slope of Cuesta de la Rija); 4) olivine basalt (crest of volcano between Amfienta and Villa); 5) picritic basalt lava (caldera, Gairia); 6) olivine basalt (upper slope of lava on Mt. Gomma); 7) red basalt; 8) trap (older lava) (Pajaro Valley); 9) feldspar pyroxenite with augite and olivine; 10) peridotite (La Oliva region); 11) basanitoid; 12) olivine basalt (Mt. Arrabales); 13) basalt with labradorite; 14) diorite with olivine; 15) olivine basalt (north of Pozo R. valley); 16) pyroxenite (Pajaro Valley); 17) gabbro with olivine; 18) essexite (Pajaro Valley); 19) essexite (Rio de Palomas Valley); 20) diorite with hornblende and biotite; 21) alkalic syenite with augite and hornblende; 22) alkalic syenite (Puerto de Rozarno); 23) syenite with aegirine; 24) quartz keratophyre (northeastern end of island); 25) comendite; 26) rhyolite.

## TABLE 32. Gran Canaria

### (Smulikowski [1945])

| Analysis No. Component | 1 % | 1 M.Q. | 2 % | 2 M.Q. | 3 % | 3 M.Q. | 4 % | 4 M.Q. | 5 % | 5 M.Q. | 6 % | 6 M.Q. | 7 % | 7 M.Q. | 8 % | 8 M.Q. |
|---|---|---|---|---|---|---|---|---|---|---|---|---|---|---|---|---|
| SiO₂ | 40.02 | 666 | 40.60 | 676 | 46.74 | 778 | 52.30 | 871 | 59.17 | 985 | 62.74 | 1045 | 65.86 | 1097 | 66.78 | 1112 |
| TiO₂ | 8.18 | 53 | 5.56 | 70 | 2.12 | 26 | 1.42 | 18 | 0.94 | 12 | 0.61 | 8 | 1.00 | 13 | 0.68 | 9 |
| Al₂O₃ | 10.82 | 106 | 13.46 | 132 | 15.18 | 149 | 19.71 | 193 | 17.23 | 169 | 17.85 | 176 | 10.35 | 102 | 11.00 | 108 |
| Fe₂O₃ | 4.58 | 29 | 4.89 | 30 | 4.19 | 26 | 2.68 | 17 | 3.69 | 23 | 1.96 | 12 | 6.89 | 43 | 3.09 | 19 |
| FeO | 8.84 | 123 | 9.00 | 125 | 4.12 | 57 | 2.59 | 36 | 1.00 | 14 | 1.33 | 18 | 0.62 | 8 | 3.15 | 43 |
| MnO | 0.18 | 3 | 0.11 | 1 | 0.18 | 3 | 0.12 | 1 | 0.18 | 3 | 0.20 | 3 | 0.25 | 4 | 0.31 | 4 |
| MgO | 14.03 | 348 | 6.30 | 156 | 5.34 | 133 | 0.85 | 21 | 0.77 | 19 | 0.72 | 17 | 2.06 | 51 | 0.38 | 10 |
| CaO | 12.34 | 220 | 14.28 | 255 | 9.08 | 162 | 4.50 | 80 | 0.89 | 16 | 1.66 | 30 | 0.78 | 14 | 0.60 | 11 |
| Na₂O | 1.47 | 24 | 2.49 | 40 | 4.69 | 76 | 8.17 | 132 | 7.33 | 118 | 6.32 | 102 | 4.53 | 73 | 5.10 | 82 |
| K₂O | 1.15 | 12 | 1.80 | 19 | 3.57 | 38 | 4.54 | 48 | 5.51 | 58 | 4.94 | 52 | 4.70 | 50 | 4.05 | 43 |
| qz | | −42 | | −44 | | −52 | | −72 | | −32 | | +2 | | +72 | | +92 |

Note: 1) limburgite-ankaramite; 2) analcime basalt; 3) mesocratic trachyte; 4) trachytite; 5) phonolite with aegirine; 6) aegirine metaphonolite; 7) pantellerite with aegirine; 8) vitrophyric pantellerite.

## TABLE 33. Tenerife

### (Smulikowski [1945]; Wolff [1931]; Gagel [1912]; Jérémine [1930])

| Analysis No. Component | 1 % | 1 M.Q. | 2 % | 2 M.Q. | 3 % | 3 M.Q. | 4 % | 4 M.Q. | 5 % | 5 M.Q. | 6 % | 6 M.Q. | 7 % | 7 M.Q. |
|---|---|---|---|---|---|---|---|---|---|---|---|---|---|---|
| SiO₂ | 40.05 | 667 | 40.86 | 680 | 41.00 | 683 | 41.57 | 693 | 41.49 | 691 | 42.45 | 707 | 42.77 | 712 |
| TiO₂ | 4.02 | 50 | 5.31 | 69 | 4.94 | 62 | 4.76 | 60 | 3.50 | 44 | 4.03 | 50 | 3.08 | 39 |
| Al₂O₃ | 3.82 | 135 | 13.40 | 131 | 16.06 | 158 | 12.36 | 122 | 16.27 | 160 | 11.77 | 116 | 15.80 | 155 |
| Fe₂O₃ | 8.47 | 53 | 7.60 | 48 | 2.73 | 17 | 5.75 | 36 | 3.08 | 19 | 3.61 | 23 | 3.34 | 21 |
| FeO | 7.47 | 104 | 5.89 | 82 | 8.89 | 124 | 8.44 | 117 | 8.57 | 119 | 9.10 | 126 | 10.85 | 151 |
| MnO | 0.18 | 3 | 0.20 | 3 | 0.18 | 3 | 0.17 | 2 | 0.45 | 6 | 0.17 | 2 | 0.18 | 3 |
| MgO | 7.57 | 189 | 7.74 | 193 | 8.05 | 201 | 7.96 | 199 | 8.97 | 222 | 12.21 | 305 | 9.04 | 224 |
| CaO | 12.00 | 214 | 11.71 | 209 | 11.62 | 207 | 13.04 | 232 | 11.70 | 209 | 11.61 | 208 | 9.77 | 174 |
| Na₂O | 3.52 | 56 | 2.20 | 35 | 2.44 | 36 | 2.82 | 45 | 3.26 | 53 | 2.33 | 37 | 3.49 | 56 |
| K₂O | 1.00 | 11 | 2.40 | 26 | 2.08 | 22 | 1.23 | 13 | 1.24 | 13 | 1.47 | 16 | 1.65 | 18 |
| qz | | −51 | | −41 | | −43 | | −44 | | −48 | | −42 | | −49 |

| Analysis No. Component | 8 % | 8 M.Q. | 9 % | 9 M.Q. | 10 % | 10 M.Q. | 11 % | 11 M.Q. | 12 % | 12 M.Q. | 13 % | 13 M.Q. | 14 % | 14 M.Q. |
|---|---|---|---|---|---|---|---|---|---|---|---|---|---|---|
| SiO₂ | 43.48 | 725 | 44.25 | 737 | 44.64 | 743 | 45.15 | 753 | 45.25 | 754 | 45.72 | 761 | 48.60 | 809 |
| TiO₂ | 3.88 | 49 | 4.00 | 50 | — | — | 4.08 | 51 | 3.94 | 49 | — | — | — | — |
| Al₂O₃ | 12.65 | 124 | 15.90 | 156 | 16.55 | 162 | 14.05 | 138 | 15.10 | 148 | 20.72 | 203 | 21.03 | 206 |
| Fe₂O₃ | 4.64 | 29 | 3.10 | 19 | 7.53 | 47 | 2.83 | 18 | 4.93 | 31 | 4.22 | 26 | 0.96 | 6 |
| FeO | 8.08 | 113 | 8.62 | 119 | 7.52 | 104 | 7.63 | 106 | 7.66 | 107 | 3.66 | 51 | 7.34 | 102 |
| MnO | 0.19 | 3 | 0.17 | 2 | — | — | 0.17 | 3 | 0.16 | 2 | — | — | — | — |
| MgO | 9.66 | 241 | 7.10 | 180 | 9.52 | 236 | 5.89 | 147 | 7.03 | 176 | 3.76 | 93 | 3.72 | 92 |
| CaO | 11.90 | 213 | 10.38 | 186 | 11.25 | 200 | 10.44 | 186 | 9.77 | 174 | 9.62 | 171 | 8.10 | 144 |
| Na₂O | 2.32 | 37 | 3.88 | 63 | 2.47 | 40 | 5.68 | 92 | 3.50 | 56 | 7.52 | 121 | 5.67 | 91 |
| K₂O | 1.45 | 15 | 1.85 | 19 | 1.34 | 14 | 1.46 | 16 | 1.82 | 19 | 1.10 | 12 | 2.26 | 24 |
| qz | | −36 | | −46 | | −38 | | −56 | | −39 | | −68 | | −48 |

160

## TABLE 33. (continued)

| Component | 15 % | 15 M.Q. | 16 % | 16 M.Q. | 17 % | 17 M.Q. | 18 % | 18 M.Q. | 19 % | 19 M.Q. | 20 % | 20 M.Q. | 21 % | 21 M.Q. |
|---|---|---|---|---|---|---|---|---|---|---|---|---|---|---|
| $SiO_2$ | 48.67 | 811 | 49.73 | 828 | 50.86 | 847 | 54.24 | 903 | 54.24 | 903 | 55.30 | 922 | 55.76 | 928 |
| $TiO_2$ | 2.12 | 26 | 0.38 | 5 | 2.29 | 29 | 1.96 | 25 | 0.51 | 6 | 1.34 | 16 | — | — |
| $Al_2O_3$ | 18.15 | 178 | 22.84 | 224 | 17.72 | 174 | 16.25 | 160 | 20.84 | 204 | 19.45 | 191 | 17.56 | 173 |
| $Fe_2O_3$ | 5.41 | 34 | 6.10 | 38 | 3.45 | 22 | 3.31 | 21 | 2.26 | 14 | 2.83 | 18 | 4.64 | 29 |
| $FeO$ | 5.03 | 70 | 5.98 | 83 | 5.71 | 80 | 3.74 | 52 | 2.09 | 29 | 2.08 | 29 | 2.09 | 29 |
| $MnO$ | 0.22 | 3 | — | — | 0.18 | 3 | 0.14 | 2 | 0.15 | 2 | 0.15 | 2 | 0.82 | 11 |
| $MgO$ | 4.08 | 102 | 3.91 | 97 | 3.44 | 85 | 2.46 | 61 | 1.21 | 30 | 1.66 | 41 | 2.76 | 67 |
| $CaO$ | 7.68 | 138 | 9.0 | 161 | 6.84 | 122 | 5.02 | 89 | 2.99 | 54 | 3.61 | 64 | 5.46 | 97 |
| $Na_2O$ | 4.13 | 67 | 1.89 | 31 | 5.35 | 86 | 6.39 | 103 | 9.22 | 148 | 7.89 | 127 | 6.82 | 110 |
| $K_2O$ | 2.24 | 23 | 1.04 | 11 | 2.45 | 26 | 2.95 | 32 | 3.84 | 40 | 4.06 | 44 | 1.42 | 15 |
| $qz$ | | −30 | | −3 | | −36 | | −33 | | −72 | | −55 | | −24 |

| Component | 22 % | 22 M.Q. | 23 % | 23 M.Q. | 24 % | 24 M.Q. | 25 % | 25 M.Q. | 26 % | 26 M.Q. | 27 % | 27 M.Q. | 28 % | 28 M.Q. |
|---|---|---|---|---|---|---|---|---|---|---|---|---|---|---|
| $SiO_2$ | 56.06 | 933 | 56.56 | 943 | 57.76 | 962 | 58.64 | 976 | 59.00 | 982 | 59.08 | 984 | 59.18 | 986 |
| $TiO_2$ | — | — | 0.60 | 8 | | | 0.19 | 3 | 0.84 | 10 | 0.80 | 10 | 0.98 | 13 |
| $Al_2O_3$ | 21.05 | 206 | 20.90 | 205 | 17.56 | 173 | 20.86 | 205 | 20.15 | 198 | 19.92 | 195 | 20.07 | 197 |
| $Fe_2O_3$ | 3.42 | 21 | 1.56 | 10 | 4.64 | 29 | 1.60 | 10 | 0.65 | 4 | 1.00 | 6 | 0.96 | 6 |
| $FeO$ | 2.14 | 30 | 1.41 | 19 | 2.09 | 29 | 2.23 | 31 | 2.93 | 41 | 2.07 | 29 | 2.48 | 35 |
| $MnO$ | — | — | 0.16 | 2 | 0.82 | 11 | 0.11 | 1 | 0.19 | 3 | 0.17 | 2 | 0.18 | 3 |
| $MgO$ | 1.69 | 42 | 0.72 | 18 | 2.76 | 68 | 0.29 | 7 | 0.67 | 16 | 0.64 | 16 | 0.50 | 12 |
| $CaO$ | 3.76 | 67 | 2.69 | 48 | 5.46 | 97 | 1.57 | 28 | 1.39 | 25 | 1.38 | 25 | 1.82 | 32 |
| $Na_2O$ | 8.44 | 136 | 7.06 | 114 | 6.82 | 110 | 9.79 | 158 | 8.45 | 136 | 9.80 | 158 | 8.65 | 140 |
| $K_2O$ | 2.67 | 29 | 4.87 | 52 | 1.42 | 15 | 4.62 | 49 | 5.49 | 58 | 4.96 | 53 | 4.95 | 53 |
| $qz$ | | −51 | | −42 | | −18 | | −70 | | −62 | | −40 | | −56 |

| Component | 29 % | 29 M.Q. | 30 % | 30 M.Q. | 31 % | 31 M.Q. | 32 % | 32 M.Q. | 33 % | 33 M.Q. |
|---|---|---|---|---|---|---|---|---|---|---|
| $SiO_2$ | 59.32 | 988 | 59.46 | 991 | 59.78 | 996 | 59.82 | 997 | 60.68 | 1011 |
| $TiO_2$ | 0.71 | 9 | 0.45 | 6 | 0.73 | 9 | 0.54 | 7 | 0.79 | 10 |
| $Al_2O_3$ | 19.65 | 192 | 19.49 | 191 | 19.12 | 187 | 18.20 | 178 | 19.03 | 186 |
| $Fe_2O_3$ | 2.41 | 15 | 1.45 | 9 | 2.95 | 19 | 2.95 | 18 | 0.57 | 4 |
| $FeO$ | 1.27 | 18 | 2.30 | 32 | 0.71 | 10 | 1.12 | 15 | 3.16 | 44 |
| $MnO$ | 0.14 | 2 | 0.24 | 3 | 0.22 | 3 | 0.15 | 2 | 0.16 | 2 |
| $MgO$ | 0.73 | 18 | 1.05 | 26 | 0.42 | 11 | 1.52 | 38 | 0.39 | 10 |
| $CaO$ | 1.42 | 25 | 1.66 | 30 | 0.86 | 15 | 0.96 | 17 | 1.68 | 30 |
| $Na_2O$ | 8.09 | 131 | 9.34 | 151 | 8.39 | 135 | 8.18 | 132 | 8.44 | 136 |
| $K_2O$ | 5.64 | 60 | 4.34 | 46 | 4.99 | 53 | 4.84 | 51 | 4.69 | 50 |
| $qz$ | | −53 | | −41 | | −46 | | −40 | | −43 |

Note: 1) analcime tephrite; 2) trachytephrite with biotite and hornblende; 3) trachytephrite; 4) trachybasalt with analcime and bowlingite; 5) basalt (Anaga Peninsula); 6) melanocratic basanitoid, rich in olivine; 7) basalt (Esperanza); 8) olivine basalt; 9) trachytephrite (Pico de Teide complex); 10) basalt; 11) analcime trachytephrite; 12) tephrite, young basaltic lava; 13) tephrite; 14) tephritoid; 15) trachyandesite (Pico de Teide basaltic complex); 16) basalt; 17) trachyandesite, containing olivine (Pico de Teide basaltic complex); 18-19) basaltic complex (Pico de Teide) — 18 is trachyandesite (Mt. Grande) and 19 is phonolite with nosean (north of Mercedes); 20) analcime phonolite with plagioclase (Pico de Teide complex); 21) trachydolerite; 22) haüyne tephrite; 23) phonolite with plagioclase; 24) trachydolerite (Fuente Agria); 25) vitrophyric phonolitoid; 26) vitrophyric phonolitoid (Pico de Teide complex); 27) phonolite obsidian; 28) phonolite (Pico de Teide basaltic complex); 29) phonolite obsidian; 30) phonolite (Pico de Teide complex); 31) foyaite (inclusion in phonolite); 32) phonolite with sodalite and nosean; 33) phonolite with haüynite (31-33 from basaltic complex of Pico de Teide).

## TABLE 34. Gomera

### (F. Wolff [1931])

| Analysis No. | 1 | | 2 | | 3 | | 4 | | 5 | | 6 | |
|---|---|---|---|---|---|---|---|---|---|---|---|---|
| Component | % | M.Q. | % | M.Q. | % | M.Q. | % | M.Q. | % | M.Q. | % | M.Q. |
| SiO$_2$ | 40.60 | 676 | 42.73 | 712 | 42.86 | 714 | 43.80 | 729 | 44.50 | 741 | 44.63 | 743 |
| TiO$_2$ | 5.95 | 74 | 5.71 | 71 | 6.16 | 77 | 4.30 | 54 | 4.18 | 53 | 4.94 | 62 |
| Al$_2$O$_3$ | 13.66 | 134 | 16.53 | 162 | 13.03 | 127 | 17.44 | 171 | 16.60 | 163 | 15.00 | 147 |
| Fe$_2$O$_3$ | 9.30 | 58 | 2.14 | 13 | 2.03 | 12 | 2.21 | 14 | 2.40 | 15 | 5.96 | 38 |
| FeO | 6.39 | 89 | 8.92 | 124 | 8.98 | 125 | 9.18 | 128 | 9.18 | 128 | 4.90 | 68 |
| MnO | | | 0.20 | 3 | 0.18 | 3 | | | | | 0.19 | 3 |
| MgO | 7.85 | 195 | 5.76 | 143 | 8.48 | 210 | 5.15 | 128 | 6.63 | 165 | 6.23 | 155 |
| CaO | 11.90 | 212 | 10.20 | 182 | 11.30 | 201 | 11.55 | 206 | 11.11 | 198 | 9.75 | 174 |
| Na$_2$O | 2.15 | 34 | 3.07 | 49 | 2.93 | 47 | 3.59 | 58 | 3.59 | 58 | 3.91 | 63 |
| K$_2$O | 1.03 | 11 | 1.48 | 16 | 1.62 | 17 | 0.75 | 8 | 1.23 | 13 | 1.48 | 16 |
| qz | | −38 | | −38 | | −26 | | −36 | | −39 | | −40 |

| Analysis No. | 7 | | 8 | | 9 | | 10 | | 11 | | 12 | |
|---|---|---|---|---|---|---|---|---|---|---|---|---|
| Component | % | M.Q. | % | M.Q. | % | M.Q. | % | M.Q. | % | M.Q. | % | M.Q. |
| SiO$_2$ | 44.98 | 749 | 46.50 | 774 | 56.50 | 941 | 57.70 | 961 | 58.32 | 971 | 60.50 | 1.007 |
| TiO$_2$ | 3.43 | 43 | 2.33 | 29 | 0.26 | 4 | 1.23 | 15 | 0.50 | 6 | 0.72 | 9 |
| Al$_2$O$_3$ | 17.24 | 169 | 18.15 | 178 | 20.22 | 198 | 19.20 | 188 | 19.41 | 190 | 19.80 | 194 |
| Fe$_2$O$_3$ | 11.16 | 70 | 7.85 | 49 | 1.33 | 8 | 1.66 | 11 | 3.21 | 20 | 1.98 | 12 |
| FeO | 4.68 | 65 | 2.52 | 35 | 1.46 | 20 | 2.52 | 35 | 1.14 | 16 | 1.15 | 16 |
| MnO | 0.22 | 3 | 0.27 | 4 | 0.23 | 3 | | | 0.27 | 4 | 0.25 | 4 |
| MgO | 1.21 | 30 | 4.25 | 106 | 0.78 | 19 | 1.56 | 39 | 0.45 | 11 | 1.06 | 26 |
| CaO | 9.69 | 173 | 10.03 | 179 | 1.05 | 19 | 4.28 | 77 | 0.74 | 13 | 2.65 | 47 |
| Na$_2$O | 5.40 | 87 | 3.46 | 56 | 9.65 | 156 | 7.50 | 121 | 5.30 | 85 | 6.60 | 106 |
| K$_2$O | 1.76 | 19 | 1.22 | 13 | 3.20 | 34 | 2.95 | 32 | 9.35 | 100 | 2.80 | 30 |
| qz | | −52 | | −25 | | −61 | | −32 | | −52 | | +4 |

| Analysis No. | 13 | | 14 | | 15 | | 16 | | 17 | |
|---|---|---|---|---|---|---|---|---|---|---|
| Component | % | M.Q. | % | M.Q. | % | M.Q. | % | M.Q. | % | M.Q. |
| SiO$_2$ | 60.50 | 1.007 | 62.13 | 1.035 | 62.51 | 1.041 | 63.51 | 1.051 | 64.71 | 1,077 |
| TiO$_2$ | 0.68 | 9 | 0.31 | 4 | 0.52 | 6 | 0.39 | 5 | 0.20 | 3 |
| Al$_2$O$_3$ | 21.50 | 211 | 16.93 | 166 | 19.40 | 190 | 17.10 | 168 | 17.60 | 173 |
| Fe$_2$O$_3$ | 2.11 | 13 | 3.14 | 19 | 1.68 | 11 | 1.71 | 11 | 2.05 | 12 |
| FeO | 0.72 | 10 | 1.87 | 26 | 0.79 | 11 | 1.59 | 22 | 1.35 | 18 |
| MnO | 0.15 | 2 | 0.11 | 1 | 0.15 | 2 | 0.40 | 6 | 0.30 | 4 |
| MgO | 2.32 | 57 | 0.68 | 17 | 1.04 | 26 | 1.05 | 26 | 0.80 | 20 |
| CaO | 0.69 | 13 | 2.17 | 39 | 1.30 | 23 | 1.04 | 19 | 1.02 | 18 |
| Na$_2$O | 6.50 | 105 | 9.47 | 153 | 8.01 | 129 | 7.21 | 116 | 7.37 | 119 |
| K$_2$O | 3.12 | 33 | 2.20 | 23 | 3.11 | 33 | 3.95 | 42 | 4.15 | 44 |
| qz | | 0 | | −21 | | −10 | | 0 | | −1 |

Note: 1) basalt (San Sebastian); 2) essexite-diabase (Calle Hermigua); 3) trachydolerite (Hermigua); 4) dolerite-basalt (San Sebastian); 5) dolerite-basalt (San Sebastian); 6) essexite-diabase (San Sebastian); 7) augite-labradorite rock (Agulo); 8) basalt (Agulo); 9) phonolite (Valle Hermoso); 10) phonolitic trachyte (Fortaleza de Chipude); 11) phonolite (La Pila); 12) aegirine phonolite (Alajerá); 13) trachyte (Alajerá); 14) aegirine phonolite (Alajerá); 15) trachyte (San Sebastian); 16) phonolite (Valle Hermoso); 17) trachyte (San Sebastian).

## TABLE 35. La Palma

### (Smulikowski [1945]; Fink [1913]; Wolff [1931])

| Analysis No. | 1 | | 2 | | 3 | | 4 | | 5 | | 6 | |
|---|---|---|---|---|---|---|---|---|---|---|---|---|
| Component | % | M.Q. | % | M.Q. | % | M.Q. | % | M.Q. | % | M.Q. | % | M.Q. |
| $SiO_2$ | 36.38 | 606 | 37.46 | 624 | 40.22 | 669 | 40.80 | 679 | 42.91 | 715 | 42.96 | 716 |
| $TiO_2$ | 4.80 | 60 | 5.22 | 65 | | | 3.44 | 43 | 3.80 | 48 | 3.55 | 45 |
| $Al_2O_3$ | 12.62 | 124 | 10.03 | 98 | 14.41 | 141 | 4.77 | 145 | 13.57 | 133 | 13.04 | 128 |
| $Fe_2O_3$ | 10.61 | 66 | 8.47 | 53 | 17.42 | 109 | 7.91 | 49 | 5.64 | 35 | 6.52 | 41 |
| FeO | 6.10 | 85 | 6.41 | 89 | 2.36 | 33 | 7.33 | 102 | 7.77 | 108 | 8.49 | 118 |
| MnO | 0.18 | 3 | 0.14 | 2 | | | | | 0.21 | 3 | 0.24 | 3 |
| MgO | 9.66 | 240 | 13.93 | 346 | 7.29 | 181 | 5.09 | 127 | 8.65 | 216 | 8.42 | 210 |
| CaO | 15.60 | 278 | 14.78 | 264 | 11.53 | 206 | 1.63 | 208 | 11.94 | 213 | 10.77 | 193 |
| $Na_2O$ | 2.74 | 44 | 2.00 | 32 | 3.94 | 64 | 4.38 | 71 | 3.54 | 57 | 3.60 | 58 |
| $K_2O$ | 0.88 | 10 | 0.76 | 8 | 1.90 | 20 | 2.14 | 22 | 1.80 | 19 | 1.43 | 15 |
| qz | | —57 | | —51 | | —62 | | —60 | | —50 | | —47 |

| Analysis No. | 7 | | 8 | | 9 | | 10 | | 11 | |
|---|---|---|---|---|---|---|---|---|---|---|
| Component | % | M.Q. | % | M.Q. | % | M.Q. | % | M.Q. | % | M.Q. |
| $SiO_2$ | 43.56 | 726 | 43.64 | 727 | 43.68 | 728 | 44.50 | 741 | 48.23 | 803 |
| $TiO_2$ | 3.83 | 48 | 4.85 | 61 | 4.04 | 50 | 1.72 | 21 | 1.90 | 24 |
| $Al_2O_3$ | 15.22 | 149 | 14.15 | 139 | 12.52 | 123 | 13.23 | 129 | 18.41 | 180 |
| $Fe_2O_3$ | 5.56 | 35 | 3.85 | 24 | 4.57 | 29 | 4.11 | 26 | 3.27 | 21 |
| FeO | 7.04 | 98 | 7.62 | 106 | 8.25 | 114 | 7.76 | 108 | 5.00 | 70 |
| MnO | 0.20 | 3 | 0.18 | 3 | 0.11 | 1 | | | | |
| MgO | 6.47 | 162 | 7.64 | 191 | 8.64 | 214 | 13.19 | 327 | 1.92 | 48 |
| CaO | 10.60 | 189 | 11.18 | 200 | 12.94 | 231 | 11.20 | 200 | 6.43 | 115 |
| $Na_2O$ | 4.96 | 80 | 4.15 | 67 | 3.04 | 49 | 1.69 | 27 | 7.77 | 125 |
| $K_2O$ | 1.87 | 20 | 1.79 | 19 | 1.37 | 15 | 0.74 | 8 | 3.16 | 34 |
| qz | | —57 | | —51 | | —41 | | —29 | | —73 |

| Analysis No. | 12 | | 13 | | 14 | | 15 | | 16 | |
|---|---|---|---|---|---|---|---|---|---|---|
| Component | % | M.Q. | % | M.Q. | % | M.Q. | % | M.Q. | % | M.Q. |
| $SiO_2$ | 48.85 | 813 | 51.02 | 849 | 51.38 | 855 | 55.40 | 922 | 68.54 | 1145 |
| $TiO_2$ | 2.30 | 29 | 2.12 | 26 | 1.45 | 18 | 0.43 | 5 | 0.13 | 2 |
| $Al_2O_3$ | 16.53 | 162 | 19.67 | 193 | 15.91 | 156 | 21.03 | 206 | 15.96 | 157 |
| $Fe_2O_3$ | 5.85 | 36 | 3.72 | 23 | 3.17 | 20 | 1.64 | 10 | 1.33 | 8 |
| FeO | 5.68 | 79 | 3.65 | 50 | 4.03 | 56 | 3.04 | 42 | 1.63 | 23 |
| MnO | | | 0.18 | 3 | — | — | — | — | — | — |
| MgO | 2.95 | 73 | 1.70 | 42 | 2.14 | 53 | 0.91 | 22 | 0.24 | 6 |
| CaO | 6.51 | 116 | 6.66 | 119 | 3.60 | 64 | 3.57 | 63 | 0.65 | 1 |
| $Na_2O$ | 5.49 | 89 | 7.00 | 113 | 6.07 | 98 | 7.64 | 123 | 6.25 | 101 |
| $K_2O$ | 2.91 | 31 | 3.51 | 37 | 3.35 | 36 | 4.42 | 47 | 4.90 | 52 |
| qz | | —47 | | —59 | | —37 | | —54 | | —49 |

Note: 1) melanocratic inclusions in lava (Antonio region); 2) melanocratic inclusions in lava (Antonio region); 3) limburgite; 4) essexite (Caldeira); 5) basanitoid; 6) basanitoid; 7) analcime basanite; 8) tephritoid; 9) basanitoid (San Martin Volcano); 10) essexite-diabase; 11) sodalite gauteite; 12) essexite (Caldeira); 13) ordanchite (Malpais); 14) trachydolerite; 15) haüynite tephrite (Campanario); 16) alkalic granite.

## TABLE 36. Madeira

### (Fink [1913]; Gagel [1912]; Smith [1930]; Wolff [1931])

| Analysis No. | 1 | | 2 | | 3 | | 4 | | 5 | | 6 | | 7 | |
|---|---|---|---|---|---|---|---|---|---|---|---|---|---|---|
| Component | % | M.Q. | % | M.Q. | % | M.Q. | % | M.Q. | % | M.Q. | % | M.Q. | % | M.Q. |
| $SiO_2$ | 40.07 | 667 | 41.43 | 690 | 41.72 | 694 | 41.96 | 699 | 42.19 | 703 | 42.37 | 705 | 42.39 | 706 |
| $TiO_2$ | 2.35 | 30 | 2.67 | 34 | 3.41 | 43 | 2.16 | 27 | 3.15 | 40 | 3.21 | 40 | 2.61 | 33 |
| $Al_2O_3$ | 8.95 | 88 | 13.18 | 129 | 11.47 | 113 | 15.85 | 156 | 13.80 | 135 | 13.29 | 130 | 15.77 | 155 |
| $Fe_2O_3$ | 4.82 | 30 | 6.95 | 44 | 4.04 | 25 | 7.64 | 48 | 5.52 | 34 | 3.79 | 24 | 5.89 | 37 |
| FeO | 7.81 | 109 | 7.31 | 102 | 10.58 | 148 | 7.24 | 101 | 8.87 | 123 | 10.24 | 142 | 8.66 | 121 |
| MnO | — | — | — | | | | | | | | | | | |
| MgO | 13.86 | 344 | 11.91 | 295 | 12.55 | 311 | 8.45 | 209 | 8.55 | 212 | 10.76 | 267 | 7.44 | 184 |
| CaO | 13.83 | 246 | 10.74 | 192 | 10.82 | 193 | 9.54 | 170 | 11.39 | 203 | 11.17 | 199 | 9.40 | 168 |
| $Na_2O$ | 1.34 | 22 | 1.60 | 26 | 2.28 | 37 | 2.05 | 33 | 2.50 | 40 | 2.94 | 47 | 2.05 | 33 |
| $K_2O$ | 0.56 | 6 | 0.93 | 10 | 1.22 | 13 | 1.17 | 13 | 1.21 | 13 | 1.17 | 13 | 1.24 | 13 |
| qz | | —36 | | —35 | | —43 | | —33 | | —31 | | —45 | | —30 |

| Analysis No. | 8 | | 9 | | 10 | | 11 | | 12 | | 13 | | 14 | |
|---|---|---|---|---|---|---|---|---|---|---|---|---|---|---|
| Component | % | M.Q. | % | M.Q. | % | M.Q. | % | M.Q. | % | M.Q. | % | M.Q. | % | M.Q. |
| $SiO_2$ | 42.40 | 706 | 42.42 | 706 | 42.71 | 711 | 43.30 | 721 | 43.79 | 729 | 43.85 | 730 | 44.40 | 739 |
| $TiO_2$ | 3.68 | 46 | 0.30 | 4 | 3.38 | 43 | 2.83 | 35 | 2.82 | 85 | 2.53 | 32 | 2.77 | 35 |
| $Al_2O_3$ | 14.19 | 139 | 1.32 | 13 | 14.62 | 143 | 14.07 | 138 | 13.73 | 134 | 12.94 | 126 | 15.40 | 151 |
| $Fe_2O_3$ | 6.14 | 38 | 4.27 | 27 | 3.12 | 19 | 5.53 | 34 | 3.37 | 21 | 2.70 | 17 | 5.20 | 33 |
| FeO | 7.69 | 107 | 6.96 | 97 | 9.34 | 130 | 7.17 | 100 | 10.20 | 142 | 10.51 | 146 | 7.81 | 109 |
| MnO | | | | | | | | | | | | | | |
| MgO | 9.02 | 224 | 40.80 | 1012 | 8.91 | 221 | 9.62 | 239 | 9.46 | 235 | 11.90 | 295 | 7.23 | 180 |
| CaO | 11.08 | 198 | 1.19 | 20 | 10.68 | 191 | 10.87 | 194 | 10.54 | 188 | 9.49 | 169 | 9.92 | 177 |
| $Na_2O$ | 2.50 | 40 | 0.72 | 11 | 3.11 | 50 | 2.41 | 39 | 2.71 | 44 | 2.42 | 39 | 2.83 | 46 |
| $K_2O$ | 1.43 | 15 | 0.45 | 4 | 1.55 | 16 | 1.12 | 12 | 1.25 | 14 | 1.06 | 12 | 1.19 | 13 |
| qz | | —39 | | —47 | | —43 | | —34 | | —43 | | —36 | | —32 |

| Analysis No. | 15 | | 16 | | 17 | | 18 | | 19 | |
|---|---|---|---|---|---|---|---|---|---|---|
| Component | % | M.Q. | % | M.Q. | % | M.Q. | % | M.Q. | % | M.Q. |
| $SiO_2$ | 44.50 | 741 | 44.86 | 747 | 45.04 | 750 | 45.69 | 761 | 46.08 | 768 |
| $TiO_2$ | 2.61 | 33 | 2.52 | 31 | 3.67 | 46 | 1.30 | 16 | 2.73 | 34 |
| $Al_2O_3$ | 13.85 | 135 | 16.18 | 159 | 16.41 | 161 | 17.02 | 167 | 17.39 | 171 |
| $Fe_2O_3$ | 3.47 | 22 | 7.22 | 45 | 6.02 | 38 | 4.59 | 29 | 10.95 | 68 |
| FeO | 9.02 | 125 | 7.10 | 99 | 7.30 | 102 | 8.52 | 118 | 2.56 | 36 |
| MnO | | | | | | | | | | |
| MgO | 11.00 | 273 | 5.34 | 133 | 3.93 | 98 | 5.62 | 139 | 2.66 | 66 |
| CaO | 10.06 | 179 | 9.95 | 178 | 11.42 | 203 | 11.31 | 201 | 8.87 | 158 |
| $Na_2O$ | 2.70 | 44 | 3.78 | 61 | 3.09 | 50 | 3.21 | 52 | 3.72 | 60 |
| $K_2O$ | 0.92 | 10 | 1.39 | 15 | 0.93 | 10 | 1.07 | 12 | 1.38 | 15 |
| qz | | —35 | | —39 | | —29 | | —32 | | —27 |

| Analysis No. | 20 | | 21 | | 22 | | 23 | | 24 | | 25 | |
|---|---|---|---|---|---|---|---|---|---|---|---|---|
| Component | % | M.Q. | % | M.Q. | % | M.Q. | % | M.Q. | % | M.Q. | % | M.Q. |
| $SiO_2$ | 46.44 | 773 | 47.70 | 794 | 49.15 | 818 | 49.34 | 822 | 49.87 | 830 | 51.78 | 863 |
| $TiO_2$ | 2.90 | 38 | 2.54 | 32 | 0.83 | 10 | 0.81 | 10 | 2.60 | 33 | 1.05 | 14 |
| $Al_2O_3$ | 16.30 | 160 | 17.32 | 170 | 17.86 | 176 | 16.50 | 162 | 14.98 | 147 | 18.68 | 183 |
| $Fe_2O_3$ | 4.82 | 30 | 5.43 | 34 | 1.07 | 7 | 7.92 | 49 | 6.17 | 39 | 6.42 | 40 |
| FeO | 7.07 | 98 | 4.71 | 65 | 10.77 | 150 | 2.41 | 33 | 4.40 | 61 | 2.77 | 39 |
| MnO | — | — | — | | 0.75 | 10 | 0.98 | 14 | | | | |
| MgO | 4.92 | 122 | 3.62 | 90 | 3.24 | 80 | 2.66 | 66 | 1.77 | 44 | 1.86 | 46 |
| CaO | 10.03 | 179 | 7.98 | 143 | 6.57 | 117 | 8.04 | 143 | 6.34 | 113 | 6.04 | 108 |
| $Na_2O$ | 3.82 | 61 | 4.21 | 68 | 5.49 | 89 | 5.20 | 84 | 5.08 | 82 | 5.53 | 89 |
| $K_2O$ | 1.44 | 23 | 2.45 | 40 | 2.29 | 24 | 3.42 | 36 | 2.04 | 21 | 2.34 | 38 |
| qz | | —38 | | —44 | | —30 | | —31 | | —23 | | —39 |

## TABLE 36. (continued)

| Analysis No. Component | 26 | | 27 | | 28 | | 29 | | 30 | |
|---|---|---|---|---|---|---|---|---|---|---|
| | % | M.Q. | % | M.Q. | % | M.Q. | % | M.Q. | % | M.Q. |
| $SiO_2$ | 52,40 | 873 | 52,47 | 874 | 52,75 | 878 | 55,54 | 925 | 57,67 | 960 |
| $TiO_2$ | 1,60 | 20 | 1,57 | 20 | 0,94 | 12 | 0,71 | 9 | 0,40 | 5 |
| $Al_2O_3$ | 19,27 | 189 | 15,84 | 155 | 18,29 | 179 | 18.20 | 178 | 19,17 | 188 |
| $Fe_2O_3$ | 4,56 | 29 | 3,30 | 21 | 4,68 | 29 | 5,92 | 37 | 4,55 | 28 |
| FeO | 3,57 | 50 | 8,42 | 117 | — | — | — | — | — | — |
| MnO | — | — | — | — | 4,03 | 60 | 1,14 | 16 | 0,99 | 14 |
| MgO | 2,03 | 51 | 1,52 | 38 | 2,15 | 54 | 1,32 | 32 | 1,22 | 30 |
| CaO | 6,68 | 120 | 5,05 | 90 | 7,39 | 132 | 5,64 | 101 | 3,94 | 70 |
| $Na_2O$ | 5,50 | 89 | 7,03 | 114 | 5,66 | 91 | 6.44 | 102 | 6,84 | 110 |
| $K_2O$ | 2,03 | 33 | 2,52 | 27 | 2,29 | 37 | 2.30 | 37 | 3,35 | 36 |
| qz | | −34 | | −29 | | −40 | | −13 | | −25 |

Note: 1) madeirite (Ribeira de Massapez); 2) basaltic trachydolerite (Calheta); 3) trachydolerite (Serrado); 4) trachydolerite (Serrado); 5) trachydolerite (Caniçal); 6) nepheline basanite (Caniçal); 7) trachydolerite (Chapanna); 8) typical spotty trachydolerite (Gran Curral); 9) lherzolitic dunite bomb; 10) nepheline basalt (Curral Bocca das Corregos); 11) porphyritic (feldspar) basalt ? (Ribeiro Frio); 12) spotty trachydolerite (nepheline basanite ?) (Rabaçal); 13) essexite porphyrite (Ribeira de Massapez); 14) trachydolerite (Curral Lombo Grande); 15) essexite melaphyre (Punta Delgada); 16) trachydolerite (Serrado); 17) essexite-diabase (Ribeira das Voltas); 18) essexite-diabase (Soca, Ribeira de Massapez); 19) trachydolerite (Serrado); 20) trachydolerite (Punta do Sol); 21) trachydolerite (Ribeiro Frio); 22) essexite (Ribeira de Massapez); 23) sodalite syenite (Soca); 24) hornblende ankerite; 25) trachytoidal trachydolerite (Ilheo, Porto da Cruz); 26) trachytoidal trachydolerite (Achada, Porto da Cruz); 27) sodalite syenite (Soca); 28) spotty trachydolerite (Ribeira de Massapez); 29) trachydolerite (Serrado); 30) trachyandesite (gauteite) (Gran Curral).

### TABLE 37. Porto Santo

| Analysis No. Component | 1 | |
|---|---|---|
| | % | M.Q. |
| $SiO_2$ | 68,79 | 1145 |
| $TiO_2$ | 16,83 | 165 |
| $Al_2O_3$ | 1,54 | 9 |
| $Fe_2O_3$ | 0.61 | 8 |
| FeO | 0.24 | 6 |
| MnO | 0.51 | 9 |
| MgO | 6.65 | 107 |
| CaO | 3,71 | 39 |
| qz | | +6 |

Note: 1) quartz bostonite (Serra de Feteira).

### TABLE 38. Cape Verde Islands

#### Santo Antão
#### [Bacelar [1932])

| Analysis No. Component | 1 | | 2 | | 3 | |
|---|---|---|---|---|---|---|
| | % | M.Q. | % | M.Q. | % | M.Q. |
| $SiO_2$ | 41,12 | 684 | 47,44 | 790 | 48,46 | 807 |
| $TiO_2$ | — | — | — | — | — | — |
| $Al_2O_3$ | 10.17 | 100 | 23,71 | 232 | 21,81 | 214 |
| $Fe_2O_3$ | 2.60 | 16 | 6,83 | 43 | 2,17 | 14 |
| FeO | 9.82 | 136 | 3,53 | 49 | 3,75 | 52 |
| MnO | — | — | — | — | — | — |
| MgO | 13.34 | 331 | 1,95 | 48 | 0,68 | 16 |
| CaO | 14,90 | 266 | 6,47 | 115 | 4,58 | 82 |
| $Na_2O$ | 6,61 | 106 | 6,40 | 103 | 8,41 | 135 |
| $K_2O$ | 2,27 | 24 | 3,34 | 35 | 5,86 | 63 |
| qz | | −74 | | −64 | | −97 |

Note: 1) limburgite (Pedra Molar); 2) tephrite (Cova); 3) leucitite (Siderão).

## TABLE 39. São Vicente

### (Bacelar [1932])

| Analysis No. Component | 1 % | 1 M.Q. | 2 % | 2 M.Q. | 3 % | 3 M.Q. | 4 % | 4 M.Q. | 5 % | 5 M.Q. |
|---|---|---|---|---|---|---|---|---|---|---|
| $SiO_2$ | 36.77 | 612 | 38.60 | 643 | 39.13 | 651 | 43.58 | 726 | 47.63 | 793 |
| $TiO_2$ | 4.07 | 51 | 4.23 | 53 | 4.98 | 63 | 3.97 | 50 | 1.39 | 18 |
| $Al_2O_3$ | 9.80 | 96 | 11.64 | 114 | 10.20 | 100 | 9.98 | 98 | 17.20 | 169 |
| $Fe_2O_3$ | 3.79 | 24 | 9.44 | 59 | 4.40 | 28 | 2.56 | 16 | 3.60 | 23 |
| $FeO$ | 7.89 | 110 | 5.33 | 74 | 8.93 | 124 | 8.78 | 122 | 8.09 | 113 |
| $MnO$ | 0.14 | 2 | 0.71 | 10 | 0.15 | 2 | 0.13 | 2 | — | — |
| $MgO$ | 14.94 | 370 | 6.23 | 155 | 12.72 | 315 | 11.50 | 285 | 6.25 | 155 |
| $CaO$ | 15.67 | 279 | 14.43 | 258 | 13.03 | 233 | 13.36 | 238 | 6.42 | 114 |
| $Na_2O$ | 2.49 | 40 | 5.43 | 88 | 3.32 | 53 | 2.28 | 37 | 4.65 | 75 |
| $K_2O$ | 0.97 | 11 | 2.49 | 27 | 0.61 | 6 | 1.71 | 18 | 1.31 | 14 |
| $qz$ | | —57 | | —78 | | —53 | | —40 | | —37 |

| Analysis No. Component | 6 % | 6 M.Q. | 7 % | 7 M.Q. | 8 % | 8 M.Q. | 9 % | 9 M.Q. | 10 % | 10 M.Q. |
|---|---|---|---|---|---|---|---|---|---|---|
| $SiO_2$ | 48.67 | 810 | 51.18 | 852 | 52.18 | 869 | 55.11 | 917 | 56.26 | 937 |
| $TiO_2$ | 2.12 | 26 | 2.40 | 30 | 1.24 | 15 | 0.17 | 2 | 0.60 | 8 |
| $Al_2O_3$ | 13.79 | 135 | 17.44 | 171 | 19.07 | 187 | 19.27 | 189 | 18.68 | 183 |
| $Fe_2O_3$ | 1.60 | 10 | 4.70 | 29 | 3.36 | 21 | 2.67 | 17 | 3.61 | 23 |
| $FeO$ | 6.98 | 97 | 4.15 | 58 | 1.30 | 18 | 1.41 | 19 | 1.87 | 26 |
| $MnO$ | 0.10 | 1 | 0.10 | 1 | 0.07 | 1 | 0.08 | 1 | 0.81 | 10 |
| $MgO$ | 6.73 | 167 | 2.87 | 71 | 1.58 | 39 | 0.05 | 1 | 0.22 | 5 |
| $CaO$ | 11.55 | 206 | 9.60 | 171 | 4.56 | 81 | 1.33 | 24 | 2.36 | 42 |
| $Na_2O$ | 2.58 | 42 | 5.84 | 94 | 4.06 | 66 | 9.11 | 147 | 7.19 | 116 |
| $K_2O$ | 0.37 | 4 | 0.44 | 4 | 5.51 | 58 | 6.54 | 69 | 5.21 | 55 |
| $qz$ | | 0 | | —27 | | —24 | | —73 | | —48 |

Note: 1) melilite ankaratrite (Mt. Verde); 2) nepheline ankaratrite (Nha Claudia); 3) nepheline ankaratrite (Mt. Amargosa); 4) basanitoid (Da Estrada Mindelo-Madeiral); 5) diorite (Porto Grande); 6) gabbro-dolerite (Mt. Cavalo); 7) diorite (Porto Grande); 8) trachyte (Mt. Fateixa); 9) sodalite phonolite (Mt. Cavalo); 10) nepheline syenite (Pedras Brancas).

## Table 40. Sal

### (Bacelar [1932]; Part [1950])

| Component | 1 % | 1 M.Q. | 2 % | 2 M.Q. | 3 % | 3 M.Q. | 4 % | 4 M.Q. | 5 % | 5 M.Q. |
|---|---|---|---|---|---|---|---|---|---|---|
| $SiO_2$ | 37.18 | 619 | 38.94 | 648 | 48.96 | 815 | 53.18 | 885 | 54.66 | 910 |
| $TiO_2$ | 4.60 | 58 | — | — | 1.96 | 25 | 1.80 | 23 | 0.81 | 10 |
| $Al_2O_3$ | 10.37 | 102 | 9.20 | 90 | 18.24 | 178 | 19.62 | 192 | 21.96 | 216 |
| $Fe_2O_3$ | 5.64 | 35 | 4.52 | 28 | 3.57 | 23 | 3.39 | 21 | 1.10 | 7 |
| $FeO$ | 7.53 | 105 | 8.48 | 118 | 4.48 | 62 | 3.57 | 50 | 2.19 | 31 |
| $MnO$ | 0.26 | 4 | — | — | 0.25 | 3 | 0.04 | 0 | 0.21 | 3 |
| $MgO$ | 12.52 | 310 | 14.78 | 366 | 1.49 | 37 | 1.40 | 35 | 0.35 | 8 |
| $CaO$ | 15.08 | 269 | 13.86 | 247 | 6.12 | 109 | 4.84 | 86 | 3.02 | 54 |
| $Na_2O$ | 2.87 | 46 | 3.44 | 55 | 6.48 | 105 | 7.45 | 120 | 8.91 | 144 |
| $K_2O$ | 0.93 | 10 | 1.26 | 14 | 3.99 | 42 | 3.84 | 40 | 5.55 | 58 |
| $qz$ | | —36 | | —59 | | —61 | | —56 | | —80 |

Note. 1) ankaratrite; 2) ankaratrite, rich in olivine; 3) nepheline monzonite; 4) nepheline monzonite; 5) nepheline phonolite.

## TABLE 41. Boa Vista

### (Bacelar [1932])

| Component | 1 % | 1 M.Q. | 2 % | 2 M.Q. | 3 % | 3 M.Q. |
|---|---|---|---|---|---|---|
| $SiO_2$ | 35.76 | 595 | 46.40 | 773 | 59.16 | 985 |
| $TiO_2$ | 3.48 | 11 | 3.62 | 45 | 1.39 | 18 |
| $Al_2O_3$ | 9.76 | 96 | 12.93 | 126 | 18.46 | 101 |
| $Fe_2O_3$ | 4.70 | 29 | 5.69 | 36 | 2.04 | 12 |
| $FeO$ | 6.48 | 90 | 6.18 | 86 | 1.99 | 28 |
| $MnO$ | 0.21 | 3 | 0.11 | 1 | 0.34 | 5 |
| $MgO$ | 16.97 | 421 | 7.50 | 186 | 0.74 | 18 |
| $CaO$ | 15.28 | 273 | 12.02 | 214 | 1.38 | 25 |
| $Na_2O$ | 3.46 | 56 | 2.67 | 43 | 7.50 | 121 |
| $K_2O$ | 1.38 | 15 | 1.10 | 12 | 4.99 | 53 |
| $qz$ | | —69 | | —25 | | —18 |

Note: 1) basalt; 2) dolerite; 3) syenite-monzonite.

## TABLE 42. Maio
### (Bacelar [1932])

| Analysis No. Component | 1 % | 1 M.Q. | 2 % | 2 M.Q. | 3 % | 3 M.Q. | 4 % | 4 M.Q. | 5 % | 5 M.Q. | 6 % | 6 M.Q. |
|---|---|---|---|---|---|---|---|---|---|---|---|---|
| $SiO_2$ | 36.16 | 602 | 41.60 | 693 | 41.94 | 698 | 44.0 | 733 | 44.49 | 741 | 50.05 | 834 |
| $TiO_2$ | — | — | 4.98 | 63 | 5.24 | 65 | 4.42 | 55 | — | — | — | — |
| $Al_2O_3$ | 6.85 | 68 | 11.88 | 117 | 12.91 | 126 | 14.54 | 142 | 22.94 | 225 | 20.98 | 2C6 |
| $Fe_2O_3$ | 6.27 | 39 | 5.07 | 32 | 4.24 | 26 | 6.51 | 41 | 7.90 | 49 | 2.12 | 13 |
| $FeO$ | 6.94 | 96 | 7.24 | 101 | 8.28 | 115 | 4.29 | 60 | 6.14 | 85 | 4.05 | 56 |
| $MnO$ | — | — | 0.12 | 1 | 0.08 | 1 | 0.13 | 2 | — | — | — | — |
| $MgO$ | 13.88 | 344 | 7.12 | 177 | 8.15 | 202 | 3.85 | 95 | 5.75 | 142 | 1.65 | 41 |
| $CaO$ | 18.88 | 336 | 13.42 | 239 | 12.78 | 228 | 11.22 | 200 | 2.96 | 53 | 4.12 | 73 |
| $Na_2O$ | 2.65 | 43 | 3.39 | 55 | 2.64 | 43 | 4.79 | 77 | 5.36 | 86 | 8.43 | 136 |
| $K_2O$ | 1.00 | 11 | 0.51 | 5 | 1.28 | 14 | 1.39 | 15 | 2.10 | 22 | 6.19 | 66 |
| *qz* | | —60 | | —40 | | —40 | | —46 | | —56 | | —96 |

Note: 1) ankaratrite, rich in olivine; 2) ankaratrite; 3) ankaratrite; 4) basalt; 5) pyroxenite.

## TABLE 43. São Tiago
### (Bacelar [1932]; Part [1950])

| Analysis No. Component | 1 % | 1 M.Q. | 2 % | 2 M.Q. | 3 % | 3 M.Q. | 4 % | 4 M.Q. | 5 % | 5 M.Q. |
|---|---|---|---|---|---|---|---|---|---|---|
| $SiO_2$ | 39.64 | 660 | 44.55 | 742 | 53.65 | 893 | 53.80 | 896 | 54.97 | 915 |
| $TiO_2$ | — | — | 3.24 | 10 | 1.30 | 16 | — | — | 0.64 | 8 |
| $Al_2O_3$ | 16.98 | 167 | 11.62 | 114 | 20.58 | 202 | 23.59 | 231 | 20.06 | 197 |
| $Fe_2O_3$ | 6.61 | 41 | 5.53 | 34 | 1.00 | 6 | 3.57 | 23 | 3.43 | 21 |
| $FeO$ | 9.31 | 129 | 5.20 | 72 | 1.36 | 19 | 1.88 | 26 | 1.25 | 18 |
| $MnO$ | — | — | 0.18 | 3 | 0.08 | 1 | — | — | 0.21 | 3 |
| $MgO$ | 6.65 | 165 | 9.40 | 233 | 1.28 | 1 | 0.87 | 21 | 0.43 | 11 |
| $CaO$ | 10.58 | 189 | 13.16 | 235 | 4.50 | 80 | 2.26 | 40 | 2.60 | 46 |
| $Na_2O$ | 5.95 | 96 | 2.19 | 35 | 6.34 | 102 | 9.05 | 146 | 8.27 | 133 |
| $K_2O$ | 3.09 | 33 | 1.86 | 20 | 5.08 | 54 | 4.77 | 51 | 5.14 | 54 |
| *qz* | | —83 | | —33 | | —43 | | —81 | | —67 |

Note: 1) teschenite (Praia); 2) essexite; 3) nepheline monzonite; 4) phonolite (Praia); 5) nepheline syenite (Pedras Brancas).

## TABLE 44. Fogo
### (Bacelar [1932])

| Component | Analysis No. 1 % | 1 M.Q. | 2 % | 2 M.Q. | 3 % | 3 M.Q. |
|---|---|---|---|---|---|---|
| $SiO_2$ | 42.18 | 703 | 42.65 | 710 | 45.04 | 750 |
| $TiO_2$ | 5.06 | 64 | — | — | — | — |
| $Al_2O_3$ | 14.01 | 137 | 15.35 | 150 | 16.04 | 157 |
| $Fe_2O_3$ | 4.58 | 29 | 6.46 | 41 | 7.10 | 44 |
| $FeO$ | 8.28 | 115 | 8.19 | 114 | 8.23 | 114 |
| $MnO$ | 0.13 | 2 | — | — | — | — |
| $MgO$ | 5.90 | 146 | 7.14 | 177 | 4.46 | 111 |
| $CaO$ | 13.16 | 235 | 11.96 | 213 | 10.19 | 182 |
| $Na_2O$ | 3.69 | 60 | 5.02 | 81 | 6.11 | 98 |
| $K_2O$ | 2.26 | 24 | 1.47 | 16 | 2.85 | 30 |
| *qz* | | —53 | | —59 | | —60 |

Note: 1) basalt; 2) porphyritic (plagioclase) basalt; 3) pyroxenite (tephrite).

## TABLE 45. Fernando de Noronha
### (Lacroix [1923]; Renard [1882]; Bull. ac. R. Belge, Vol. 3)

| Component | Analysis No. 1 % | 1 M.Q. | 2 % | 2 M.Q. |
|---|---|---|---|---|
| $SiO_2$ | 39.96 | 665 | 42.24 | 703 |
| $TiO_2$ | 3.03 | 38 | — | — |
| $Al_2O_3$ | 9.75 | 96 | 20.15 | 198 |
| $Fe_2O_3$ | 5.98 | 38 | 12.17 | 76 |
| $FeO$ | 7.61 | 106 | 4.07 | 57 |
| $MnO$ | — | — | — | — |
| $MgO$ | 12.95 | 321 | 5.22 | 130 |
| $CaO$ | 14.04 | 250 | 6.15 | 110 |
| $Na_2O$ | 2.86 | 46 | 6.10 | 98 |
| $K_2O$ | 0.94 | 10 | 0.35 | 4 |
| *qz* | | —51 | | —56 |

Note: 1) ankaratrite; 2) basaltic basanite.

# LITERATURE CITED

Ampferer, O., "Über das Bewegungsbild von Faltengebirgen," Jahrb. K. K. Geol. R. A., Vol. 56, Wien (1906).

Ampferer, O., "Gedanken zum Bewegungsbild des Atlantisches Raumes," Sitzber. Akad. Wiss. Wien, Mat. Kl. 1, 150, Nos. 1-2, pp. 97-114 (1941).

Ampferer, O., "Vergleich der tectonischen Virsamkeit von Contraction und Unterströmung," Mitt. Geol. Ges. Wien (1942).

Artem'ev, M. E., "Some distributional patterns of isostatic anomalies in the region of Alpine fold structures of Europe," Izv. Akad. Nauk SSSR, Ser. Geol., 9:58-66 (1962).

Artem'ev, M. E., "Gravity anomalies and the seismicity of Western Europe and the Mediterranean," Izv. Akad. Nauk SSSR, Ser. Geofiz., No. 2, pp. 309-317 (1963).

Artem'ev, M. E., "Disturbance of isostatic equilibrium and the direction of vertical movements of the Earth's crust," Dokl. Akad. Nauk SSSR, 160(5):1151-1154 (1965).

Artem'ev, M. E., Isostatic Gravity Anomalies and Some Problems of Their Geologic Interpretation, Izd. Nauka (1966).

Artyushkov, E. V., Physics of the Earth, No. 8 (1968).

Bacelar, B. J., "A Geologia do Archipelago de Cabo Verde," Com. Serv. Geol. Portugal, 18:1-275 (1932).

Bailey, E. B., Thomas, H. H., et al., Tertiary and Post-Tertiary Geology of Mull, Loch Aline, and Oban, Geol. Surv. Scotland Mem. (1924).

Balakina, L. M., Vvedenskaya, A. V., Misharina, L. A., and Shirokova, V. I., "The stress state at earthquake foci and the field of elastic stresses in the Earth," Physics of the Earth, Izv. Akad. Nauk SSSR (1967).

Barth, T. F. W., The Volcanic Geology, Hot Springs, and Geysers of Iceland, Carnegie Inst. of Washington, Publ. 587 (1950).

Belousov, V. V., "The geologic structure and development of oceanic basins," Izv. Akad. Nauk SSSR, Ser. Geol., No. 3, pp. 3-18 (1955).

Belousov, V. V., Fundamental Questions on Geotectonics, Gosgeoltekhizdat (1962).

Belousov, V. V., The Earth's Crust and Continental Mantle, Mezhduvedom, Geofiz. Kom., Izd. Nauka ($1966_1$).

Belousov, V. V., "Possible depth conditions of magmatism," Sov. Geol., No. 4, pp. 8-25 ($1966_2$).

Belousov, V. V., "Some questions on development of the Earth's crust and the upper mantle of the oceans," Geotektonika, No. 1, pp. 3-14 (1967).

Belousov, V. V., and Rudich, E. M., "The place of island arcs in development of the structure of the Earth," Sov. Geol., No. 10, pp. 3-23 (1960).

Bilibin, Yu. A., "Questions on the metallogenetic evolution of geosynclinal zones," Izv. Akad. Nauk SSSR, Geol., No. 4 (1948).

Bilibin, Yu. A., Metallogenic Provinces and Metallogenic Epochs, Gosgeoltekhizdat (1955).

Black, G. P., "The Tertiary volcanic succession of the Isle of Rum, Inverness-shire," Trans. Edinburgh Geol. Soc., 15:39-51 (1952).

Blot, C., "Origine profonde des séismes superficiels et des éruptions volcaniques," Assoc. Séism. etc., Sér. A, Travaux Scient. 23, Toulouse, pp. 103-121 (1964).

Blot, C., and Priam, R., "Volcanisme et séismicité dans l'Archipel des Nouvelles Hébrides," Bull. Volcanol. (Napoli), 26:167-180 (1963).

Boese, W., "Petrographische Untersuchungen an jungvulkanischen Ergussgesteinen von São Thomé und Fernando Poo," N. Jahrb. f. Miner., Suppl. Vol. 34, 259 pp. (1912).

Boucart, J., and Jérémine, E., "Fuerteventura," Bull. Volcanol., Sér. 2, No. 4 (1938).

Bowen, N. L., "Lavas of the African Rift Valleys and Their Tectonic Setting," Am. Jour. Sci., 35A:19-33 (1938).

Bultitude, R.J., and Green, D. H., "Experimental study at high pressures on the origin of olivine nephelinite, etc.," Earth and Plan. Sci. Letters, Vol. 3 (1967).

Cargill, H. K., Hawkes, L., and Ledeboer, J. A., "The major intrusions of southeast Iceland," Quart. J. Geol. Soc. London, 84:505-539 (1928).

Carstens, H., Cristobalite Trachytes of Jan Mayen, Oslo (1961).

Challis, G. A., "High-temperature contact metamorphism at the Red Hills ultramafic intrusion, Wairau Valley, New Zealand," J. Petrol., 6(3):395-419 (1965).

Clark, S. P., and Ringwood, A. E., "Density distribution and constitution of the mantle," Rev. Geophys., 2(1):35-88 (1964).

Daly, R. A., Igneous Rocks and the Depths of the Earth, McGraw-Hill (1933).

de Assunçao, C. T., 23rd International Geological Congress, Vol. 2 (1958).

de Saint Ours, J., Études Géologiques dans l'Extrême Nord de Madagascar et l'Archipel des Comores, Serv. Géol. Rep. Malgache, Tananarive (1960), 272 pp.

Eaton, J. P., and Murata, K. J., "How volcanoes grow," Science, 132(3432):925-938 (1960).

Egorov, L. S., Gol'dburt, T. L., and Shikhorina, K. M., "Geology and petrography of the magmatic rocks of the Gulya intrusion," in: The Gulya Intrusion of Ultramafic Alkalic Rocks, Tr. Nauchn. Issled. Inst. Geol. Arktiki, 122:3-115 (1961).

Eliseev, N. A., Structural Petrology, Leningrad (1953), 309 pp.

Engel, A. E. J., and Engel, C. G., "Composition of basalts from the Mid-Atlantic Ridge," Science, 144(3624):1330-1333 (1964$_1$).

Engel, A. E. J., and Engel, C. G., "Igneous rocks of the East Pacific Rise," Science, 146(3643): 9-17 (1964$_2$).

Engel, C. G., and Engel, A. E. J., " Basalts dredged from the Northeast Pacific Ocean," Science, 140 (3753):1321-1324 (1963).

Épshtein, E. M., Anikeeva, L. I., and Mikhailova, A. F., "Metasomatic rocks and phologopite content of the Gulya intrusion," in: The Gulya Intrusion of Ultramafic Alkalic Rocks, Tr. Nauchn. Issled. Inst. Geol. Arktiki, No. 122 pp. 116-272 (1961).

Fedotov, S. A., and Kuzin, I. P., "Velocity profile of the upper mantle in the region of the Kurile Islands," Izv. Akad. Nauk SSSR, Ser. Geofiz., No. 5, pp. 670-686 (1963).

Fink, L., "Die Gesteine der Inseln Madeira und Porto Santo," Zeitschr. Deutsch. Geol. Ges., Vol. 65, No. 4 (1913).

Frolova, N. V., "Conditions of sedimentation during the Archean," Tr. Irkutsk. Univ., Ser. Geol., No. 2 (1951).

Furon, R., The Geology of Africa, Oliver, Edinburgh and London (1963).

Gagel, C., "Studien über den Aufbau und die Gesteine Madeiras," Zeitschr. Deutsch. Geol. Ges., 64:344-491 (1912).

Gorshkov, G. S., "Some questions on theoretical vulcanology," Izv. Akad. Nauk SSSR, Ser. Geol., 11:21-27 (1958).

Goryachev, A. V., Basic Laws of Tectonic Development in the Kurile-Kamchatka Zone, Izd. Nauka (1966), 235 pp.

Green, D. H., "The petrogenesis of the high-temperature peridotite intrusion in the Lizard area, Cornwall," J. Petrol., 5(1):134-188 (1964).

Green, D. H., and Ringwood, A. E., "An experimental investigation of the gabbro to eclogite transformation and its petrologic applications," in: Petrology of the Upper Mantle, Department of Geophysics and Geochemistry, Australian National University, Publ. No. 444 (1966$_1$), pp. 1-61.

Green, D. H., and Ringwood, A. E., "Origin of the calc-alkaline igneous rock suite," in: Petrology of the Upper Mantle, Department of Geophysics and Geochemistry, Australian National University, Publ. No. 444 (1966$_2$), pp. 62-117.

Green, D. H., and Ringwood, A. E., "The genesis of the basaltic magmas," in: Petrology of the Upper Mantle, Department of Geophysics and Geochemistry, Australian National University, Publ. No. 444 (1966$_3$).

Gzovskii, M. V., Problems of Magmatism and Tectonophysics, Data of the First Vulcanological Conference (Mater. 1-go Vulkanol. Soveshch.), Izd. Akad. Nauk SSSR (1962).

Gzovskii, M. V., "Tectonophysics and the problem of the origin of magmas of different chemical composition," in: Problems of Magma and the Origin of Igneous Rocks, Izd. Akad. Nauk SSSR, pp. 194-209 (1963).

Hausen, H., On the Geology of Fuerteventura (Canary Islands), Helsinki (1959).

Heezen, B. S., Tharp, M., and Ewing, M., The Floors of the Ocean: I. North Atlantic, Geol. Soc. America Spec. Paper 65 (1959).

Hobbs, H., "Origin of the lavas of the Pacific Region," 7th Pacif. Sci. Congr. II, Geology, pp. 346-357 (1953).

Holmes, A., "The basaltic rocks of the Arctic Region," Min. Mag., 18:180-223 (1918).

Holmes, A., "Basaltic lavas of South Kivu, Belgian Congo," Geol. Mag., 2:89-101 (1940).

Holmes, A., and Harwood, H. F., The Volcanic Area of Bufumbira, Part 2, Geol. Surv. Uganda, Mem. III (1937), 300 pp.

Hoppe, H. J. "Chemische und mikroskopische Untersuchungen an isländischen Gesteine," Chem. Erde, Vol. 2 (1938).

Izokh, É. P., Ultrabasite–Gabbro–Granite Rock Series and the Formation of High-Alumina Granites, Sib. Otd. AN SSSR (1965), 138 pp.

Jérémine, E., Composition Chimique et Mineralogique de la Roche de Pic de Teide, Soc. Franc. Mineral. (1878-1928), Paris (1930).

Kapp, H., "Zur Petrologie der Subvulkane zwischen Mesters Vig und Antarctic Havn (Ost Grönland)," Medd. om Grönland, Vol. 153, No. 2, 203 pp. (1960).

Kheraskov, N. P., Some General Laws on the Makeup and Structural Development of the Earth's Crust, Tr. Geol. Inst. Akad. Nauk SSSR, Vol. 91 (1963), 118 pp.

Khrenov, P. M., Komarov, Yu. V., Bukharov, A. A., Gordienko, I. V., Kiselev, A. I., and Lobanov, M. P., "Volcanic-plutonic belts in the southern part of eastern Siberia," Dokl. Akad. Nauk SSSR, 160(6):1388-1391 (1965).

Khrenov, P. M., Komarov, Yu. V., and Sherman, S. I., Structural-Tectonic Aspects of Volcanic-Plutonic Belts (on the Basis of the Southern Part of Eastern Siberia), in: Summaries of a Symposium: Volcanic-Plutonic Formations and Their Ore Potential, Izd. Nauka (1965).

Kiselev, A. I., and Saltykovskii, A. Ya., Scheme of Mesozoic Vulcanism of Western Transbaikalia, Byul. Mosk. Obshchestva Ispytatelei Prirody, Otd. Geol. (1967).

Knipper, A. L., "Features of the formation of anticlines with serpentinite cores (northern Akerinskii zone of the Little Caucasus), Byul. Mosk. Obshchestva Ispytatelei Prirody, Otd. Geol., 10(2):46-58 (1965).

Komarov, Yu. V., "Scheme of Lower Mesozoic magmatism of the western Transbaikalian mobile zone," Geol. i Geofiz., No. 11, pp. 3-14 (1960).

Komarov, Yu. V., and Khrenov, P. M., "The type of development of continental Mesozoic structures in eastern Asia," Dokl. Akad. Nauk SSSR, 151(4):911-914 (1963).

Komarov, Yu. V., Odintsov, M. M., and Khrenov, P. M., "Features of Mesozoic continental structures and vulcanism in the interior parts of Asia," Dokl. Sov. Geologov (Problem 4), pp. 263-274 (1964).

Kraus, E., Die Entwicklungsgeshichte der Kontinente und Ozeane, Akad. Verl., Berlin (1959) (partially translated in: Problems of Continental Drift, INL, pp. 64-127 (1963).

Kuznetsov, Yu. A., The Principal Types of Magmatic Rocks, Izd. Nedra (1964), 387 pp.

Kuznetsov, Yu. A., "The principal forms of granitic magmatism and the mechanism of forming granitoidal bodies," Geol. i Geofiz., No. 6, pp. 3-15 (1966).

Lacroix, A., La Minéralogie de Madagascar, Paris (1923).

Le Pichon, X., Hontz, R. E., Drake, C. L., and Nafe, J. E., "Crustal structure of midocean ridges: I. Seismic measurements," J. Geophys. Res., 70(2):319-339 (1965).

Luk, A. A., "Structure of the upper mantle of the Earth along the Pamir-Lena profile," Sov. Geol., No. 2, pp. 106-117 (1966).

Lyubimova, E. A., "The temperature gradient in the upper shells of the Earth and the possibility of investigating the low-velocity channel," Izv. Akad. Nauk SSSR, Ser. Geofiz., No. 12 (1959).

Lyustikh, E. N., "Hypotheses of thalassogenesis and crustal blocks of the earth," Izv. Akad. Nauk SSSR, Ser. Geofiz., No. 11, pp. 1542-1549 (1959).

Lyustikh, E. N., "Hypothesis of differentiation of the earth shell and geotectonic generalizations," Sov. Geol., No. 6, pp. 28-52 (1961).

Macdonald, G. A., Petrography of the Island of Hawaii, U. S. Geol. Survey Prof. Paper 214D (1949₁), pp. 51-96.

Macdonald, G. A., "Hawaiian petrographic province," Geol. Soc. Amer. Bull., 60(10):1541-1596 (1949₂).

Macdonald, G. A., and Katsura, R., "Chemical composition of Hawaiian lavas," J. Petrol., 5(1):82-133 (1964).

Magnitskii, V. A., "Zone refining as a mechanism for formation of the Earth's crust," Izv. Akad. Nauk SSSR, Ser. Geol., No. 11, pp. 3-8 (1964).

Magnitskii, V. A., Internal Structure and Physics of the Earth, Izd. Nedra (1965), 379 pp.

Mohr, P. A., The Geology of Ethiopia, University College, Addis Ababa (1960), 268 pp.

Muratov, M. V., "Problem of the origin of oceanic basins," Byul. Mosk. Obshchestva Ispytatelei Prirody, Otd. Geol., 32:55-70 (1957).

Muratov, M. V., "The tectonic structure and position of Iceland," Izv. Vyssh. Uch. Zav., Geol. i Razvedka, 12:1 \-29 (1961).

Nagibina, M. S., "New data on the tectonics of the Mongolia—Okhotsk belt," Byul. Mosk. Obshchestva Ispytatelei Prirody, Otd. Geol., 33(3):3-22 (1958).

Nagibina, M. S., "History of the structural development of the Mongolia—Okhotsk belt," Reports of Soviet Geologists at the 21st Session of the International Geological Congress, Probl. 18, Izd. Akad. Nauk SSSR (1960).

Noe-Nygaard, A., "Chemical composition of tholeiite basalts from the Wyville-Thomson Ridge belt," Nature, 212(5059):272-273 (1966).

Noe-Nygaard, A., and Rasmussen, J., "The making of the basalt plateau of the Faeroes," International Geol. Congress, Mexico, Vol. 11, pp. 399-407 (1957).

Pakiser, L. C., "Structure of the crust and upper mantle in the western United States," J. Geophys. Res., 68(20):5747-5756 (1963).

Part, G. M., "Occurrence of nepheline monzonite and allied types in the Cabo Verde Archipelago," Geol. Mag., 87(6):421-426 (1950).

Patterson, E. M., "A petrochemical study of the Tertiary lavas of northeast Ireland," Geochim. et Cosmochim. Acta, Vol. 5/6, pp. 283-299 (1952).

Patterson, E. M., and Swaine, D. J., "A petrochemical study of Tertiary tholeiitic basalts: the middle lavas of the Antrim Plateau," Geochim. et Cosmochim. Acta, Vol. 8, pp. 173-181 (1955).

Peive, A. V., and Sinitsyn, V. M., "Some basic questions in the study of geosynclines," Izv. Akad. Nauk SSSR, Ser. Geol., No. 4, pp. 28-51 (1950).

Peterschmitt, E., "Quelques données nouvelles sur les séismes profonds de la mer Tyrrhéni-
    enne," Ann. Geofis. (Rome), 9(3):305-334 (1956).

Petrushevskii, B. A., Questions on the Geologic History and Tectonics of Eastern Asia, Izd.
    Nauka (1964).

Powers, H. A., "Composition and origin of basaltic magma of the Hawaiian Islands," Geochim.
    et Cosmochim. Acta, 7:77-107 (1955).

Rezanov, I. A., "Structure of the earth's crust in platform regions," Byul. Mosk. Obshchestva
    Ispytatelei Prirody, Otd. Geol., 37(1):25-42 (1962).

Richey, J. E., The Tertiary Volcanic Districts. British Regional Geology, Scotland, Edinburgh
    (1935).

Ringwood, A. E., "A model for the upper mantle," J. Geophys. Res., No. 11 (1962).

Ringwood, A. E., and Green, D. H., "An experimental investigation of the gabbro-eclogite trans-
    formation and some geophysical implications," in: Petrology of the Upper Mantle, De-
    partment of Geophysics and Geochemistry, Australian National University, Publ. No. 444
    (1966).

Saltykovskii, A. Ya., "Possible initial average composition of magmas in southwestern Trans-
    baikalia," Izv. Akad. Nauk SSSR, Ser. Geol., No. 12, pp. 53-59 (1966).

Sheinmann, Yu. M., "Remarks on the classification of the structure of continents," Izv. Akad.
    Nauk SSSR, Ser. Geol., No. 3, pp. 19-35 (1955).

Sheinmann, Yu. M., "Some distributional patterns of volcanic phenomena on platforms," Tr.
    Vses. Aerogeol. Tresta, 2, Collection of Regional Geology (Sb. po Regional'noi Geol.),
    pp. 136-157 (1956).

Sheinmann, Yu. M., "The place of the Atlantic and Indian Oceans in formation of earth struc-
    ture," Dokl. Akad. Nauk SSSR, 119(4):779-781 (1958).

Sheinmann, Yu. M., "The Mohorovičić discontinuity, the depth of magma generation, and the
    distribution of ultramafic rocks," Sov. Geol., No. 8, pp. 31-44 (1961).

Sheinmann, Yu. M., "A mechanism of formation of oceanic magma," in: The Crust of the
    Pacific Basin, Geophysical Monograph No. 6, Am. Geophys. Union, Publ. No. 1035,
    pp. 181-186 (1962).

Sheinmann, Yu. M., "Tectonic conditions of magma formation," in: Problems of Magma and
    the Origin of Igneous Rocks, Izd. AN SSSR, pp. 183-193 (1963$_1$).

Sheinmann, Yu. M., "Are there juvenile granites?" Sov. Geol., No. 1, pp. 61-69 (1963$_2$).

Sheinmann, Yu. M., "The time required for restructuring continental crust to oceanic," Izv.
    Akad Nauk SSSR, Ser. Geol., No. 1 (1964$_1$).

Sheinmann, Yu. M., "Magmas and the geosynclinal process. Types of earth crust and magma,"
    in: Tectonics, Magmatism, and Distributional Patterns of Ore Deposits, Izd. Nauka,
    pp. 102-119 (1964$_2$).

Sheinmann, Yu. M., "Ways of studying the composition of the mantle," Byul. Mosk. Obshchestva
    Ispytotelei Prirody, Otd. Geol., 39(4):3-20 (1964$_3$).

Sheinmann, Yu. M., "Duration of the earth-crust transformation on the basis of data for the
    northern Atlantic," Tectonophysics, 1(5):377-383 (1964$_4$).

Sheinmann, Yu. M., "One of the features of midocean ridges," Geotektonika, No. 4, pp. 106-108
    (1965$_1$).

Sheinmann, Yu. M., Sov. Geol., No. 8 (1965$_2$).

Sheinmann, Yu. M., Izv. Akad. Nauk SSSR, Ser. Geol., No. 4 (1970).

Sheinmann, Yu. M., Apel'tsin, F. R., and Nechaeva, E. A., Alkalic Intrusions, Their Distribution
    and Associated Mineralization, Geology of Rare-Element Deposits (Geologiya Mestorozhd.
    Redk. Élementov), Nos. 12-13, Gosgeoltekhizdat (1961), 178 pp.

Sigvaldson, G. E., "Das Liparitvorkommen der Móscaradshnjúkar auf Island," Betr. Miner.
    und Petr., 6(2):100-107 (1958).

Smith, W., "Quest," British Museum Natural History Shakelton Rowett Expedition, Reports of 1921-1922, London (1930).

Smulikowski, K., "Contributions à la petrographie des iles Canariés," Arch. Mineral. [Towarzystwa Warszawkiego], Warsowie, Vol. 15 (1945).

Speranskaya, I. M., Ignimbrites in Volcanogenic Sequences on the North Shore of the Sea of Okhotsk and Questions Concerning Their Origin, Tr. Lab. Vulkanol., Vol. 20, Izd. AN SSSR (1961).

Stark, J. T., and Hay, R. L., Geology and Petrology of Volcanic Rocks of the Truk Islands, East Caroline Islands, U. S. Geol. Survey Prof. Paper, No. 409 (1963), 40 pp.

Steinmann, G., "Geologische Beobachtungen in den Alpen, Pt. 2: Die Schardtsche Überfaltungs-theorie und die geologische Bedeutung der Tiefseeabsätze und der ophiolitischen Massen-gesteine," Berichte Naturforsch. Gesellsch. Freiburg, Vol. 16, pp. 18-67 (1905).

Steinmann, G., "Die ophiolitischen Zonen in der mediterranien Kettengebirgen," Comptes Rendus, 14e Congr. Intern. Géol., Madrid, pp. 637-668 (1926).

Stille, H., Zur Frage der Herkunft der Magmen, Berlin (1940$_1$).

Stille, H., Einführung in den Bau Amerikas, Borntraeger, Berlin (1940$_2$).

Talwani, M., Le Pichon, X., and Ewing, M., "Crustal structure of midocean ridges: 2, Com-puted model from gravity and seismic refraction data," J. Geophys. Res., 70(2):341-352 (1965).

Thorarinsson, S., "The eruption of Mt. Hekla in 1947-48," Bull. Volkanol., Ser. 2, 10:157-168 (1950).

Tryggvason, E., "Das Skjalbreid gebiet auf Island, eine petrographische Studie," Geol. Inst. Uppsala Univ., Sect. B, Vol. 30 (1943).

Tryggvason, E., On the Stratigraphy of the Sog Valley in SW Iceland, Acta Nat. Island., Vol. 1, No. 10 (1950).

Tsuboi, K., "Petrographical investigations of some volcanic rocks from the South Sea Islands, Palau, Yap, and Saipan," Jap. J. Geol. Geogr., 9(3-4):201-212 (1932).

Turner, F. J., and Verhoogen, J., Igneous and Metamorphic Petrology, 2nd Ed., McGraw-Hill, New York (1960).

Tyrrell, G. W., "The petrography of Jan Mayen," Trans. Roy. Soc. Edinburgh, 54(3):745-765 (1926).

Tyrrell, G. W., "Flood basalts and fissure eruptions," Bull. Volcanol., Ser. 2, No. 1, pp. 89-111 (1937).

Tyrrell, G. W., "Petrography of igneous rocks from the Vatajökull region, Iceland, collected by F. W. Anderson," Trans. Roy. Soc. Edinburgh, 61:3 (1949).

Tyrrell, G. W., and Peacock, M. A., "The petrology of Iceland," Trans. Roy. Soc. Edinburgh, 55:1 (1926).

Uffen, R. J., "On the origin of rock magma," J. Geophys. Res., 64(1):117-122 (1959).

Uffen, R. J., and Jessop, A. M., "The stress release hypothesis of magma formation," Bull. Volcanol., 26(5):57-66 (1963).

Ustiev, E. K., "The Okhotsk tectonic—magmatic belt and some problems connected with it," Sov. Geol., No. 3 (1959).

Ustiev, E. K., "The Okhotsk structural belt and problems of volcanic-plutonic formations," in: Problems of Magma and the Origin of Igneous Rocks, Izv. AN SSSR (1963$_1$), pp. 161-182.

Ustiev, E. K., "Problems of vulcanism-plutonism, Volcanic-plutonic formations," Izv. Akad. Nauk SSSR, Ser. Geol., No. 12, pp. 3-30 (1963$_2$).

Ustiev, E. K., "Composition of parental magmas on the basis of the Cretaceous and Paleogene rocks of the Okhotsk volcanic belt," Izv. Akad. Nauk SSSR, Ser. Geol., No. 3 (1965).

Vinogradov, A. P., "Meteorites and the earth's crust," Izv. Akad. Nauk SSSR, Ser. Geol., Vol. 10, pp. 5-27 (1959).

Vinogradov, A. P., "The origin of the material of the earth's crust," Geokhimiya, 1:3-29 (1961).

Vinogradov, A. P., "The origin of the earth's shells," Izv. Akad. Nauk SSSR, Ser. Geol., No. 11 (1962).

Wager, L. R., and Deer, W. A., "The petrology of the Skaergaard intrusion, Kangerdlugssuag, East Greenland," Medd. om Grönland, Vol. 105, No. 4 (1939).

Walker, F., "The geology of the Shiant Isles," Q. J. Geol. Soc. London, 86:355-398 (1930).

Walker, F., and Davidson, C. T., "A contribution to the geology of the Faeroes," Trans. Roy. Soc. Edinburgh, Vol. 58, III (Session 1935-36), pp. 869-897 (1936).

Washington, H. S., Chemical Analyses of Igneous Rocks, U. S. Geol. Survey Prof. Paper 99 (1917).

Washington, H. S., "Isostasy and rock density," Bull. Geol. Soc. Am. 33(2):375-410 (1922).

Wyllie, P. J., and Drever, H. J., "The petrology of a picrite sill on the island of Soai (Hebrides)," Trans. Roy. Soc. Edinburgh, Vol. 65, No. 8 (1961-1962).

Willis, B., and Wills, S., "Eruptivity and mountain building," Bull. Geol. Soc. Am., 52(10):1643-1684 (1941).

Winchell, H., "Honolulu series, Oahu, Hawaii," Bull. Geol. Soc. Am., 58:1-48 (1947).

Wolff, F., Der Vulkanismus, Vol. 2, Pt. 2, Stuttgart (1931).

Yagi, K., "Petrochemistry of the alkalic rocks of Panape Island, western Pacific Ocean," Rept. 21st Session Norden, International Geol. Congress XIII, pp. 108-122 (1960).

Yoder, H. S., Jr., and Tilley, C. E., "Origin of basalt magmas: an experimental study of natural and synthetic rock systems," J. Petrol., 3(3):342-532 (1962).

Zavaritskii, A. N., An Introduction to Petrochemistry, Izd. AN SSSR (1944), 324 pp.